Developments in Sedimentology
Volume 66

Mid-Latitude Slope Deposits (Cover Beds)

Developments in Sedimentology

Volume 66

Mid-Latitude Slope Deposits (Cover Beds)

Editors

Arno Kleber

TU Dresden, Faculty of Environmental Sciences,
Department of Geosciences

Birgit Terhorst

Julius-Maximilians-Universität Würzburg,
Institut für Geographie und Geologie

ELSEVIER

AMSTERDAM • BOSTON • HEIDELBERG • LONDON • NEW YORK • OXFORD
PARIS • SAN DIEGO • SAN FRANCISCO • SINGAPORE • SYDNEY • TOKYO

Elsevier
Radarweg 29, PO Box 211, 1000 AE Amsterdam, The Netherlands
The Boulevard, Langford Lane, Kidlington, Oxford OX5 1GB, UK
30 Corporate Drive, Suite 400, Burlington, MA 01803, USA
525 B Street, Suite 1900, San Diego, CA 92101-4495, USA

First edition **2013**

British Library Cataloguing in Publication Data
A catalogue record for this book is available from the British Library

Library of Congress Cataloging-in-Publication Data
A catalog record for this book is available from the Library of Congress

Printed and bound by CPI Group (UK) Ltd, Croydon, CR0 4YY
Transferred to digital print 2013

ISBN: 978-0-444-53118-6
ISSN: 0070-4571

Contents

Contributors

Bibus, E. (229), Universität Tübingen, Geographisches Institut, Tübingen, Germany

Bullmann, H. (9), Alfred-Kästner-Straße 28b, 04275 Leipzig, Germany

Chifflard, P. (127), University of Natural Resources and Life Sciences, Vienna, University of Vienna, Danube University Krems, WasserCluster Lunz, Lunz am See, Austria

Damm, B. (153), University of Vechta, Institute of Structural Research and Planning in Rural Areas, Applied Physical Geography, Vechta, Germany

Dietze, M. (9), Technische Universität Dresden, Faculty of Environmental Sciences, Institute of Geography, Dresden, Germany

Felix-Henningsen, P. (9), Universität Gießen, Institut für Bodenkunde, Giessen, Germany

Frühauf, M. (95), Universität Halle-Wittenberg, Institut für Geowissenschaften, Halle, Germany

Hübner, R. (127), Technische Universität Dresden, Faculty of Environmental Sciences, Institute of Geography, Institute of Geography, Dresden, Germany

Heinrich, J. (9), Universtität Leipzig, Fakultät für Physik und Geowissenschaften Institut für Geographie, Leipzig, Germany

Heller, K. (127), Technische Universität Dresden, Faculty of Environmental Sciences, Institute of Geography, Dresden, Germany

Hülle, D. (9), Universität Köln, Geographisches Institut, Köln, Germany

Kleber, A. (1, 9, 95, 127, 171, 229, 253), Technische Universität Dresden Department of Forest, Earth and Hydro Sciences, Geosciences Dresden, Germany

Leopold, M. (9, 171), The University of Western Australia, School of Earth and Environment, Perth, Australia

Lorz, C. (95), Weihenstephan-Triesdorf University of Applied Sciences, Faculty Woods and Forestry, Freising, Germany

Mailänder, R. (95), Eidgenössisches Nuklearsicherheitsinspektorat, Brugg, Switzerland

Müller, S. (9), Universität Frankfurt, Institut für Physische Geographie, Frankfurt am Main, Germany

Moldenhauer, K.-M. (127), Universität Bayreuth, Fachrichtung Geowissenschaften, Bayreuth, Germany

Ottner, F. (153), University of Natural Resources and Life Sciences, Institute of Applied Geology, Vienna, Austria

Phillips, J.D. (95), University of Kentucky, Department of Geography, Lexington, USA

Raab, T. (9), Brandenburgische Technische Universität Cottbus, Fakultät für Umweltwissenschaften und Verfahrenstechnik, Cottbus, Germany

Sauer, D. (9), Technische Universität Dresden, Faculty of Environmental Sciences, Institute of Geography, Dresden, Germany

Scholten, T. (9), Universität Tübingen, Geographisches Institut, Tübingen, Germany

Spies, E.-D. (9), Landesamt für Geologie und Bergbau Rheinland-Pfalz, Mainz, Germany

Terhorst, B. (1, 9, 153, 229, 253), Institut für Geographie und Geologie, Julius-Maximilians-Universität Würzburg, Würzburg, Germany

Thiemeyer, H. (9), Universität Frankfurt, Institut für Physische Geographie, Frankfurt am Main, Germany

Völkel, J. (171), Technische Universität München, Center of Life and Food Sciences Weihenstephan, Freising, Germany

Vonlanthen, C. (171), Amt für Wald und Landschaft, Kanton Obwalden, Sarnen, Switzerland

Preface

The idea for this book originated in Summer 2003, when, at the INQUA Congress in Reno, Nevada, A.K. met A.J. van Loon, series editor of the renowned book series "Developments in Sedimentology." A.K. proposed editing a book on certain types of slope deposits that have been in the focus of German research for decades; this research had, however, not made its deserved way into the international dispute, mainly because of the language barrier. It was around this time that discussion of the "critical-zone concept" grew rapidly. It was evident to us that slope deposits have the potential to play a decisive role in this dispute so that contributing an alternative view on the materials near the surface might have a significant impact. Because of the continued scientific interest in the critical zone, this still holds true today. A.K. drew up an initial table of contents for the book, which streamlined the discussions between him and the series editor during the conference. Although the table of contents underwent many changes during the preparation of the book, its final version is quite close to the original concept.

Two renowned researchers in the field, H. Veit and J. Völkel, helped shape the structure of the book and potential authors were identified. Unfortunately, both were unable to commit to the book due to changes in the university system at the time. A.K. and B.T., motivated by Arno Semmel, who believed in the project and kept the idea alive, agreed to share the burden (and the fun) of finalizing this book. B.T. organized a session at the European Geosciences Union General Assembly at Vienna in 2010 where the main authors of several chapters presented the concepts for their respective chapters.

We are indebted to A.J. van Loon, who patiently assisted and supported the evolution of this book, and for his final review of the entire book. We are also grateful for thorough reviews of the chapters provided by several peers.

We wholeheartedly dedicate this book to Arno Semmel, who for decades had been the major researcher and the nestor of the field of science presented here. Despite being unable to contribute directly due to ill health, he always followed with great interest the evolution of the book. Until he passed away on October 10, 2010, he encouraged us editors to stay on track because of the great importance he felt this book would have for the particular field of science that had filled much of his life. He had a tremendous influence on the scientific development of both editors as a mentor and a critical commentator, as he had on almost all the authors of this book, several of whom have been his direct scholars.

Arno Kleber
Birgit Terhorst

Chapter 1

Introduction

A. Kleber and B. Terhorst

1.1 SCOPE OF THE BOOK

Slopes, be they steep or almost insignificantly inclined, are the single most important landform on Earth, as they occupy some 90% of all ice-free land (Huggett and Huggett, 2011). Accordingly, slopes are of particular interest to mankind because most human activities take place upon them. Therefore, slopes have been in the human scope probably from the onset of human intelligence and in the focus of scientific interest from antiquity (e.g., Aristotle; see Lyell, 1832).

Modern scientific research on slopes, especially on processes acting on slopes, started with W.M. Davis, G.K. Gilbert, and W. Penck at around the turn to the twentieth century (Selby, 1993). Since then, a vast amount of literature on slopes, the materials they consist of, and the processes acting on them have arisen (reviewed by De Wolf, 1988; Selby, 1993). A large number of processes have been identified, namely, creep, fall, flow, heave, erosion by overland flow, slide, etc.; accordingly, various types of deposits are being distinguished, ascribed to these different processes and triggered by a manifold of factor combinations. These are all so well accepted that they are referred to in every geomorphologic textbook (e.g., most recently, Huggett and Huggett, 2011). According to this literature, one may differentiate between reallocated hillslope materials, the slope deposits, and *in situ* materials, derived from weathering of the local bedrock, which have been disturbed in their structure by bio- or pedoturbation at most.

However, starting mainly in the 1960s, a different view on some of these materials, especially on materials regarded as having remained *in situ*, came into being in Central Europe. Thereby, many of these materials were identified as reallocated, too. Because of their widespread nature, often without much small-scale variation, they are difficult to be kept apart from *in situ* materials. This insight has only sparsely found its way into the international literature yet, though a huge amount of literature has arisen on these deposits as well; but its major part is published rather dispersed in German language, and there are only a few review articles in English (Kleber, 1992a, 1997; Raab et al., 2007; Semmel and Terhorst, 2010).

Developments in Sedimentology, Vol. 66. http://dx.doi.org/10.1016/B978-0-444-53118-6.00001-5

Therefore, the main scope of this book is to provide to a broad geoscientific readership a comprehensive review of the knowledge of these particular types of slope deposits, in the following referred to as "cover beds," which knowledge has been accumulated over the past 50 years.

The book is regionally confined to the mid-latitudes because the major findings have been made in Central Europe and because there already is impressive knowledge on the layering of the subsurface from other climatic zones (Paton et al., 1995; Semmel, 1991c), whereas evidence from other mid-latitude areas is still scarce.

1.2 STRUCTURE OF THE BOOK

The principal structure of this book is to build upon the basics of cover-bed research several aspects of application regarding ecological implications of these deposits, then attempting to get a broader regional perspective, and finally utilizing cover beds for relative dating.

The most extensive Chapter 2 embraces cover beds in the subdued mountains of Central Europe, the area where this type of research has started and where discussion on the matter is still by far the most vivid. The various types of layers are introduced as to their properties and their distribution patterns. Based on this, problems in the classification and distinction of these layers are discussed, followed by statistical analyses of layer properties. Building upon this, the genesis of the various layers and their ages is discussed in the following sections. The chapter closes with a selection of smaller scale studies to demonstrate the regional variability of the discussed phenomena.

The following Chapters 3–5 comprise important aspects of the application of the cover-bed concept in various ecological disciplines. The most evident effect of layered parent materials is on soil formation and soil properties. As soils could not have evolved by far as deep as they actually are during the Holocene without relying on preexisting crushed matter, the widespread occurrence of cover beds rather than bare rock at the onset of most soil formation is crucial for the existence of rather mature soils. Not that evident, but ecologically very important though, is that cover beds decisively affect vadose zone water paths in slope environments as depicted in Chapter 4. As there has been done by far less work on this issue, this chapter mainly portrays some case studies that examined these relationships to some detail. Chapter 5 stresses a third important applied aspect, the tendency of slope failures to occur in, and adapt in their structure to, the layered subsurface.

Because there is strong evidence that such deposits are widespread in the mid-latitudes, Chapter 6 leaves the regional scope of the previous chapters behind and attempts transferring the concept of cover beds to regions other than Central Europe subdued mountains. These are basins and lowlands (in Russia, Turkey, and the western USA) and high mountains (European Alps and Rocky Mountains) of the mid-latitudes.

As another application of the concept, Chapter 7 utilizes cover beds for relative dating purposes, analyzing the relationship between cover beds and underlying landforms or landforms that have established after cover-bed deposition. The examples stem from Central Europe subdued mountains and the western USA. The foundations for understanding these applications of cover-bed research have been laid in Chapters 2 and 6.

There is ongoing, vivid discussion on these issues. Thus, the concluding Chapter 8 mainly focuses on open questions and future research demands in the context of cover beds.

More than 90% of the book review and critically discuss existing, mostly German literature; the remaining parts contain unpublished research accompanied by original data.

1.3 TERMINOLOGY

Daniels and Hammer (1992: 78) defined slope deposits as materials derived from upslope through erosion processes, possibly combined with frost heave or bioturbation, which merge into alluvium on foot slopes. However, this definition does not explicitly include admixed loess that stems not from an earlier upslope loess deposition but from incorporation of eolian matter during dislocation, which loess may be a constitutive component of such deposits. Also, such deposits may overlie alluvium rather than merge into it. Kleber (1990, 1992a) coined the term "cover bed" to include loess-bearing deposits not covered by previous definitions. Thus, cover beds may and often do contain loess material. Furthermore, he defined cover beds as allochthonous deposits (cf. the "mobile zone" of Ollier and Pain, 1996) occurring on surfaces of varying inclination and resulting from processes such as solifluction not limited to linear discharges or local rock failures. Therefore, cover beds typically have a wide distribution with relatively uniform properties; this important property is expressed by the word "cover" in this term. Thus, there are benefits in using the term "cover bed," which is defined as follows:

Cover beds are deposits that formed by unconcentrated dislocation processes chiefly from upslope materials, but that may contain admixed eolian matter. They cover undifferentiated slopes to a large extent or even completely rather than being restricted to drainage ways, linear discharges, or local failures. Cover beds typically consist of several layers separated by disconformities.

Readers may be aware that there is usage of the term cover bed in New Zealand, too, which encompasses a broader meaning, that is, colluvia, pure loesses, and tephra layers overlying some other material such as alluvial terraces or tills (e.g., Suggate and Almond, 2005). Both definitions have in common the widespread, covering character of the described deposits. Similar deposits to those addressed in this book have been termed "stratified slope deposits" by

De Wolf (1988) and "solifluction sheets" by Harrison (2002). The major difference is that their definitions did not include admixed loess, which, on the other hand, was included in the definition of "pseudo-grèzes", whereas the more popular French term "grèzes litées" usually is limited to different spatial distributions and finer clasts (De Wolf, 1988; Ozouf et al., 1995; also see the adoption of the term to some German slope deposits by Karte, 1983) compared to typical cover beds as described in this book. Furthermore, the assumed processes of their dislocation (De Wolf, 1988) are different. However, none of these other approaches has led to a systematic perception of successions of such layers.

Almost all cover beds addressed in this book are assumed to be periglacial in origin. Therefore, we usually omit the qualifier "periglacial", rather we explicitly label diverging cases.

As cover-bed successions typically comprise more layers than one, the identification of layers is a major task in the study of cover beds. We restrict the term "layer" to geologic bodies, not using it for soil horizons. Layers may be distinguished by the fact that they are separated by unconformities, separating two strata of different ages, which indicates discontinuous sediment deposition. Usually, the layers comprising a cover-bed succession have been deposited approximately parallel to one another, so that they are separated by so-called disconformities, which fact often complicates the identification of layers.

A disconformity is the interface between two parallel-bedded materials, which are internally uniform and which are distinct from each other in terms of pattern of texture, fabric, or mineralogy (Arnold, 1968). This distinctness is abrupt, that is, the change is evident over a vertical distance of at least a few (typically <5) centimeters, if the fine fractions are used for defining the disconformity, and by the diameter of the mean clast, if the clast content is used for defining.

Lithological discontinuity (used as a synonym to disconformity in this book; see Schaetzl and Anderson, 2005) is a diagnostic property of soils as it may provide considerable changes in particle size distribution or mineralogy, representing differences in parent material within a soil. A lithological discontinuity requires one or more of the following (WRB, 2006) characteristics:

- an abrupt change in particle size distribution except for a sole change in clay content[1];
- a relative change of >20% in the ratios of coarse, medium, and fine sand;
- clasts with a lithology different from the underlying continuous rock;
- a layer with clasts without weathering rinds on top of a layer containing such rinds;

1. Because differences in the clay fraction could be of pedogenic rather than geologic origin, clay content is typically not used as a criterion; however, where pedogenic clay formation is slow (Colman, 1982; Chadwick and Davis, 1990), a marked change in clay content may well indicate a disconformity.

- a change between materials with angular and with rounded clasts;
- abrupt color changes not resulting from pedogenesis; or
- striking differences in size and shape of resistant minerals between adjacent layers (revealed by micromorphology or mineralogy).

Besides these criteria, disconformities may also be identified by intercalated paleosols, which indicate some time of subaerial exposure before the deposition of the overlying strata, or by considerable changes in sediment structures (e.g., orientation of clast length axes), which hint at changed processes of sedimentation. Stone lines (thin layers with concentrations of rock fragments) are also occasionally used as indicators of disconformities.

As a caveat, even obvious evidence of a disconformity according to the given criteria is not an assured proof of the existence of a chronological hiatus or a change in deposition processes because the dislocation of cover beds may have been laminar so that just older changes in the composition of the subsurface originally located upslope might have been allocated during their formation (see the book cover photo with a stone line carried along by laminar dislocation); furthermore, parts of dislocating layers may have been disproportionated during transport so that a disconformity may have formed within a moving deposit. To overcome such caveats, the tracing of disconformities over large distances in exposures or via systematic augering is needed, which, indeed, has often been utilized in many of the studies cited in this book. Disconformities described in this book were identified in the field and often later controlled by laboratory methods (e.g., grain-size or mineralogical analyses).

Soils in this book are mostly described according to Soil Taxonomy (SSS, 2010); in a few sections outlined in the introduction to the respective chapters, their classification is after the World Reference Base for Soil Resources (WRB, 2006). Since soil classification is a vast scientific field per se, only short descriptions of taxonomic terminology can be provided in footnotes for readers unfamiliar with soil classification, which certainly cannot replace a correct definition as given in the above references. Even worse, most soil classification in this book except where stated otherwise is tentative because the original classification often was in different classification systems (mostly the German system; AGB, 2005).

Because there is no internationally accepted terminology on cover beds, this book uses approximate translations of the original German terms (following Kleber, 1992a). This causes a somewhat unusual usage of the term "layer." If not used with a qualifier (upper, intermediate, basal), this term is defined as a single, uniform, undivided geologic entity. However, the various types of cover beds are also designated as layers with qualifiers added; because these actually are types of layers and not necessarily single geologic units, occasionally separation of such a layer in several sublayers may occur. The book strictly adheres to the established acronyms for the various types of layers as introduced by AGB (1994), though they are not self-explaining.

1.4 HISTORY OF IDEAS

Schönhals (1957), Priehäuser (1958), and Büdel (1959) were among the first in Germany to describe slope deposits as consisting of distinct layers. Systematic research on slope deposits in subdued mountains started with the apprehension that soils typically have not formed from bedrock through *in situ* weathering; rather, they have formed from slope deposits, which are occasionally mixed with loess, and which are typically very widespread and rather uniform. A conspicuous relationship between these deposits and their layering, and the resultant soils and their horizons, respectively, was soon recognized. At this stage, Schilling and Wiefel (1962) and Schwanecke (1966, 1970) in East Germany and Semmel (1964, 1968a) in West Germany described layer successions comprising various members with different properties and distributions. These authors differed widely in their interpretation of the layers' ages, of their horizontal distribution, and in their terminology (see Stahr, 1979; Völkel, 1995; Völkel et al., 2002a). Subsequent contributors introduced additional, incompatible terms and definitions for the layers (Altermann et al., 2008; Hoffmann, 1970, 1977; Kopp, 1970). Soon after the German reunification, Altermann (1990, 1993) and Frühauf (1990a, 1991) attempted resolving the discrepancies between the viewpoints in the former parts of the country. On this basis, the German Soil Science Society revised its "key to soil mapping" by inclusion of those deposits (AGB, 1994, 2005; Sabel, 1989; Schilling and Spies, 1991; Wittmann, 1991). With this standardization, it became possible to summarize the knowledge of the time regarding slope deposits covering the slopes in Germany (Kleber, 1992a).

Three different approaches characterize the modern phase of the study of cover beds: (1) a broadened focus to further geoecological relationships such as heavy metal contents of soils from cover beds (Kleber et al., 1998b; Lorz, 1996; Sabel, 1989) and consequences of the subcutaneous slope internal structures for the vadose zone flow of water and, with it, contaminants (Chifflard et al., 2008; Heller, 2012; Kleber and Schellenberger 1998; Moldenhauer, 1993); (2) an attempt to transfer the concept of cover beds to other regions of the mid-latitudes (Bussemer, 2002; Kleber, 1990, 1997, 1999b; Leopold et al., 2008a,b); and (3) the use of new dating techniques to relatively and numerically date cover beds (Hülle et al., 2009; Völkel and Mahr, 1997, 2001; Völkel and Leopold, 2001). Altogether, a lot of knowledge has accumulated regarding properties, genesis, and the ages of cover beds and their geoecological significance—issues that will be addressed in later chapters of this book.

This all has been achieved within the disciplines of geomorphology and pedology, which were well connected in the early years when mainly geomorphologists became hired for soil surveying. In neighboring scientific disciplines such as geology and edaphologically oriented soil science, these findings were not unequivocally accepted. In soil science, receipt slowly started evolving not till the inclusion of cover beds in the German soil classification system (AGB, 1994). However, in geology, the concept still has not achieved a good reputation. In our opinion, there are two reasons for this slow establishing of the concept, which both have to do with the fact that the research on cover beds was from its very beginning focused on applied issues, while neglecting some basic inquiries. The first problem is the claim of an almost ubiquitous layer of constant thickness mainly regardless of topographic position and other external factors (see Section 2.6.3); this is difficult to accept, as geomorphic processes are generally believed to depend on topography—it is more plausible to suspect soil formation or bioturbation to be the cause of this astonishing property of the subsurface. The second problem is that stratigraphic significance was often claimed or implicated for the particular layers in a cover-bed succession. This book addresses both issues, proposing a solution for the former problem and clearly disapproving the existence of a confined chronology of the layers in Central Europe, at least at the present stage of knowledge. Cover beds are classified by composition and distribution, to name the most important criteria, not by chronostratigraphic means.

1.5 COVER BEDS IN THE CONTEXT OF THE "EARTH'S CRITICAL-ZONE" CONCEPT

Earth's surface is literally the fundament of all human uses of landscape. However, human influence is not restricted to the surface; rather there is much interaction with the near-surface materials, for example, in terms of nutrients and water delivered from there to crops and trees or of pollution introduced into the subsurface. For basic research, surface and subsurface are of equal interest because they make up the major interface for all Earth-surface processes.

This importance of the near-surface materials—with slope deposits being among the most widespread of these materials—has been acknowledged since decades, but this has gained much progress through the "Earth's critical zone" concept. This concept attempts integrating all ecological interdependencies from the top of the canopy layer down to and including the active phreatic zone in a holistic way. It views the subsurface as the interface between the solid materials Earth is composed of and its fluid envelopes (atmosphere, open water bodies). This is where the coevolution of landforms, soils, and biota takes place, which, on their part, affect one another as well as the critical zone as a whole through various feedback mechanisms (Brantley et al., 2006). Accordingly, this concept crosses discipline boundaries, involving essentially all fields of earth and life sciences (Brantley et al., 2007). The phenomena and processes in the critical zone are acknowledged to be crucial for sustaining life on the planet (Rasmussen et al., 2010). This book presents a close look to the composition and structure of the critical zone's solid materials on slopes and to some of the aspects regarding their role as an interface. It does not put forward an alternative concept but a concretization of the substrates and structures in the core of the critical zone.

The biomantle concept may be considered as a part of the current critical-zone concept (Johnson and Lin, 2006), although it is somewhat older (Johnson, 1990). The biomantle is defined as the upper part of the soil, which is chiefly a product of the activity of biota, where bioturbation is a dominant process in the formation of soil properties. The major advantage of this concept is its focus on the impact of organisms on near-surface materials, which had often been neglected previously. However, it is often assumed that bioturbation has even produced the "epidermal" upper part of the soils, with other processes being subsidiary at most (Johnson, 1990; Johnson et al., 2005; Paton et al., 1995; Schaetzl and Anderson, 2005).

It is well accepted among cover-bed researchers that the uppermost cover bed (the so-called upper layer) of Central Europe has been modified by the action of fauna (including man), flora, and microorganisms during the more than 10,000 years since its deposition (Frühauf, 1991; Russow and Heinrich, 2001). The interdependency between layered subsurface and vegetation has already been demonstrated by Heinrich (1991), who analyzed effects of a strong storm event that threw trees that had developed less deep rooting on nutrient-rich threefold cover-bed successions, whereas deeper rooting trees on meager twofold successions remained staying alive. However, turbation by biota alone is rarely able to explicate all differences between the materials closest to the surface and those beneath, especially if clast contents, stable-mineral composition, or other properties—outlined above to discriminate disconformities—diverge remarkably; or, if primary sediment features such as clast orientation have been preserved. Biota as many other processes of pedogenesis rather may well adapt to, and thereby accentuate und reinforce, preexisting boundaries within the soils, but not more—at least in the mid-latitudes this book focuses on.

Though known for long (e.g., Yaalon and Ganor, 1973), the addition of eolian matter to soils derived mainly from other materials may be considered as one aspect of current critical-zone research (Derry and Chadwick, 2007). This addition is often understood as a quasi-continuous process (e.g., Birkeland, 1999) although Chadwick and Davis (1990) showed that eolian addition may also have occurred in pulses. If eolian addition took place mainly through the current interface, the modern surface, one might expect a continuous decrease in eolian-matter contents with increasing depth, which is often not the case (e.g., Kleber, 2011). Rather, eolian addition may come from older, reworked materials, may be syngenetic with the deposition of the respective sediment, or may, indeed, be admixed later.

Chapter 2

Subdued Mountains of Central Europe

A. Kleber, B. Terhorst, H. Bullmann, D. Hülle, M. Leopold, S. Müller, T. Raab, D. Sauer, T. Scholten, M. Dietze, P. Felix-Henningsen, J. Heinrich, E.-D. Spies and H. Thiemeyer

2.1 INTRODUCTION

A. Kleber, B. Terhorst

This chapter addresses cover beds in the subdued mountains[1] of Central Europe. Subdued mountains are uplands or mountain ranges that are not sufficiently high to reach the timberline. Though usually comprising gently inclined slopes bordering wide valleys, they may occasionally rise to steep slopes. However, bedrock outcrops fostering steep walls are rare.

These areas are of special interest with regard to the development of the cover-bed concept because all early findings on the matter, and most since then, were reported from there. Therefore, plenty of literature and partially contrasting opinions have derived from these mountains. However, though or just because of the diverse history of ideas, there still is need for clarification regarding terminology, genesis, and, especially, ages of cover beds.

In general, periglacial cover beds in subdued mountains may be divided into three main types. One type, containing merely eolian matter (i.e., loess), is referred to as allochthonous slope deposit. This book does not address allochthonous, that is, pure, unmixed eolian deposits, which already are in the scope of a plethora of literature. The second type is referred to as autochthonous and is dominated by material which has been reworked from underlying and up-slope rocks. The third—mixed—type includes both local components and loess. Accordingly, these latter two types of cover beds experience a local aspect attributed to onsite geological materials, though to a varying degree.

In the following, we will address the various types and subtypes as they typically occur in Central European subdued mountain ranges, starting with a

1. The English term was first used by W.M. Davis (1899). The German translation is "Mittelgebirge."

Developments in Sedimentology, Vol. 66. http://dx.doi.org/10.1016/B978-0-444-53118-6.00002-7

TABLE 2.1 Cover-Bed Terminology with Emphasis on Literature in English Language

Semmel (1964)	AGB (1994, 2005)	Kleber (1992a, 1997)	Völkel et al. (2001)[a]	Lorz and Phillips (2006)	Terhorst (2007)
Deckschichten	Periglaziale Lagen	Cover beds	Periglacial slope deposits	Cover beds	Periglacial cover beds
Deckschutt	Hauptlage (LH)**	Upper layer	Upper head	Upper cover bed	Upper periglacial cover bed
Mittelschutt	Mittellage (LM)	Intermediate layer	Middle head	Middle cover bed	Middle periglacial cover bed
Basisschut	Basislage (LB)	Basal layer	Lower head	Lower cover bed	Basal periglacial cover bed

[a]See also Raab et al. (2007).
**established acronyms used throughout this book

classification based upon sediment properties, which is the most important criterion in classification.

The classification used in this volume follows the German soil survey manual (AGB, 1994, 2005), which uses the terms "Basislage," "Mittellage," and "Hauptlage," that is, basal, intermediate, and upper layer, respectively (Table 2.1). However, we preclude their "Oberlage," which is described as a rock-fall deposit very locally found at high elevations at the foot of outcrops of solid rock, and which is, though without much proof, believed to be periglacial in origin (AGB, 2005; see also the discussions in Emmerich, 1994 and Semmel, 2002a). Such material may be compared to the French "éboulis ordonnés" in their classical definition (De Wolf, 1988). The preclusion of rock-fall material is based on the restricted local occurrence of this type of deposit, which prevents qualifying it as a cover bed.

The sediment properties of these various deposits will be discussed in Section 2.2. AGB (1994, 2005) provided information on the distribution and the thickness of specific layers used in particular cases as additional classifiers for layers. Following this approach, these criteria are dealt with thereafter (Section 2.3). This is concluded with some remarks on potential confusion between the described layers, especially in the field, and how to avoid such confusion (Section 2.4). The criteria discussed until then will be mainly qualitative. Therefore, a section on quantitative aspects of layer properties and distribution will evaluate properties and distribution by means of statistics (Section 2.5). These pieces of information provide the basis for a discussion on the genesis of the layers (Section 2.6), emphasizing processes, their reconstruction, and some modern analogs for each layer type individually, first, and then with regard to mutual relationships between the layers (Section 2.6.4). Subsequently, we will focus on the age of the layers, first on relative-age criteria (Section 2.7.1) and then on attempts to date cover beds numerically (Section 2.7.2). Finally, the last section of this chapter depicts some of the even more complicated regional peculiarities (Section 2.8). Figure 2.1 lists the major study areas discussed in this section.

FIGURE 2.1 Map of Central Europe depicting the study areas of this section (without numbering in parentheses), in the subsequent Chapters 3–5 (numbering in parentheses). (For color version of this figure, the reader is referred to the online version of this chapter.)

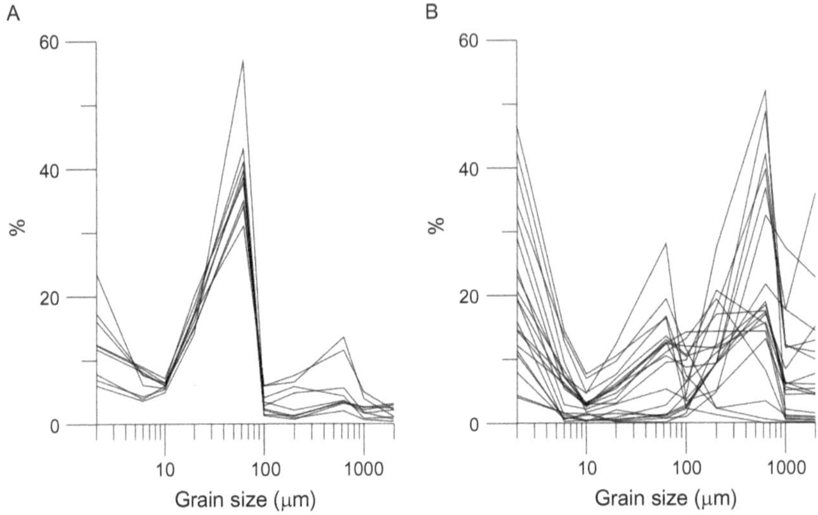

FIGURE 2.2 Combined grain-size distributions of cover beds in the Tharandt Forest, Saxony, based on 48 analyses of the fine material from closely spaced soil profiles upon various types of bedrock. All loessic layers have a similar grain-size distribution with a distinct peak in the coarse-silt fraction and only slight differences in the other textural classes. The basal layers differ markedly, depending on the type of bedrock. Basal layers are considerably more influenced by rock type than loessic layers (*reproduced from Dietze, 2006*). (A) Loess-enriched upper and intermediate layers. (B) Basal layers.

All statements on layer properties, except for the eolian admixture, take for granted that local rocks are able to deliver appropriate materials: referring to typical grain sizes implies reference to the rock types from which these grains are derived.

2.2 SEDIMENTARY PROPERTIES OF LAYERS

A. Kleber, M. Dietze, B. Terhorst

2.2.1 The Basal Layer

The basal layer (LB[2]) is derived from local materials, that is, from rock either directly underlying the respective site or outcropping farther up-slope. Rock fragments have been delivered to the layer, after weathering and erosion of the parent rock, by dislocation.

Basal layers are typically unsorted (Fig. 2.2). All grain sizes present in the local rocks, as well as rock fragments of various sizes, mostly derived from frost

2. This book uses the established acronyms for the various types of layers as introduced by AGB (Arbeitsgemeinschaft Boden) (1994) throughout.

shattering, may be incorporated. Rarely, there are layers or lenses consisting of sorted, often well-bedded, mainly sandy matter within the basal layer intercalated between coarse rock fragments or at its top. The basal layers usually contain more clasts than the other layers, but not necessarily larger ones than the upper layer. However, where material from old, deeply weathered saprolites had remained present until the basal layers formed, these saprolites rather than the bedrock may dominate the layer properties (Felix-Henningsen et al., 1991; Sauer and Felix-Henningsen, 2006).

If present, the clasts' longitudinal axes are well oriented parallel to the slope inclination. If rock fragments are platy and frequent, imbricate (also known as tegular) structures are often observed.

Basal layers do not show a loess component in the field, except for occasional silt caps on rock fragments (Thalheim and Fiedler, 1990), which occur mainly in basal layers from rock types delivering very coarse fragments such as gneiss (see Fig. 4.2). However, laboratory analyses commonly provide evidence of some minor amounts of loess. Dietze (2006), for instance, identified mica minerals through X-ray diffraction in basal layers on a basalt plateau, with the basalt being free of mica. These minerals were found in the clay and fine-silt fractions, so that they may be interpreted as far-transported eolian components; loess-enriched layers, in contrast, typically comprise coarse silt.

Almost all basal layers are remarkably compacted and show bulk densities around and beyond 1.8 g cm^{-3} (Table 2.2, also see Table 4.1 in Chapter 4).

Basal layers may consist of more than one sublayer, with the particular subordinate layers being appreciably different or quite similar, depending on upslope variability in rock composition (Altermann et al., 1977; Sauer, 2002; Sauer and Felix-Henningsen, 2006; Semmel, 1968; Stahr, 1979; see also Fig. 2.3).

2.2.2 The Intermediate Layer

By definition, intermediate layers consist of local debris mainly derived from basal layers with an addition of eolian particles. Therefore, there usually is a marked increase in silt, particularly in coarse-silt (Semmel, 1998a) content in an intermediate compared to the underlying basal layer. This does, obviously, not hold true for basal layers derived from bedrock that delivers large quantities of silt, such as many shales and other metamorphic rocks. Most intermediate layers are claimed to have, on average, a higher loess component than overlying upper layers (e.g., AGB, 2005; Altermann, 1993). Furthermore, intermediate layers are often stated to typically contain less coarse clasts than the underlying and overlying layers (AGB, 1994). However, these differences between upper and intermediate layers are not robust enough to withstand statistic tests (Scholten, 2003; see Section 2.5.5).

The contents of loess in intermediate layers are variable and mainly controlled by elevation and distance to sources (Friedrich, 1996), which were mainly alluvial plains of braided rivers. In addition, the topographic factors that

TABLE 2.2 Bulk Densities and Other Data Derived from Cover Beds in a Profile in the Frankenwald Mountains, North Bavaria

Depth (cm)	Horizon	Cover bed	Color (dry)	Color (moist)	Texture	Clasts (vol.%)	Density (g cm^{-3})	pH ($CaCl_2$)
3	A	LH	10YR 4/3	10YR 3/1	Sandy loam	50	n.d.	2.93
44	AE	LH	2.5YR 6/4	2.5YR 4/2	Sandy loam	40	1.0[a]	3.52
85	2Bt	LM	2.5YR 6/6	2.5YR 5/4	Sandy loam	20	1.5[a]	4.21
114	3Bwg	LM	2.5YR 6/4	2.5YR 5/4	Sandy loam	70	1.6[a]	4.28
153	4Bg1	LB	2.5YR 6/2	2.5YR 4/2	Loamy sand	95	1.8[a]	4.08
190	4Bg2	LB	n.d.	n.d.	n.a.	100	1.9[a]	n.d.
215	4C	LB	n.d.	n.d.	n.a.	100	2.0[a]	n.d.
335+	4Bwg	LB	n.d.	n.d.	n.a.	100	1.5[a]	n.d.

Pedogenetically, this is a complex profile with at least two levels of imprint by soil water (see also Section 4.3.2). The bulk densities of the layers are characteristic, however, very low in the upper layer, moderate in the bipartite intermediate layer, and high in the basal layers. In the deepest part of the profile, the bulk density seems to decrease again. It is unclear whether this is a primary sediment property or due to rinsing out of fines by percolating slope water (*modified from Audorff et al., 2007*).

n.a., not applicable; n.d., not determined; LH, upper layer; LM, intermediate layer; LB, basal layer.

[a]The lower part was cored by motor auger, which prevented laboratory-based measurement of bulk densities. Values were assessed by field methods (AGB, 2005) using measured ones for calibration.

FIGURE 2.3 Stacked basal layers beneath an upper layer in the Steigerwald Mountains, NW Bavaria. The local bedrock is characterized by alternating sandstones and shales. The clay-rich (darker) and sand-rich (lighter) basal layers exactly reflect this up-slope alternation (*photo by A. Kleber*). Geological designations: LH = upper layer; LB = basal layer; kmo = bedrock (Late Triassic sandstone, saprolized). Ruler is 2 m. (For color version of this figure, the reader is referred to the online version of this chapter.)

principally influence the occurrence of intermediate layers (see Section 2.3.2) are also important for its loess content (Friedrich, 1996), though the correlation is not always very distinct. Furthermore, the composition of the intermediate layer varies due to differences in eolian sedimentation rates. In case of low sedimentation rates, the relative share in local debris is enhanced, whereas increased loess sedimentation causes a reduced share in clasts (cf. Sauer, 2004). Another variation in intermediate-layer sediment properties may be derived from reworked, older paleosol material (Bibus, 1985; Bibus et al., 1991; Terhorst and Felix-Henningsen, 2010).

The orientation of rock fragments in intermediate layers may vary largely, with a tendency for longitudinal axes being aligned parallel to the slope, though often not or not as clearly expressed as in the basal layer (Völkel, 1995; Völkel and Raab, 1999). Mainly platy rock fragments (e.g., shale, limestone) show such an alignment, but it may also been observed from gneiss fragments.

FIGURE 2.4 Stacked intermediate layers between the upper and the basal layer, Franconian fault-block Mountains, NE Bavaria. The disconformity within the intermediate layers is located approximately where the ruler (20 × 20 cm) is. It is well visible due to an abrupt color change from more brownish (above) to lighter with whitish, wedge-shaped inclusions (*photo by A. Kleber*). Geological designations: LH, LB as in Fig. 2.3, LM = intermediate layer, So2 = bedrock (Early Triassic sandstone of loose consistency with purple cast). Soil horizons after Soil Taxonomy (SSS, 2010). (For interpretation of the references to color in this figure legend, the reader is referred to the online version of this chapter.)

Intermediate layers typically have bulk densities between those of the basal and upper layers in the range of 1.4–1.6 g cm^{-3} (Table 2.2). Intermediate layers sometimes exhibit a platy fabric.

Intermediate layers may consist of more than one sublayer as well (cf. Sauer, 2002). Such units are mostly similar to one another (Fig. 2.4). Often the clast content is somewhat higher, the deeper the layer is positioned in a stacked intermediate-layer succession (Kleber et al., 1998b; Völkel, 1995).

2.2.3 The Upper Layer

The upper layer is mainly defined in terms of thickness, its position in the profile, and its spatial distribution (Section 2.3.3), whereas its sedimentological properties are similar to those of the intermediate layer; that is, coarse silt is typically enriched compared to basal layers and underlying rocks. The upper layer contains a mixture of local, bedrock-derived material and eolian fines. The loess content depends on the same parameters as in intermediate layers.

In areas covered by tephra of the Late Glacial eruption (12.9 ka, Schmincke et al., 1999) of Lake Laach Volcano (Laacher See, cf. Fig. 2.1), traces of tephra may be found in upper layers. Distal to Lake Laach, heavy-mineral analysis is needed to identify tephra constituents (Table 2.3), and the search for tephra constituents will entirely fail farther away. In proximal areas, in contrast, tephra is often recognized in the field, mainly by the allophone-field test according to Fieldes and Perrott (1966) as applied by Sauer (2002) or by the so-called greasing effect, which is assumed to be due to clay minerals derived from volcanic glasses, such as allophanes and imogolites; through greasing, soils set water free under moderate pressure even under apparently dry condition. However, there are other constituents in soils that may produce the same effect, as greasing attributed to poorly crystallized iron oxides occurs in mountains far away from tephra input (Mahr, 1998; Mahr and Völkel, 2000). Very close to Lake Laach, tephra may form a discrete layer just beneath the upper layer (see Fig. 2.23 later in this chapter).

Upper layers usually contain coarse clasts. Several authors claim that they are typically richer in clasts than underlying intermediate layers (see Section 2.5.5).

TABLE 2.3 Heavy Minerals in Periglacial Cover Beds of the Spessart Mountains, Hesse (*Modified from Semmel, 2005*)

	Upper layer	Intermediate layer	First basal layer	Second basal layer
Pyroxene	15	3	–	–
Brown hornblende	45	1	–	–
Sphene (titanite)	5	–	–	–
Garnet	2	3	–	–
Rutile	2	1	3	3
Topaz	2	–	–	–
Tourmaline	17	47	79	84
Zircon	12	45	18	13

Numbers are in percentages; –: not contained. Beyond field evidence and marked differences in grain size (Semmel, 2005), the minerals diagnostic for Lake Laach tephra (printed in italics, see also Poetsch, 1975) allow to differentiate between upper layer, intermediate layer (which contains just a few, probably translocated grains), and two basal layers, derived from up-slope sandstone and underlying mudstone, respectively.

A striking feature of numerous upper layers is that they contain—proportionally to the amount that the source rock delivered—more coarse to very coarse clasts than fine or medium ones (Völkel, 1995), a feature that has been observed to be formed currently in active stone pavements (Dietze and Kleber, 2011).

Upper layers are sedimentologically uniform in most cases but may display some concentration of clasts in their lower or—rarely—upper parts so that they show some grading in the clast content. Rock fragments directly above the lower boundary of the upper layer are explained by sorting/sinking processes in the former active layer, or they are interpreted as a residue of older periglacial layers or stone pavements (Kösel, 1996).

In general, the occurrence and orientation of clasts vary due to geological and geomorphological conditions. The clasts' longitudinal axes may be oriented parallel to the slope or parallel to a fossil paleotopography (Bibus, 1985; Bibus et al., 1991; Mailänder and Veit, 2001). Their orientation may be decisively influenced by cryoturbation processes as well. Furthermore, in the case of remarkably random orientation of clasts, influence of humans or bioturbation cannot be excluded.

One of the most conspicuous attributes of upper layers is their very low bulk density of usually $<1.2\ g\ cm^{-3}$, often around or below $1.0\ g\ cm^{-3}$ (Table 2.2), making them ideal substrates for plant roots. This property is so typical that it is often regarded as diagnostic for upper layers.

2.3 DISTRIBUTION AND THICKNESS OF LAYERS

A. Kleber, T. Scholten

2.3.1 The Basal Layer

Basal layers are widespread in subdued mountains of Central Europe. They cover essentially all relief of subdued mountains far down into the basins and are rarely missing in any position of the relief. However, there are a few exceptions: basal layers are often absent or just very thin on pronounced slope knickpoints, that is, where plateaus bend into upper slopes. Occasionally, basal layers are also absent on plateaus or on other wide, very flat relief.

Many sources describe occasional thickening of basal layers (Sauer, 2002; Sauer and Felix-Henningsen, 2006; Semmel, 1968; Völkel, 1995). Information on layer thickness is often difficult to obtain, because 1-m hand-coring augers cannot reach the base of such thick deposits (see Pfeiffer, 2002). As seen from Fig. 2.3, the basal layer may comprise more than one layer, each typically having a thickness around 50 cm (Scholten, 2003), which fact may distort coring results if the second basal layer is confused with *in situ* saprolite (cf. Dietze and Kleber, 2010). Thus, one may expect basal layers being typically underestimated in the literature in terms of thickness. Digging or motor augering may yield more accurate data but comes with much additional expenditure. With the advent of geophysical sounding (Gerber et al., 2007; Kleber and

Schellenberger, 1998; Sauer and Felix-Henningsen, 2004; Völkel et al., 2001), there is a chance of improvement in the data situation in the near future. Basal layers tend to smoothen any preexisting relief, that is, they infill rough parts of older landscapes. This often makes their real thickness pretty unpredictable. However, most often these thickened sections are found in a few, typical positions of the relief: they prevail in areas of transition from steeper slope sections to foot slopes, on slope flattenings, and in slope depressions (slope depressions are often tagged as "slope hollows" in the literature), where thicknesses up to 6 m have been reported. Many slope depressions once were deeper than they are at present, with the basal layer veiling much of the former relief (see Figs. 2.5 and 2.6).

2.3.2 The Intermediate Layer

The intermediate layer is the least far distributed of the three periglacial cover beds. Its distribution is part of the diagnostic criteria for this layer according to AGB (1994, 2005): thereafter, intermediate layers are restricted to where they were sheltered from later erosion, that is, to positions of the relief where there was least overland runoff which could have eroded the deposit. This claim has been adopted by almost the entire literature without critical reflection and will be discussed in Section 2.6.2. A typical situation, where intermediate layers occur, is depicted in Figs. 2.5 and 2.6. The relief is a slope depression. Wherever the topography provides slope depressions without perennial channels, they are the most likely sites to find intermediate layers (Fig. 2.7). In many cases, the intermediate layer occurs in relation to paleodepressions or paleoconcavities on slopes, which have successively been filled during periglacial times. These geomorphological positions are often not easily detected in the present relief. In slope depressions with deeply incised channels, which are usually attributed to man-induced soil erosion (Semmel, 1996a), intermediate layers have been eroded just in these channels.

The thickness of intermediate layers usually reaches not more than 50 cm. However, some intermediate layers are divided in sublayers and may be more than 1 m thick then (Kleber and Stingl, 2000; Sauer, 2002; see Fig. 2.4).

2.3.3 The Upper Layer

Upper layers cover essentially the entire landscape, provided they have not been eroded since their deposition. Bedrock outcrops and valley floors are the only landscape elements where they typically are absent.

The most important diagnostic property of upper layers is their astonishing constant thickness of 50 ± 15 cm (2σ). This thickness is not apparently influenced by slope angle, slope aspect, or relative position within the relief (Bibus, 1985). Hence, this layer essentially is a parallel veil of the landscape.

The only known relationship between layer thickness and any of the common triggers of geomorphic processes is that the layer thickness correlates

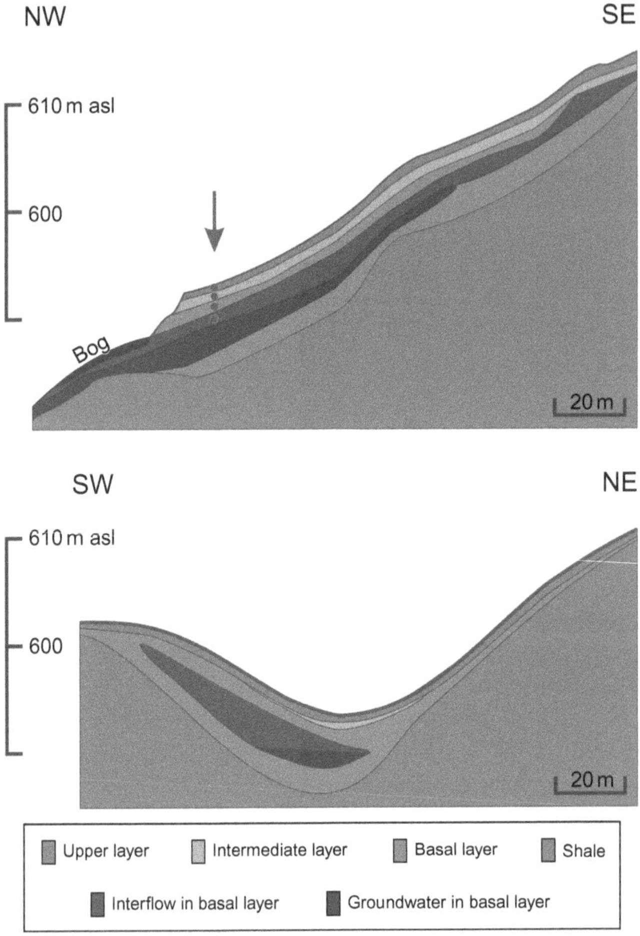

FIGURE 2.5 Thickness variation of layers in a slope depression, revealing steepness of a buried landscape that existed before the deposition of the basal layer, Frankenwald Mountains, North Bavaria (*modified from Kleber and Schellenberger, 1999*). Longitudinal (top) and cross (bottom) section from motor augering and geoelectrical sounding. Note: exaggeration of heights pretends a closed depression in the old relief, which actually does not exist. The components of the hydrological system also depicted in this figure will be referred to in Section 4.3.2. (For color version of this figure, the reader is referred to the online version of this chapter.)

somewhat with its texture (see Section 2.5.4). This means that upper layers are usually a bit thicker if the substratum is sandy (typically 60 cm), and somewhat less thick on silt- (50 cm) or clay- (40 cm) dominated materials (Scholten, 2003; Semmel, 1968; Semmel and Plass, 1965). Since silt is a diagnostic component of upper layers; extreme values, which might be expected in pure sands or clays, are not encountered.

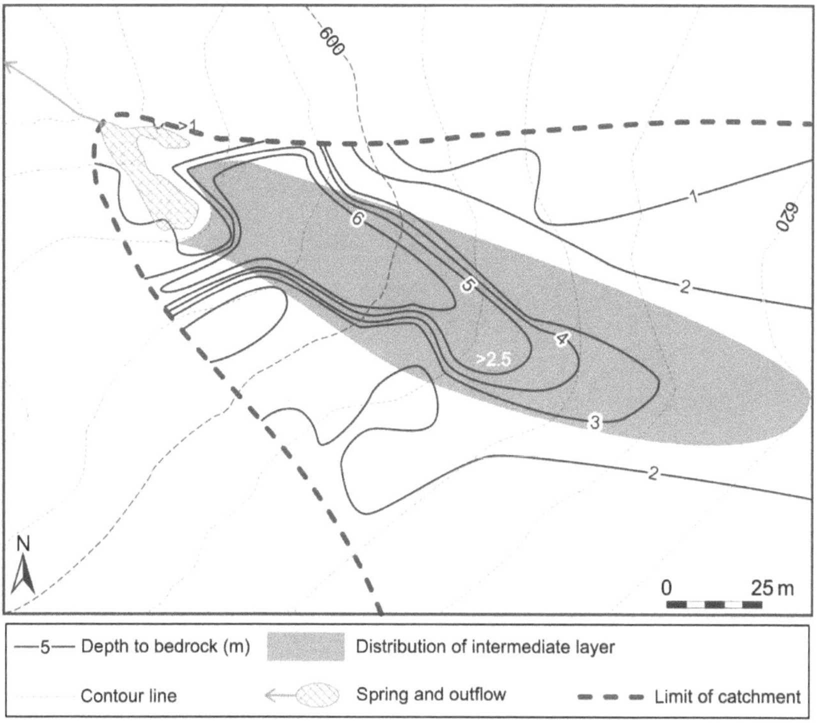

FIGURE 2.6 Intermediate- and basal-layer distribution in a slope depression, Frankenwald Mountains, North Bavaria. The figure depicts both layers in the central part of a small spring catchment, which is formed as a slope depression (*modified from Kleber and Schellenberger, 1998, 1999; Kleber et al., 1998b*). The basal-layer thickness is given in iso-lines; thicknesses of up to 4 m were determined by a coring motor auger, those exceeding 4 m are derived from geoelectrical sounding. For the intermediate layer, not the thickness but just the distribution is given because the thickness does not exceed 50 cm even in the center of the slope depression.

2.3.4 Overview of the Distribution of Periglacial Cover Beds

An attempt to simplify the complex distribution patterns of periglacial cover beds as they are found in most subdued mountains of Central Europe is given with Fig. 2.8. Though rather crude, this scheme may serve as a useful first approximation to the spreading of cover beds. It depicts not only the thickening of basal layers in foot-slope positions but also the smoothening of any older relief, which used to be rougher before the basal layer was deposited. Another typical attribute of the distribution of basal layers is their absence near crest lines and at sharp, convex slope bends.

The distribution of intermediate layers is controlled by relief forms. Concave slopes, especially concave with regard to plan curvature (i.e., slope depressions, difficult to depict in a two-dimensional figure like this), are most appropriate deposition centers for intermediate layers. However, there is also

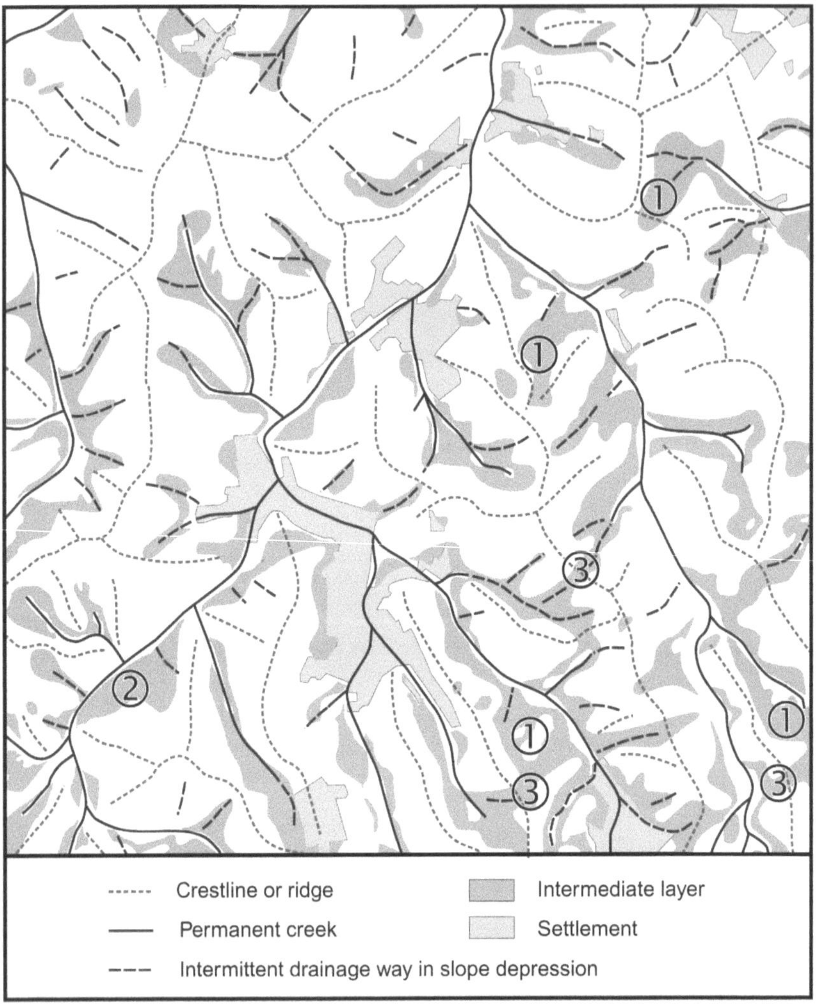

FIGURE 2.7 Intermediate-layer distribution dependent on landforms in the Oberreifenberg area, Hesse, depending on landforms. Gray = intermediate-layer distribution (*modified from Friedrich, 1996, map 1*). By far the most occurrences of intermediate layers are located in slope depressions, mostly regardless of the slope aspect, with some exceptions: Larger occurrences of intermediate layers outside slope depressions are mostly on east-facing slopes. The only occurrence on a west-facing slope is a foot-slope area which, however, is close to crestlines in the west that may have acted as wind shield. Intermediate layers usually do not occur on crestlines or ridges, except where two occurrences align from two sides crossing a low, intersecting saddle. (For color version of this figure, the reader is referred to the online version of this chapter.)

a tendency of intermediate layers to be more frequent on lee than on luff slopes (see Section 2.6.2). Upper layers are distributed even farther than basal layers, but they do not remarkably vary in thickness.

Because of the relationships between relief parameters and the spatial distribution of cover beds in mid-latitude subdued mountain ranges, which are

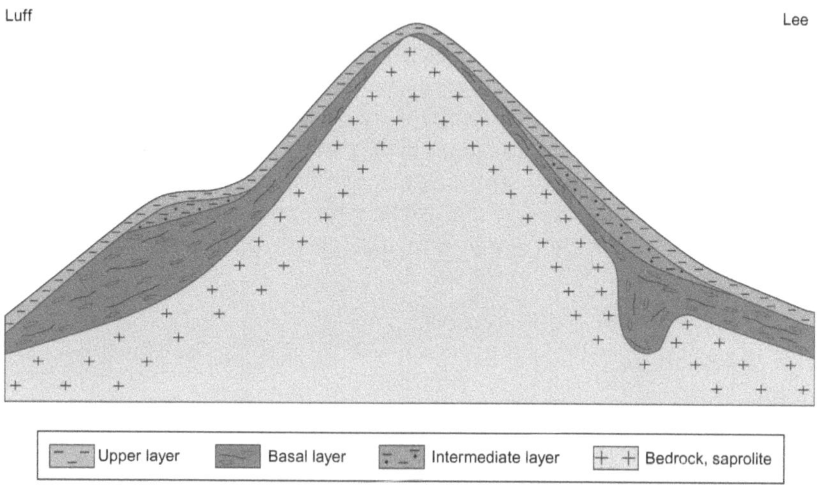

FIGURE 2.8 Schematic distribution of cover beds in Central European subdued mountains. The figure is not to scale because it is valid at various scales: ridges of several tens of meters relative height may exhibit essentially the same distribution patterns as entire mountain ranges do. (For color version of this figure, the reader is referred to the online version of this chapter.)

obvious at least on the landscape scale, modeling is a promising approach to the forecasting of cover-bed distribution. Most studies agree that the properties of periglacial cover beds are highly correlated to characteristics of underlying rocks. The evaluation and processing of digital elevation models and of geological maps thus allow for GIS-supported modeling of the spatial distribution and the properties of cover beds. First attempts have been undertaken by Friedrich (1996) and Scholten (Scholten, 2003; Scholten et al., 2001, 2002). The authors derived algorithms that describe this correlation in a statistically significant way, based on geomorphometric relief parameters and the petrographic composition of the underlying rocks (Scholten, 2003; see Section 2.4). The derived relationships and combination rules permitted rather precise predictions of thickness, texture, and clast content of cover beds in some Central European mountain ranges. Calculations of this relief/geology model exhibited deviations of ±10 cm for mean total thicknesses. Predictions of texture and clast content of the multilayered cover beds deviated from measured values by less than 30%. Dietze and Kleber (2010) included a modified flow model into cover-bed prediction to account for the fact that clasts in cover beds have been dislocated from their original position.

Another important step in cover-bed research is the introduction of nondestructive geophysical techniques that allow high precision as well as coverage of large areas. Sauer and Felix-Henningsen (2004) and Gerber et al. (2010) used ground-penetrating radar as a tool for nondestructive soil-depth and cover-bed mapping and underpinned the close link between the spatial distribution of cover beds and slope morphology in the Lahn-Dill Mountains (Hesse). Similarly, Kleber and Schellenberger (1998) used geoelectrical sounding in the Frankenwald Mountains. The distribution of cover beds interpreted from these data clearly follows the slope curvature: especially intermediate layers and anthropogenic layers appear linked to landform parameters, as they are connected to concave parts of the slope, whereas upper and basal layers cover most landscape positions mainly independent from topographical factors. However, cover beds are generally thinning on convex slope sections, whereas an increased thickness of especially the basal layer is typical at footslope positions where profiles with multiple basal layers are common (see Section 2.3.1).

2.4 CLASSIFICATION ISSUES

M. Leopold, A. Kleber

2.4.1 Discrimination Among Pleistocene Cover Beds

Van Steijn et al. (1995) discussed various types of periglacial deposits and provided criteria for their discrimination. In Central European subdued mountains, most of these types are not common, so that we may restrict our discussion to classification difficulties of a very few deposits, which are not even necessarily periglacial. However, an important source of confusion involves basal layers and glacial till, which may have several properties in common. The measurement of clast orientations has turned out to be an excellent tool for discriminating among the two (see Section 2.8.1, in particular Fig. 2.28).

Since the definitions of the various layers have some weaknesses, as outlined in the previous two sections, an obvious problem is keeping layers apart from one another. These weaknesses appear inevitable at the current stage of knowledge, considering that the layers are diamictites, which fact induces a wide range of physical, chemical, and mineralogical properties.

At first sight, distinguishing between basal layers and layers with an admixed loess component, "loessic layers," appears to be the most trivial discrimination. However, this may be complicated if the basal layer contains remarkable amounts of silt delivered from bedrock, such as some shales or gneisses do, so that the diagnostic loess constituents may not be determined via texture examination. Neither field-derived texture estimates nor grain-size analysis—usually the means of first choice for distinguishing these two layer types—is able to assist then. In such cases, it needs to be assured whether the upper of, let's say, two layers is a loessic intermediate or just a silty basal layer. This often demands for more elaborate laboratory analyses. Especially, heavy-mineral analyses of very fine sand have turned out effective (Table 2.4), though there are regions where the local loess component has a similar mineralogical composition as the bedrock (Völkel, 1995). Technically more demanding, the silt fraction may be analyzed as to its heavy-mineral contents (Thiemeyer and Veit, 1993), or clay-mineral analyses may be utilized to identify eolian constituents (Dietze and Kleber, 2010; Völkel, 1995; Völkel et al., 2002a).

TABLE 2.4 Heavy Minerals in Cover Beds in the Frankenwald Mountains, North Bavaria (*Modified from Kleber et al., 1998b*)

Mineral	LH	LM	LB
Augite	–	2.3	–
Brookite	–	0.7	–
Chlorite	67.2	57.4	81.2
Epidote	1.7	2.3	–
Green hornblende	24.7	25.3	2.1
Muscovite	4.6	7.6	16.2
Titanite	0.6	2.2	0.5
Tourmaline	0.6	–	–
Zircon	0.6	1.5	–
Others	–	0.7	–
Σ Grains	174	131	191

Numbers are in percentages. The profile is the same as in Table 2.2. The underlying shale delivers large amounts of silt. Therefore, eolian influx cannot be ascertained from the silt content but calls for laboratory testing. Heavy-mineral analysis is often the most appropriate approach. The only minerals (just a few of the individual bedrock-derived grains have sufficient specific weight at all) known from local rocks are chlorite and muscovite (Dahlke, 1984). The basal layer, indeed, contains almost no other minerals. Very small amounts of other minerals may be due to later contamination. The intermediate layer shows a mineralogical variety which cannot be explained by the composition of the underlying rocks in the area. Since it also differs from the upper layer, the results indicate that each layer has a different source contributing to its composition, with the major hiatus being between the basal and the intermediate layer. See also Table 2.3 for heavy-mineral analyses of a profile with Lake Laach components.

However, there is a field approach, which has proven a valuable tool in identifying intermediate layers, though it clearly involves the danger of circular reasoning. Many silt-bearing types of bedrock, especially those belonging to the crystalline basement, constitute soils which are rather acid from their early beginning. Thus, they never reached a pH range high enough to allow for clay translocation (see Birkeland, 1999). Loess, on the other hand, typically contains minerals having a larger buffering capacity. Thus, loess admixed to other parent materials may raise the pH value of an incipient soil to a higher level than bedrock alone could do. For some time—until soil acidification ceases the process—this level of pH may enable clay translocation. If so, many soils have developed argillic properties, though sometimes extremely weak. In the field, clay cutans are archives of this pedogenic process. They allow conclusions regarding the former buffering component, even if the present soil is much more acid. Clay cutans in a substrate that would be acid without loess constituents have turned out to be valuable indicators, though their absence does not necessarily disprove loess admixture.

Keeping intermediate and upper layers apart is also simple—at least in theory. As their properties are similar or—at the extreme—even almost identical (see Sections 2.2.2, 2.2.3, and 2.5.5; Völkel, 1993, 1994), the intermediate layer is just the second loess-bearing layer in a succession starting from the surface. Actually, this simple approach often fails, because the upper layer (1) may be absent due to soil erosion or other human intervention; (2) may be overlain by anthropogenic soil sediment attributable to up-slope soil erosion (see subsequent section), which may be confused with an upper layer, giving rise to erroneously classifying the upper as an intermediate layer; or (3) may be entirely substituted by anthropogenic soil sediment. Evidently, a major task is the identification of anthropogenic influence and especially soil-erosion sediments to clarify which is the upper layer if there is one at all. Some references even argue (4) that the upper layer may be split into two rather similar layers (e.g., Kösel, 1996; Rohdenburg, 1968), but this would imply that there are independent handles to keep upper layers apart from intermediate ones with certainty, which is actually not the case.

2.4.2 Anthropogenic Layers

2.4.2.1 Genesis and Classification of Anthropogenic Layers

Anthropogenic layers are created by processes that are either directly or indirectly linked to human activity. Processes directly linked include excavation and re-deposition of soils and sediments during construction works in and around settlement areas. These sediments do not qualify as, and usually cannot be confused with, cover beds because of their restricted spatial distribution, though they have an increasing proportion in the overall global budget of allocated soil material (Hooke, 2000; Zalasiewicz et al., 2008).

However, soil-erosion sediments have many properties in common with periglacial cover beds, which fact may cause confusion. Modern agriculture currently forms anthropogenic layers throughout the world by plowing fields, which causes tillage erosion (Imeson et al., 1980; Kwaad and Mücher, 1979). Besides plowing, artificial deposition of material, mainly for manure processes, is known from vineyards or in areas with long-term application of plaggen manure deposition (Blume and Leinweber, 2004). Due to the long settlement history in Central Europe, anthropogenic layers may vary in age by several thousands of years (e.g., Bork et al., 1998; Giessübel, 1977; Leopold, 2003; Niller, 1998), though they certainly are younger than the periglacial cover beds.

Processes indirectly linked with human activity establish the preconditions to allow gravity, water, and wind to either initiate erosion or to intensify effects of human activities such as settling, clearing, mining, grazing, or farming. All these activities open landscapes and destroy covering vegetation and, thus, the protection from natural erosion (Verstraeten et al., 2009).

As the genesis of anthropogenic layers is linked to anthropogenic settlement and the use of the surrounding nature, they occur throughout Central Europe. Depending on historic and prehistoric land-use patterns, the spatial distribution of anthropogenic layers may be notably different from the modern one (Houben, 2011). Often they are associated with archeological findings such as ceramics, agricultural tools, and accumulations of charcoal pieces. However, especially in forested areas of the lower elevated subdued mountains, old land-use techniques like grazing cattle in the forests or collecting organic matter, mainly litter, from forest soils produced anthropogenic layers without accompanying archeological remains (Dotterweich, 2008; Raab and Völkel, 2005; Semmel, 1993a; Stolz and Grunert, 2010). These activities set the necessary preconditions for erosion and the formation of anthropogenic layers even in nowadays densely vegetated forests. Mainly finer fractions were eroded from the upper layer and deposited on top of the periglacial layers. Figure 2.9 gives an example of a 2500-year-old

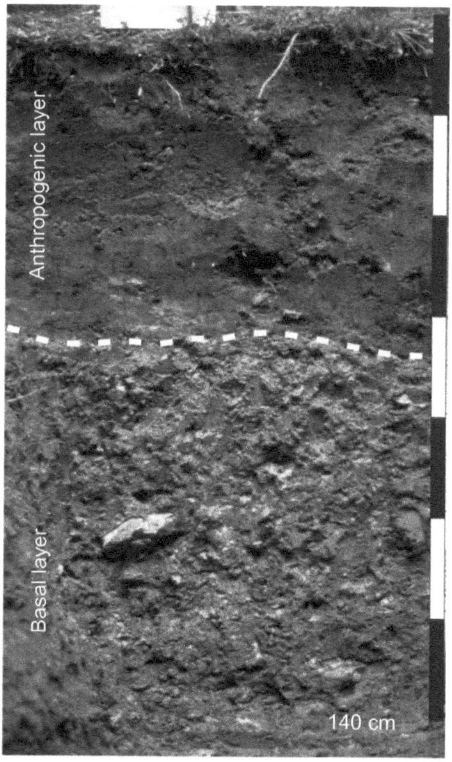

FIGURE 2.9 Anthropogenic layer on top of a well-developed basal layer in the Vils River Valley, Germany. Note the sharp sedimentary boundary between the two layers at 65 cm. Radiocarbon ages of integrated charcoal pieces in the anthropogenic layer yielded an age of around 2.5 ka (Raab, 2005) (*picture by M. Leopold*). (For color version of this figure, the reader is referred to the online version of this chapter.)

anthropogenic layer on top of a well-developed basal layer. Both sediments are separated by a pronounced disconformity, indicating their different geneses.

2.4.2.2 Properties of Anthropogenic Layers

Sedimentological properties of anthropogenic layers derived from soil erosion are strongly connected with substrate properties of their source area. Anthropogenic soil sediments often may be correlated with up-slope source materials. As these sediment sources may vary in texture so do the anthropogenic layers. They may have a fine to coarse texture and may contain clasts even up to boulder size. However, as surface runoff is the main mode of transportation, large clasts are very rare, and a sand- or silt-rich texture is much more common.

Many studies provide evidence for re-deposition of eroded soil horizons in an upside-down sequence (e.g., Bork et al., 1998; Lang and Hönscheidt, 1999; Leopold, 2003; Niller, 1998). Former A–B–C successions of soil horizons in up-slope areas may be reverted to C-B-A-equivalent soil sediments at foot- and toe-slope positions by soil erosion (Fig. 2.10). Due to the water-driven character of the processes and to interruptions of the processes by agrotechnical means, anthropogenic layers often become stratified during sedimentation. As they rarely

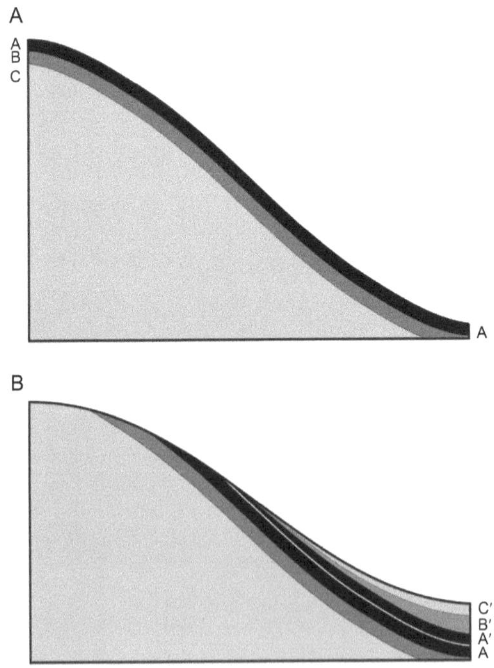

FIGURE 2.10 Schematic cross-section of anthropogenic layers reworked by soil erosion. (a) Original situation with normal soil-horizon sequence A–B–C; (b) situation after reworking with up-slope truncation of soil and deposition of A′ material derived from the A horizon eroded first, B′ from B, and C′ from C eroded last.

accumulated by a single, large event but rather by many smaller ones, individual sedimentary strata typically are rather thin. However, over the decades, bioturbation and subsequent plowing may destroy the original stratification, and anthropogenic layers thus become transformed into an unsorted and nonstratified sediment (see French, 1992), often quite similar to certain cover beds.

The different modes of transportation and re-deposition along slopes, including temporary caching, induce changes in texture distribution depending on slope position. Sediment erodibility in the source area plays an important role. The different volumes of anthropogenic layers may be the result of considerable differences in erodibility of the various sediments sources (see Butler, 1959; Pesci and Richter, 1996). However, the kind of vegetation of the depositional area also plays a crucial role. A rough deposition surface like a dense grassland is much more effective in storing such sediments than a smooth one such as a partly degenerated grassland due to pasture (Leopold and Völkel, 2007).

Textural data may be used to identify facies of anthropogenic layers and provide clues on their origin (Fig. 2.11). The differences of four facies, which are all derived from one particular loess source, cannot fully be explained by different soil horizons as sources as suggested in Fig. 2.10. This holds especially true for the clayey (no. III) and very sandy facies (no. IV). These facies may be explained by differences either in event intensities or in properties of depositional areas. Heavy slope-wash events may winnow fine from coarser material and convey the former into ephemeral gullies (Boardman, 1992), leaving on the slope the latter, constituting facies no. IV. During a series of weak events, coarse material may be deposited on the foot slope, whereas the finer and lighter

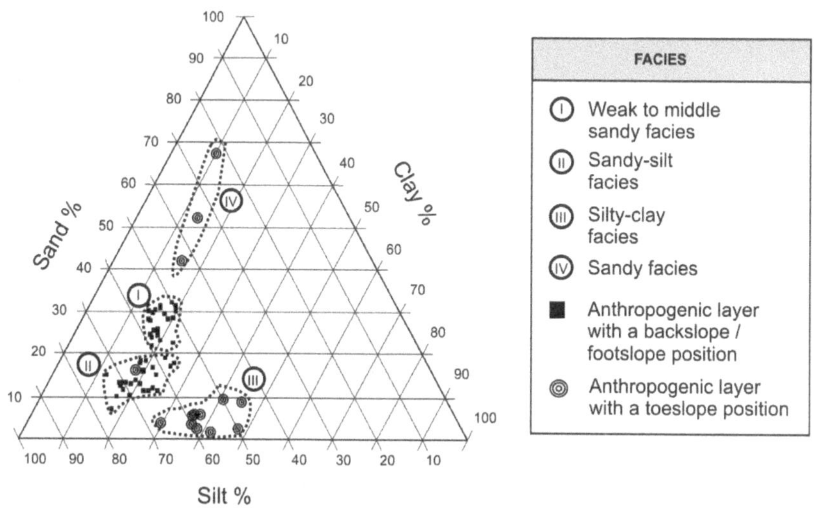

FIGURE 2.11 Triangular diagram of the texture distribution of anthropogenic soil sediments at a loess slope, Loess Rolling Hills, SE Bavaria, Germany. Note the various facies caused by different depositional and weather situations during their genesis (*modified from Leopold and Völkel, 2007*).

materials get transported to the toe slope. However, the textural separation could also be caused by a rougher surface at the foot slope which catches all the coarse sediments even during heavy slope wash.

Anthropogenic layers may be found wherever there is some sloping relief. They dominantly cover foot and toe slopes but, depending on the predepositional relief, occasionally fill former depressions on mid- and up-slope positions. Niller (1998) introduced a cascade-like erosion and deposition model for anthropogenic layers, which is crucial to know for any dating approach to these layers (Fuchs and Lang, 2009). Whenever the depositional material reaches the floodplains of the rivers, it may interfinger with fluvial deposits and sets new challenges in dating (Fuchs et al., 2010).

2.4.2.3 Discrimination of Anthropogenic Layers from Periglacial Cover Beds

At a study site of about 0.16 km^2 in the headwaters of the River Dill catchment near Giessen (Hesse), a mainly forested area, the spatial distribution patterns of sediment indicate small-scale alternations of upper layers and anthropogenic layers at different relief positions (Fröhlich et al., 2005). This is mainly indicated by artifacts and remnants of charcoal in scattered soil pits as well as by a variety of anthropogenic molds that correspond to the historic land-use system, which have developed already in Medieval times. The size of the uphill watershed area correlates positively with the thickness of the anthropogenic layer on a particular slope in their study area. Slope curvatures correlate with local mass balances in terms of divergence of sediment flux and acceleration and deceleration of mass dislocation. At map scales of around 1:25,000, no influence of the relief on the upper-layer thickness may be observed. However, the study of Fröhlich et al. (2005) suggests that historic land use may strongly influence small-scale patterns of the spatial distribution of upper and anthropogenic layers. Their findings clearly indicate the importance of identifying anthropogenic layers for any unequivocal classification of cover beds.

Cultivation leaves traces beyond sediments, which may be utilized to recover that there was agriculture at all. The most common indicators are agrarian terraces, which usually develop where two fields or plots had been separated by, for instance, a hedge for some time. Accumulation of anthropogenic soil sediment was enhanced up-slope and soil erosion down-slope of this barrier. This induced a small terrace-like terrain modification, which may be well evident in the field. Unfortunately, many of these terraces in modern forests stem from Medieval or even earlier times and, thus, are partially leveled and recognized only after meticulous surveillance. Forest-border terraces are signs of former agriculture more easily to identify (though less frequent): After cultivation, soil erosion removed the upper materials from the field and exposes deeper layers, whereas in the nearby forest, the surface remained protected (Kleber et al., 1998a). In firm materials, the resulting step is usually steep and well noticeable. Another frequent, easily identified indicator of former agriculture are gullies carved into the forest soils, especially into cover beds (see Semmel, 1996a; Thiemeyer, 1988). Usually, these formed at the time when farming was abandoned, though this indicator is not unambiguous, especially in higher elevated subdued mountains characterized by a subcontinental to continental climate, where vegetation disturbances due to needle ice may produce similar effects (Beierkuhnlein and Kleber, 1999). If all these indicators are not found, overprint of the landscape due to historical agriculture seems unlikely. Otherwise, profiles must be examined thoroughly as to the occurrence of anthropogenic soil sediment.

Anthropogenic layers may show a similar loose consistency and may have the same thickness as upper layers. Thus, both are easily confused. Usually, anthropogenic soil sediments show better sorting than upper layers. However, plowing at a time when the anthropogenic soil sediment was thinner may have brought clasts into higher horizons, rendering this indicator uncertain. Anthropogenic soil-sediment materials have undergone nutrient withdrawal though agriculture at a time when nutrients were rarely re-supplied through fertilization, provided cultivation lasted sufficiently long. Therefore, pedogenesis in anthropogenic soil sediment has often developed remarkable podzolization, whereas uncultivated cover beds usually have sufficient buffering capacity to prevent this from happening. Thus, pedogenesis may often serve as an indicator for discrimination. However, this criterion does not apply to all anthropogenic soil sediments. Semmel (2002a) described podzolization as well as weak clay translocation in anthropogenic soil sediment. Frequently, anthropogenic layers are spangled with stains and tinsels of charcoal, whereas natural charcoal usually is identified by different distribution patterns. However, charcoal has also been found in clearly identified periglacial cover beds under a forest of intensive use since centuries, probably accompanied by charcoal burning (Dietze, 2006); apparently, charcoal may be washed into cover beds, and syn-depositional admixture to anthropogenic layers is not the only incorporation path for charcoal.

If all these criteria fail to discriminate anthropogenic layers from periglacial cover beds, intense mapping of the distribution and thickness of the respective sediment may provide clues, because anthropogenic soil sediments typically change in thickness depending on slope inclination and former parcel boundaries, and they wedge out somewhere up-slope in the domain of soil erosion. There is no average thickness of anthropogenic layers, but in general they become thicker along the slope and reach their maximum thickness at the transition from the foot to the toe slope, varying from several decimeters up to several meters depending on the sediment source area and the overall duration of their formation. This often is a major difference to the rather constant upper-layer thickness and may, thus, be used to differentiate these layers in the field. In areas with thick anthropogenic layers at the base of a slope, up-slope occurrences of periglacial cover beds often show considerable truncation due to soil erosion. Often, sedimentation of the anthropogenic layer prograded backward from foot slopes to higher areas so that soil sediments became deposited on top of eroded, truncated cover-bed profiles and may substitute entire layers or layer successions.

2.5 STATISTICAL APPROACH TO LAYER PROPERTIES AND DISTRIBUTION

T. Scholten, M. Dietze, A. Kleber

2.5.1 Approach

Scholten (2003) made an attempt at forecasting layer properties and thicknesses of periglacial cover beds using a GIS-aided approach (see also Behrens, 2003; Dietze and Kleber, 2010). For this analysis, he could utilize field and lab data of more than 900 soil pits from three subdued mountain ranges: the Harz, Solling, and Lahn-Dill Mountains (Fig. 2.1). The Harz Mountains are underlain by metamorphic and magmatic rocks. The profiles included in the study comprise an elevational range from about 170 to 1150 m asl. The Solling Mountains are underlain by sedimentary, mainly psammitic but also some pelitic rock types. The profiles are elevated between 70 and 530 m asl. The Lahn-Dill Mountains mainly consist of weakly metamorphic shale, with the profiles being elevated between about 150 and 680 m asl. A major difference between the study areas was that in the two first-mentioned mountains, profiles from all relief units were included, whereas in the third one the focus was on steeper slopes, which thus are overrepresented in this area.

Scholten (2003) presented characteristic measures and relationships between sedimentological data and relief/bedrock parameters, using correlation and variance analysis. Here we report some of the results in the form of statistical data such as arithmetic mean, minimum, maximum, and standard deviation (Tables 2.5–2.7). Furthermore, we discuss correlation results that are significant ($p < 0.01$) and of which the Pearson's correlation coefficient (r) is better than 0.5, except where stated otherwise (then r is given in parenthesis). The same holds true for results of variance analyses: significant effect sizes of η^2 (expressed as explanation of the variance in %) better than approx. 0.3 are reported. The morphographical indices referred to are—besides self-explaining ones—"convexity index," a measure of the radius of slope curvature, "topographic moisture index," a measure for the hydrologically contributing area (see Dietrich and Montgomery, 1998), and the "topographic height," which is the height above the associated valley floor. For correlation and variance analyses, data from the Harz and Solling Mountains have been pooled.

2.5.2 Basal-Layer Properties

The average basal layer in the study areas is approx. 50 cm thick. Thereby it is usually the thickest of the periglacial layers. However, with a standard deviation of 20 cm or more, it is also the most variable one regarding thickness. The average basal layer in the eastern Harz Mountains has a 50% clast content (estimated

TABLE 2.5 Statistical Measures of Layers, Eastern Harz Mountains (*Modified from Scholten, 2003*)

Parameter	*n*	Min.	Max.	Mean	Std. deviation
Basal layer					
Thickness (cm)	470	10	160	56	24
Clay content (%)	408	3	77	18	9
Silt content (%)	408	11	78	42	13
Sand content (%)	408	3	80	40	14
Clast content (%)	470	5	90	49	20
Intermediate layer					
Thickness (cm)	259	10	95	45	17
Clay content (%)	227	8	42	23	7
Silt content (%)	227	26	83	61	9
Sand content (%)	227	0	58	15	9
Clast content (%)	259	0	45	9	7
Upper layer					
Thickness (cm)	472	20	70	44	10
Clay content (%)	408	5	40	18	6
Silt content (%)	408	19	85	63	10
Sand content (%)	408	1	70	18	10
Clast content (%)	472	0	55	11	10
All layers					
Total thickness (cm)	483	35	250	121	35

n, number of cases; min., minimum; max., maximum; mean, arithmetic mean; std., standard.

by volume in the field, see Birkeland, 1999). Sand and silt comprise some 40% each, and clay less than 20% (Tables 2.5 and 2.6). In the Lahn-Dill Mountains, the basal layers are thinner and richer in clasts, probably reflecting the steeper slopes where the profiles were positioned (Table 2.7).

Basal-layer thickness correlates negatively with convexity ($r = -0.31$) and marginally with some other morphometric indices (topographic moisture index and topographic height, all $r \leq 0.2$). Most variance in basal-layer thickness is explained by the topographic moisture index, topographic height, and curvature with the largest effect size (35%).

With regard to field evidence (Section 2.3.1), one may doubt whether each profile utilized for thickness determination in this study was dug sufficiently

TABLE 2.6 Statistical Measures of Layers, Eastern Solling Mountains (*Modified from Scholten, 2003*)

Parameter	*n*	Min.	Max.	Mean	Std. deviation
Basal layer					
Thickness (cm)	142	10	95	48	20
Clay content (%)	142	3	35	12	8
Silt content (%)	142	18	66	30	13
Sand content (%)	142	18	80	59	17
Clast content (%)	142	1	90	57	25
Intermediate layer					
Thickness (cm)	66	5	85	38	22
Clay content (%)	66	9	24	19	3
Silt content (%)	66	25	79	69	11
Sand content (%)	66	8	65	13	11
Clast content (%)	66	1	70	10	12
Upper layer					
Thickness (cm)	195	25	75	53	9
Clay content (%)	195	4	24	13	3
Silt content (%)	195	23	79	63	18
Sand content (%)	195	8	70	25	18
Clast content (%)	195	1	40	6	7
All layers					
Total thickness (cm)	156	35	208	117	33

See Table 2.5 for explanation.

deep to identify stacked basal layers (see Fig. 2.3 with layers deeper in the profile than typically reached by auger or hand digging). To rule this out, profiles that did not reach at least to the base of the first basal layer have been excluded from this study. However, this approach probably underestimates total basal-layer depths decisively but is able to provide clues on the uppermost basal layer's thickness at least.

As one may expect based on the fact that basal layers are mainly controlled by local rocks, average texture data are very scattered (minimum and maximum values) and highly variable (standard deviation). Scattered as well, the clay

TABLE 2.7 Statistical Measures of Layers, Lahn-Dill Mountains (*Modified from Scholten, 2003*)

Parameter	*n*	Min.	Max.	Mean	Std. deviation
Basal layer					
Thickness (cm)	27	6	85	31	19
Clay content (%)	30	1	39	20	9
Silt content (%)	30	13	62	35	14
Sand content (%)	30	23	68	44	13
Clast content (%)	28	3	96	61	25
Intermediate layer					
Thickness (cm)	10	15	46	29	11
Clay content (%)	11	9	36	22	7
Silt content (%)	11	20	55	41	11
Sand content (%)	11	20	71	38	16
Clast content (%)	11	10	83	50	24
Upper layer					
Thickness (cm)	37	14	75	41	17
Clay content (%)	37	13	33	22	5
Silt content (%)	37	27	62	44	9
Sand content (%)	37	19	51	34	8
Clast content (%)	37	9	75	46	17
All layers					
Total thickness (cm)	32	18	145	72	34

See Table 2.5 for explanation.

content is some exception from this in that its deviation is small; this is probably due to the fact that the study areas did not include clay-rich rock types or a large number of reworked soil residues from the Tertiary. Although they are rare, there are basal layers with a high, even dominating, silt content. Keeping basal layers apart from intermediate layers may be intricate in such cases. The sand content varies between the study areas, reflecting the varying portion of sandstones in the areas. On average, clast content in basal layers comprises about half the volume of the substrate, though with a high variability. In this respect, it outranges the other layers by far. Texture data are essentially not correlated significantly with topographic indices.

2.5.3 Intermediate-Layer Properties

Intermediate layers may be very thin or even absent but also reach almost 1 m in thickness in this study (Scholten, 2003). The thickness variability of intermediate layers is of the same order as that of basal layers, with the average around 40 cm. However, including profiles without intermediate layers would result in an average thickness of 22 cm, that is, much less. The intermediate-layer thickness correlates with some morphographic indices (positively with the slope aspect and the topographic height, and negatively with the inclination and the convexity: all $r \leq 0.32$). Its thickness variance is mostly explained by the slope aspect, preferring east-facing slopes, and by the slope inclination (37% and 32%, respectively), furthermore, topographic moisture index, and topographic height, though with smaller effect sizes of 25% each.

Most intermediate layers have a little clay, averaging around 20%, and even less sand in most areas. However, they are rich in silt, which makes up more than 60% in the first two areas under study and 40% in the third one. Their clast content is typically the lowest of all profiles. Texture data do not or little correlate with morphographic indices; correlations between convexity and sand content ($r = 0.3$), slope inclination and clast content ($r = 0.21$), and absolute elevation and, negatively, silt content ($r = -0.26$) are the most striking ones. Accordingly, variances of texture data are not or only to a small extent (26% explanation of clast content variance by inclination) explained by morphographic indices.

In the Lahn-Dill Mountains with steeper slopes under study, the intermediate layers are thinner and remarkably richer in clasts, and the relative proportion of profiles without an intermediate layer in the total number of profiles is higher. Furthermore, the silt content is considerably lower (Table 2.7).

2.5.4 Upper-Layer Properties

Upper-layer thicknesses typically average between 40 and 55 cm in the three study areas (Tables 2.5–2.7). Standard deviations of their thicknesses are remarkably small, corroborating the general conclusion on thickness variability (Section 2.3.3). The thicknesses of the upper layers are not explained by the slope aspect ($r < 1.7$; see Section 2.6.3). The same holds true for correlations with other morphographic indices ($r \leq 1.4$).

Calculations show a relatively close relationship between the upper-layer thickness and the texture of the substrate. Thirty-eight percent of the variation in thickness of the upper layer may be explained by the particle-size distribution, with sandy and sandy-loamy material causing a higher thickness of the upper layer compared to clayey material. This might reflect the influence of heat capacity, which is lower for sand than for clay, on thawing processes (Scholten, 2003).

The clay and sand content is typically low, whereas silt is the textural class with the highest proportion, only slightly excelled by clasts in the Lahn-Dill Mountains. The clay and silt content of upper layers decreases with absolute elevation ($r = -0.27$ and -0.34, respectively), whereas clasts increase ($r = 0.3$). The sand content correlates positively ($r = 0.34$) and the silt content negatively ($r = -0.31$) with the topographic height.

2.5.5 Overall View on Statistical Results

Mostly, layers have to be looked at separately, because their properties do not have many common controls. However, some exist and some comparison of layers among one another will become important in later chapters.

Averaged out, basal layers always have the largest content in clasts—in cases five times that of other layers. Furthermore, they often are the thickest layer in a succession. The silt content in intermediate and upper layers does not differ significantly, disqualifying this property as a diagnostic one (see Section 2.2.2). In two areas, they are marginally higher in intermediate layers, but in the third one, the exact reverse holds true. The main difference between these areas is the average slope inclination.

Another property often brought forward as a diagnostic difference between intermediate and upper layers is the clast content (see Section 2.2.2). Data do not clearly support this: the average clast content differs only insignificantly between the two layers, being higher in upper layers in one area and vice versa in another. Another, statistically not tested, observation is that the clast content in intermediate layers increases with elevation, often outranging upper layers above some 600 m asl. Thus, upper and intermediate layers may not be kept apart with certainty on the basis of sediment properties. Probably for this reason, Schilling and Wiefel (1962) subsumed both deposits under one common term: "Hauptfolge," which might translate as "main succession."

The clay content in intermediate layers exceeds that in upper layers by 5% points on average.

The total thicknesses of layer successions are at least as variable as the thicknesses of single layers. However, basal layers are the major contributor to the total thickness, that is, the thicker the basal layer, the thicker the entire cover-bed succession ($r = 0.6$). The average profile depth is 120 cm, but only around half that value on the steeper slopes of the Lahn-Dill Mountains. The maximum total thickness observed in all profiles is 250 cm, the minimum around 20 cm (at a site probably affected by soil erosion).

Another quantitative clue on layer properties stems from correlation analysis comparing layers with one another (Scholten, 2003). This allows for concluding on how closely certain properties in one layer are related to the same property in another cover bed. As a result of these analyses from the Harz and Solling Mountains, the sand content of upper layers correlates with those of intermediate layers ($r > 0.5$) but less, though still significant, with those of basal layers ($r < 3.8$). The same holds true for silt with almost the same strength of correlation ($r > 0.45$ and < 0.3, respectively). The clay content is uncorrelated among different layers.

Correlations of clast contents vary remarkably between the areas of study: in the first two areas, clasts in the upper layers correlate significantly with those in intermediate layers ($r=0.58$). However, in the Lahn-Dill Mountains, this is insignificant. Rather, there is a correlation between upper and basal layers regarding clasts ($r=0.51$).

2.6 GENESIS OF COVER BEDS

A. Kleber, S. Müller, B. Terhorst, H. Thiemeyer

2.6.1 The Basal Layer

There is not much dispute concerning the processes that form basal layers. The most important is believed to be gelifluction, indicated by clast orientation or even imbricate structure (Section 2.2.1; also see Bertran and Texier, 1999). Gelifluction is mainly laminar flow (better: sliding) with the longitudinal axes of the clasts aligned during sliding by friction of finer substrates sliding faster around rock fragments (Matsuoka, 2001) due to shear stress (Van Steijn et al., 1995). We will have a closer look at gelifluction in Section 2.6.3. The upper and lower boundaries of the basal layer are undulated occasionally, showing small cryoturbation structures as involutions or lenses of sorted material (Bibus, 1985; see Fig. 2.16 later in this chapter).

Some basal layers contain well laminated and sorted, mostly sandy intercalations (Section 2.2.1). This is ascribed to the effect of slope wash when the layer formed (see Bertran and Texier, 1999). The finding that these fluvial layers are intercalated and preserved within deposits of which the main driver was gelifluction indicates that gelifluction was laminar rather than turbulent. Another indicator suggesting that gelifluction, which formed the basal layer, was laminar is provided by the properties of stacked basal layers closely depending on up-slope bedrock variations. This feature would not have persisted turbulent dislocation (Fig. 2.12). Turbulent flow would result in mixing of cosmogenic nuclides. Two basal layers were analyzed as to ^{10}Be. Nuclide concentrations were found to be lower with increasing height in the respective profile, though differences were small and some were within the error bars (Schaller et al., 2002). This finding may be ascribed to the effect that material with progressively lower ^{10}Be concentrations from up-slope successively built up the basal layers without evident mixing.

Silt caps are found within basal layers typically on upper sides of rock fragments. Some authors take these caps for syngenetic features due to internal wash processes (Semmel, 1968), precluding the classification of those deposits as basal layers if the silt is of eolian origin. Bertran and Texier (1999) also ascribe this feature to lateral water flow during dislocation of the substrates. However, they fail to explain why blank undersides of stones are always connected with those silt caps. Silt caps are often observed in modern rocky

FIGURE 2.12 Laminar character of dislocation in cover beds, Little Cottonwood Canyon, Utah, USA. Though not from Central Europe, it is an impressive example of laminar dislocation of upslope matter (stone line) and its incorporation into a cover bed (*photo by A. Kleber*). (For color version of this figure, the reader is referred to the online version of this chapter.)

materials, for example, in rock-fall deposits (own observation in the Colorado Front Range), where they are accompanied by accumulation of humic substances on otherwise blank rock undersides (thin organic layers did certainly not persist in basal layers until our days). Lateral wash within the material would hardly prefer upper sides and wipe undersides, and lateral wash may often be ruled out in modern rockfall. Accordingly, silt caps are attributed to postdepositional internal wash by seeping waters by several authors (Birkeland, 1999; Richter et al., 1970; Völkel, 1995; Völkel et al., 2002a). Thus, they should not be confused with pre- or syn-sedimentary loess enrichment in an intermediate layer: The distribution of these loess derivates within basal layers is noticeably different from that in intermediate and upper layers, where the loess component appears finely dispersed.

The distribution of basal layers (Section 2.3.1) appears to provide insight into their postdepositional fate. Because they often are absent on sharply convex upper slope shoulders, one may speculate that they were exposed to erosion. This cannot have occurred after the last glaciation, as the upper layer, which likely is more susceptible for this to happen, has survived. Thus, this finding would call for slope wash acting after basal- and before upper-layer deposition. However, basal layers are sometimes missing on flat crests, too, whose positions are not likely to have been affected by periglacial slope wash. Furthermore, basal layers are often especially thick in slope depressions (Fig. 2.5), which most likely would have concentrated overland flow on slopes. This may lead to the conclusion that the basal-layer distribution indicates thickness variations directly after formation of the deposit, except for those materials incorporated into later layers. As there is no significant dependence of the basal-layer distribution from the slope aspect, this likely has not been caused by local permafrost variability. Thus, it may be assumed that the processes that formed basal layers were similar to those ascribed to the upper layer (see Section 2.6.3), with the major difference that the duration of their formation may have been much longer. This is supported by the finding of Scholten (2003) that multilayered basal layers typically have similar thicknesses of each sublayer as the upper layer.

If so, the compactness of basal layers is not well compatible with this assumed genesis. However, a high bulk density is a diagnostic property of basal layers. This effect may be due to densification and clogging of pores during sliding and clast re-orientation, that is, re-packing of constituents (Scholten, 2003). However, there is no modern analog to this proposed process: see Bertran and Texier (1999), who described a high porosity in primary gelifluctional sediments, and Williams (1957), Williams and Smith (1989), Berthling et al. (2000), and Harris et al. (2008b). Then again, fragipan-like properties were explained by the growth of segregated ice during permafrost aggradation and subsequent compaction by ice segregation (Payton, 1993; Van Vliet-Lanoë and Langohr, 1981). If so, the compactness of basal layers indicates past permafrost, which idea is corroborated by ice-wedge casts, which may sometimes be found in basal layers (Kleber, 2004b; Kleber et al., 1998b; Semmel, 1968, 2005; Semmel and Terhorst, 2010). The idea is that compactness is not a genuine property of basal layers but came into being after their formation as a post-depositional overprint.

2.6.2 The Intermediate Layer

The main process behind the formation of intermediate layers again is gelifluction. There is no evidence of a principal changeover of process regimes from a gelifluction- to an eolian-dominated regime. Rather, the eolian aspect of the layer formation is an addendum, that is, incorporation of eolian fines to the substrates. Furthermore, there even is reasoning that the eolian addition did not happen during, but (at least in parts) before the formation of the intermediate layer (see Fig. 2.13 and Section 2.6.4). On the other hand, upper layers are at least on steeper slopes in a way closer to basal layers than to intermediate ones regarding some properties (Section 2.6.3), so that intermediate layers apparently own some unique character. As with the basal layer, the upper and lower boundaries of some intermediate layers are undulated, due to cryoturbation.

Two main issues connected with intermediate layers are their concentration to slope depressions on luff-ward slopes and the overall asymmetry of their distribution within landscapes (Section 2.3.2).

On luff slopes, intermediate layers typically are found in slope depressions mostly regardless of the slope aspect or other morphometric criteria (Sections 2.3.2 and 2.5.3). Frequently, it has been claimed that intermediate layers are bound to shelter from erosion (e.g., AGB, 1994, 2005; Altermann et al., 1995; Rohdenburg, 1989; Schönhals, 1996; Semmel and Terhorst, 2010).

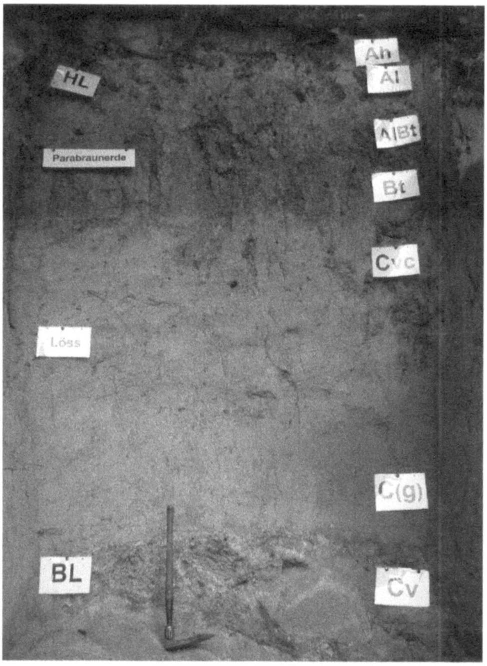

FIGURE 2.13 Loess deposition took place after the formation of the basal layer (sandstone fragments mixed with marl, indicated by "Cv") and before the intermediate layer ("Bt") formed, Vienna Forest (*photo by B. Damm*). (For color version of this figure, the reader is referred to the online version of this chapter.)

This would call for an episode of enhanced erosion between the deposition of intermediate and upper layers. However, slope depressions concentrate the water flow rather than divert it (Anderson and Burt, 1978, 1990), even if they are not occupied by permanent creeks. Accordingly, slope depressions are sites of enhanced susceptibility to erosion: it is the slope depressions where overland runoff may be observed in forests during strong rainfall, not straight slopes or areas around water divides (see Section 4.3). Thus, intermediate layers on luff slopes are concentrated to the major slope drainage areas rather than to sheltered positions. Obviously, the important criterion for discriminating intermediate layers—with a restriction to sheltered positions—does not hold on closer view. Only in the very central parts of some slope depressions, intermediate (as well as upper) layers may be eroded locally.

Then the question arises as to the reasons for the observed distribution pattern of intermediate layers. If it is not erosion (except for reworking of the uppermost decimeters into the upper layer, see Section 2.6.4) that was responsible for this pattern, differences in deposition are the likely cause. At first glance, there is a simple answer: slope depressions are sediment traps that simply collect material derived from the surrounding steeper slopes. This is certainly valid but cannot explain the differences between luff and lee slopes.

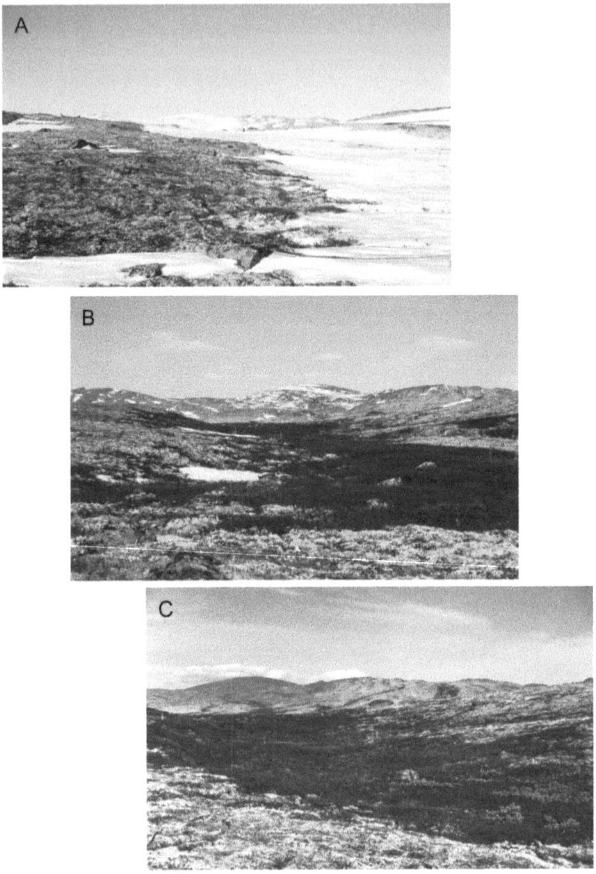

FIGURE 2.14 Snow and vegetation development in a periglacial environment, northern Norway. The photographs of a wide slope depression were taken during a single year (*pictures by J. Löffler, by permission*). In December, the distribution of snow is already remarkably tied to the slope depression, indicating that this is the preferred position to collect wind-blown snow and prevent it from further wind action (Löffler, 2002). (A) In April, most snow has vanished from the slopes, but there still is plenty in the slope depression; (B) in June, snow in the slope depression has thawed also, leaving room for vegetation growth, which is supported probably by the thermal shelter that the thick snow had provided over the winter and by the increased water supply in the slope depression; (C) this dependence of vegetation growth from topography becomes even more evident in August. (For color version of this figure, the reader is referred to the online version of this chapter.)

Two more processes, probably going hand in hand, may be inferred from Fig. 2.14. Since snow distribution is usually triggered by the wind field, the same may hold true for loess deposition. At high wind velocity, loess and snow are transported in suspension mainly (Pomeroya and Maleb, 1992), probably producing similar distribution patterns. In fact, terrain roughness is a major contributor to loess distribution patterns (Goossens, 2001). Additionally, surface roughness is decisively influenced by vegetation density and height.

Both factors add up at the same locations, the slope depressions. Thus, there is a strong model to explain the importance of these landscape elements for loess accumulation and, thus, for intermediate-layer thickness, but it is not shelter. On lee slopes, the ridge itself is the most important factor regarding surface roughness, rendering local topographic and/or vegetation differences less influential.

In many areas, intermediate layers are more widespread on east-facing than on west-facing slopes (Sections 2.3.2 and 2.5.3). Völkel (1995) provoked a debate on how this may be interpreted in terms of paleowind directions by discussing whether loess preferably accumulates on luff or lee slopes. This discussion is made possible by the fact that reconstructions of glacial wind tracks are ambiguous, as geomorphic evidences, which call for westerly winds in most of Central Europe, conflict with paleoclimate modeling results calling for easterly winds at least in northern Central Europe (Meyer and Kottmeier, 1989). However, more recent modeling attempts have reduced most of these discrepancies in favor of prevailing westerly winds during Pleniglacial and Late Glacial times (Isarin et al., 1997; Renssen et al., 2007; see Jost et al., 2005 for general limitations of paleoclimate modeling in Europe). Renssen et al. (2007) summarized plenty of evidence for prevailing geomorphologically effective westerly wind directions from northern Central Europe, though sometimes with a secondary eastern component. Eitel (2003) presented a case study corroborating westerly winds in southwestern Central Europe.

Though preferred deposition of eolian fines in the lee of mountain ranges appears accepted in most of the literature, wind-channel aided results of Goossens (Goossens, 1988, 2001, 2006; Goossens and Offer, 1997; Goossens et al., 2000) and Zufall et al. (1999a,b) raise some doubt, because these authors found preferred accumulation of fine eolian particles on luff rather than lee slopes in their experiments. Based on their findings, they argued that reconstruction of paleowind directions from loess distribution may be seriously flawed, and that the actual wind directions might have been exactly the opposite than found with the usual reconstructions. This would cast even more doubt on the current paleowind reconstructions.

However, additional instruments are available for modern reconstructions. In front of the Laurentide ice sheet, which was thicker than the Scandinavian one and thus would have influenced wind directions even more than the latter, geochemical and mineralogical imprints facilitate the exact determination of source areas. The results render older paleoclimate models false in this respect and corroborate the original reconstructions of westerly paleowinds, compatible with the classical view on distribution patterns (Bettis et al., 2003; Grimley, 2000; Muhs et al., 2008). Similar results stem from the comparison of spatial thicknesses (Bettis et al., 2003; Rousseau et al., 2007) and from the reconstruction of grain-size winnowing (Nugteren and Vandenberghe, 2004; Stevens and Lu, 2009). Most models predict lee-slope deposition (Hoffmann et al., 2008; Mason et al., 1999) and agree in this respect with distribution patterns of ancient loess (Hughes et al., 2009; Mason et al., 1999) and with measured modern dust deposition patterns (Hoffmann et al., 2008).

There may be some flaws in transferring wind-channel results to the periglacial past:

- The experiments resulted in that the dust accumulation is highest on luff slopes of obstacles (Goossens, 1988; Zufall et al., 1999b) and that the deflation is highest there as well (Goossens et al., 2000), mainly depending on drag velocity and material supply. Since the experiments were undertaken with slow to moderate wind speeds; high velocities—as they are assumed responsible for loess transportation in glacial times (Hoffmann et al., 2008)—may have re-entrained matter previously deposited on luff slopes. Only rough surfaces could then have hindered re-allocation, again pointing at slope depressions with their presumably higher and denser vegetation.
- Material will become scant in nature during a lasting storm event (Hesse and McTainsh, 2003; Smalley, 1990), causing the wind in the domain of its highest velocity to erode where it may have accumulated before. Constant material supply is a prerequisite of simulated luff accumulation (Goossens, 1988).
- Particle sizes used in the experiments were usually rather small (mostly $\leq$30 μm on average), compared to typical periglacial loess (Fang et al., 1999; Pye, 1987; Smalley, 1990). Goossens (1988) found support for accumulation of matter on windward slopes for those smaller particles, whereas he explicitly stated the opposite for sand, and he leaves the problem of coarse silt open. Zobeck et al. (1999; in contrast Goossens, 2006) found a major break in transport mechanisms of particles coarser than approx. 50 μm. Furthermore, fine dust is often transported as aggregates (Qiang et al., 2010), which would settle similar to coarser grains; this was not achieved in the experiments (Goossens, 2005b).
- Channel experiments did not consider fluvial processes (Offer and Goossens, 1995) due to precipitation, which is typical for climbing air masses as dye temperature is reached. At first sight, rain fall on the luff side would increase scavenging of particles. However, these would not necessarily accumulate upon reaching ground because the particles suspended in rain water would be transported by overland flow until they may settle in an associated valley floor or in the next vegetated slope depression.

Despite the previous conclusion that there probably is no evidence of a major erosional episode after the formation of the intermediate layer, at least not regularly, one must be aware that the modern distribution and thickness of this layer do most likely not equal the original ones. This is mainly because the upper half meter of the intermediate layer may have been reworked into the upper layer, cutting the upper decimeters of the intermediate layer, and the intermediate layer may have been entirely incorporated and, thus, is no more discernible as such, if it was thinner than this half meter.

2.6.3 The Upper Layer

In geology, usually the youngest deposit is the easiest deciphered regarding genesis, because it has lost least information through, for example, diagenesis or truncation. This does not hold true for upper layers, because they have been affected by surficial processes such as pedogenesis, bioturbation, uprooting by wind throw, and other, mainly anthropogenic geomorphic action during the Holocene (Frühauf, 1991; Göbel, 1977; Mahr, 1998; Russow and Heinrich, 2001) resulting in continuous mixing of much of the surficial materials in upper layers until modern times as derived from ^{10}Be measurements (Schaller et al., 2002). Thus, several assertions on the genesis of upper layers must rely on just the supposition that some sediment properties, as they may be observed nowadays, have been formed during deposition. The effects of overprint by the aforementioned processes, on the other hand, are usually well discernible in the field (Frühauf, 1991; Meyers and McSweeney, 1995; Russow and Heinrich, 2001).

One such property is the loose consistency of upper layers. Another assumption is that the upper layer has been formed by gelifluction and has approximately maintained its original thickness. However, the latter assumption is corroborated by the undisputed pre-Holocene age of the layer (Section 2.7.1.4) and especially by its unchanged properties buried under Late Glacial peat (Völkel and Mahr, 1997). Especially, the thickness is by most authors interpreted as a genuine sediment property, because it is supported by mineralogical and textural differences to underlying layers, and occasionally by clast orientation and gradual changes in the clast content. In the following, we will consequently treat the upper-layer thickness as an original property.

The most astonishing property of the typical upper layer is its essentially constant thickness. This is difficult to understand and to accept: it seems contradictory to every insight into fluid- or gravity-driven geomorphic processes, which are assumed (1) to vary with the slope angle and (2) to develop catenary thickness relationships, that is, become stripped where erosion prevails and thickened on accumulation.

Late Glacial climatic and geomorphic conditions have been compared to modern ones in subpolar regions (e.g., Semmel, 1985). This comparison, however, may fail because insolation undoubtedly differed considerably between such contrasting latitudes even if glacial Central Europe is compared to the modern Arctic (Kleber, 2004a). The Alps of eastern Tyrol, Austria, at around 2650 m asl, may represent a closer approximation to the Late Glacial climate: the mean annual temperature around −2.1 °C (Jaesche, 1999) resembles inferred Younger Dryas temperatures (Renssen et al., 2002) in Central European subdued mountains, assuming lapse rates were then alike modern ones. Alpine January temperatures (mean −11 °C, Jaesche, 1999) were matched even at low elevation in Central Europe during the Younger Dryas (Renssen et al., 2002). In this Alpine area, a measuring plot was installed upon a silt-rich substrate. This plot is not underlain by permafrost within at least the uppermost

10 m, as derived from temperature measurements at the snow base (Rennert, 1991) and corroborated by geophysical sounding (L. Schrott, Salzburg University, personal communication). Markers were installed in this plot in 1985, which were surveyed tachymetrically from a fix nearby rock outcrop each year, before several of them were dug out 4 years later (Emmerich, 1990). Some markers were 10-cm long wooden rods, which had been piled one on top of the other into a bore hole. Typically, rod piles that had undergone surficial displacement of, on average, more than 5 cm per year exhibited a similar picture (Fig. 2.15): their maximum dislocation occurred near the surface, whereas dislocation decreased slowly with depth. Only at 40–70 cm, typically at 45–50 cm, a sharp drop in displacement was observed. Below this drop, only minor displacement was observed. Only with the surface velocity exceeding 25 cm per year, deeper dislocation was observed. This indicates that gelifluction of silt-rich material (comparable to the upper layer with its admixed eolian matter) under modern high Alpine conditions is largely a slide process above an approx. 50 cm deep slipping plane, whereas evidence of frost creep is poor (Veit et al., 1995). These observations are not sheer coincidence due to the short period of observation, insofar as the assumed sliding plane corresponds to a disconformity recognized in the field in most cases (Veit et al., 1995). Evidently, such periglacial dislocations with an average annual surface velocity of 5–25 cm a^{-1} result in a deposit very similar to the upper layer regarding depth and sediment properties such as orientation of the clasts' longitudinal axes. Laboratory experiments support the finding of a slipping plane at shallow depth (Harris and

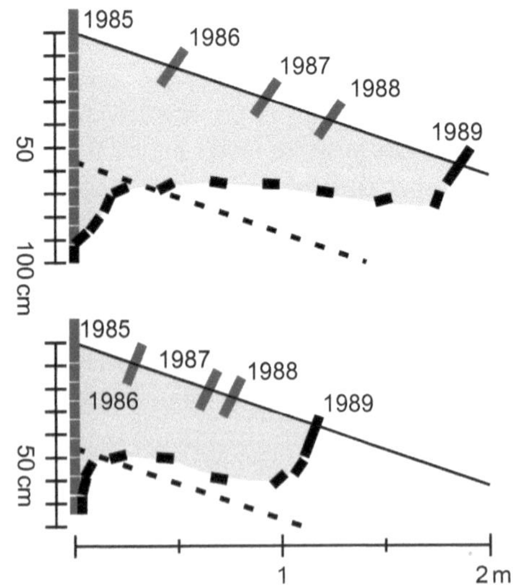

FIGURE 2.15 Gelifluction profiles from rod piles implemented in 1985 (dark gray), annually measured at the surface, and dug out in 1989 (black) from a measuring plot in eastern Tyrol, Austria, at 2650 m asl (see Fig. 6.1 for location) (*modified from Emmerich, 1990; Veit et al., 1995*). The thin line indicates the approximate position of the sliding plane.

Davies, 2000; Harris et al., 2003), and other field observations do as well (e.g., Matsumoto and Ishikawa, 2002; Matsuoka, 2001). However, Harris et al. (2008a) and Thomas et al. (2009) report that they produced and modeled slipping planes (partially at 1 m depth) only under simulated permafrost conditions. Transferring these results to periglacial conditions may be problematic because the different effects are mainly based on icing pressure between two ice fronts, which may in nature be realized in seasonally frozen ground but also as soon as re-freezing occurs during the thawing period. In fact, Harris et al. (2008b: fig. 12d) themselves depict a measured slipping plane even from the assumed outer limit of active gelifluction.

Accompanying the measurement of velocities and soil temperatures (Veit et al., 1995), the soil water content was determined later by Jaesche (Jaesche, 1999; Jaesche et al., 2002, 2003) in the same and comparable nearby plots. Some of the results appear essential for the understanding of upper-layer dynamics: (1) the slipping plane develops at around the same depth where the last occurrence of frozen soil is found at the end of the coldest and least snowy winters, indicating that dislocation is bound chiefly to near-saturation by water above perching soil ice and abates quickly after thawing (cf. Harris et al., 2008b; Williams and Smith, 1989). Warmer and snowier winters likely restrict dislocation to a depth closer to the surface, explaining part of the velocity increase at shallower depths. (2) Gelifluction under Alpine conditions does essentially not depend on slope inclination. This is mainly due to direct intake of water that is favored on flatter slopes compared to steeper ones, whose effect counteracts the inclination-triggered velocity increase that would be expected otherwise (Jaesche, 1999). This effect is influenced neither by the slope aspect, which triggers the exact timing of the process via insolation differences, but essentially not its draft, nor by the vegetation density (Jaesche, 1999). Instead, (3) the depth of dislocation is influenced by the substrate texture (Veit et al., 1995). From this, it may be concluded that gelifluction is able to produce deposits under certain circumstances that astonishingly remind of the upper layer. One of these prerequisites is the lack of permafrost in favor of seasonally frozen ground.

Interpreting the constant thickness of the upper layer as the depth of a permafrost-related active layer (e.g., Scholten, 2003; Semmel, 1985; Semmel and Terhorst, 2010) fails for several reasons: the thaw depth of permafrost is triggered by many factors, which influence one another in a complex manner; these are (atmosphere-driven) soil temperature, heat transfer through layers of snow, vegetation (ACIA, 2004), thickness of the organic soil (Marsh et al., 2009), water content (Anisimov et al., 2002), and texture (Nicolsky et al., 2009), together leading to a pronounced spatial variability of thaw depth in complex terrain (Anisimov et al., 2002). It has been known for a long time that the thickness variability of active layers is remarkably higher than the depth to the last soil ice of seasonally frozen ground (Black, 1954). Most triggers hinge on topography, especially the snow pattern (Oelke et al., 2003) and on the temperature, which depends among others strongly on aspect, elevation (Anisimov et al., 2002; Ridefelt et al., 2009), and vegetation (Löffler, 2002). Nowinski et al. (2010) demonstrated via snow depth manipulation that the thaw depth may change from 30 to >80 cm, depending solely on snow-cover variation. According to Stendel et al. (2007) and Iijima et al. (2010), the thickness of the active layer may reach almost 2 m locally. All these authors describe highly variable thaw depths (in contrast to Semmel, 1985). Thus, a deposit is barely imaginable that formed within an active layer but was not significantly triggered by the terrain. Furthermore, Semmel (1994b, 2002b) and Semmel and Terhorst (2010) addressed a major divergence between the spatial reconstruction of permafrost based on the upper-layer distribution and palynological reconstruction of the Late Glacial climate in Central Europe (see Bos, 2001; and the summary by Litt and Stebich, 1999).

On the other hand, there is evidence of permafrost shortly preceding the time when upper layers formed presumably, described mainly from the basin of Neuwied, mid-western Germany (Frechen and Rosauer, 1959; Semmel, 1985; Semmel and Terhorst, 2010). This is provided by ice-wedge casts penetrating into Lake Laach tephra, which were ejected during the Late Glacial, that is, Allerød, times (12.9 ka, e.g., Schmincke et al., 1999). Black (1976) already pointed out that individual wedge casts do not necessarily indicate continuous permafrost; rather, they may indicate local, sporadic permafrost induced by favorable, site-specific conditions. In the case of the relatively small and isolated Neuwied Basin, these conditions could have been provided by sinking cold air that became trapped and pooled. Isarin (1997) (based on permafrost indicators) and Renssen and Vandenberghe (2003) (adding a modeling approach) exclude even discontinuous permafrost during the Younger Dryas for most of southern Germany and continuous permafrost for Central Europe in its entirety. There is evidence of discontinuous permafrost during the Younger Dryas for the highest regions of the Ardenne mountains (Pissart and Juvigné, 1980). This also renders continuous permafrost conditions at that time at lower elevation and latitude unlikely. The upper layer, however, is equally distributed in southern and northern German subdued mountains, regardless of altitude.

Thus, evidence of Younger Dryas permafrost is poor. Assumptions of its existence are mainly based upon hypotheses on operating processes, although Benedict (1976) already has emphasized that none of the gelifluction processes depends on the presence of permafrost. However, as was illustrated in the aforementioned Alpine example, gelifluctional features, resembling the upper layer better than those in subpolar regions (Matsuoka, 2001; Semmel, 1985), may actually form without underlying permafrost. It should be denoted that the typical thickness of the upper layer equals the depth of contemporary gelifluction-induced dislocation measured in nonpermafrost areas worldwide (Harris et al., 2008a; Matsuoka, 2001). The best fit between observed processes in periglacial environments and upper-layer properties (thickness and continuous occurrence) is thus suggested to be just seasonally frozen ground driving gelifluction.

Völkel (1995) held selective removal of finer-grained substrate, including relatively small clasts, responsible for the fact that many upper layers contain unproportionally more coarse to very coarse than small clasts (Section 2.2.3). Similarly, Sauer (2002) assumed a widespread erosive episode before the upper layer formed, causing coarse clasts to be enriched at the former surface, which later constituted the clast surplus in the upper layer. She considered a then proportionally thicker intermediate layer, so that there was a sufficient amount of loess-rich substrate remaining after this erosional event to become incorporated into the later upper layer (see Section 2.6.4). However, there may be other processes able to achieve surficial clast enrichment; one process is the formation of desert pavements, which act as dust collectors while clasts are maintained at the surface (McFadden et al., 1998; Pelletier et al., 2007). The formation of desert pavements has been observed in modern dry-periglacial environments of the High Andes (Dietze et al., 2013; Dietze and Kleber, 2011). However, this process is not yet well understood on its part (Dietze et al., 2010). There is first evidence that stone lines in Central European cover beds may be interpreted as former desert pavements (Dietze and Roettig, 2012). Similarly, surficial enrichment of clasts by frost heave is common in periglacial environments (French, 2007). All these processes may have enriched clasts at or close to a former surface, which were incorporated into the formation of the upper layer thereafter. Furthermore, inheritance from deeper layers, originally situated upslope of observed profiles, may also account for many cases of clast enrichment in upper layers (see Section 2.6.4).

Detailed mapping and mineral analyses through microscopy and X-ray diffraction around an isolated basalt outcrop (Buchhübel, Tharandt Forest, Saxony) yielded evidence regarding transportation distances within layers outside of former runoff courses. Thereafter, minerals and rock fragments derived from the basalt were transported over 170 m from their source, whereas dislocation during the formation of the upper layer added only another 20 m (Dietze, 2006), probably precluding fluvial processes as main further distribution agents (Dietze, 2006; Dietze and Kleber, 2010).

2.6.4 Incorporation of Substrates

2.6.4.1 Incorporation Processes

Gelifluction, eolian deposition, and, as accessories, surface wash and cryoturbation are commonly agreed to be the key processes involved in the formation of cover beds. Cryoturbation is (1) the process made responsible for the incorporation of older, underlying materials into a younger layer (which will be addressed later in this section) and (2) the process assumed to be responsible for the admixture of eolian matter to substrates that are dislocated by gelifluction; in a sense, cryoturbation is contended to be the “missing link” between the major two processes: gelifluction and eolian deposition. Cryoturbation is assumed to mix loess into the substrate during or shortly after the formation of a layer. Gelifluction is usually identified through the orientation of the clasts' longitudinal axes in the direction of inclination. Cryoturbation, on the other hand, tends to tilt these axes toward a vertical direction (French, 2007). In fact, there are cases where clast axes in basal layers are aligned to ice-wedge casts (Kleber and Schellenberger, 1998) or to cryoturbation pockets in intermediate layers (Fig. 2.16). However, if cryoturbation was a major contributor to cover-bed formation, there are antagonistic trends to align clasts' longitudinal axes in incompatible directions. Cover beds would demonstrate chaotic axe orientations. However, though many upper layers show such random structures, usually attributed to postdepositional bioturbation rather than cryoturbation (Frühauf, 1991; Russow and Heinrich, 2001), most basal, intermediate, and some upper layers do not express chaotic structures but a down-slope clast orientation. Therefore, it has to be concluded that cryoturbation did not play a decisive, dominant role in their formation. Rather, their structure predominantly stems from gelifluction.

If so, the question arises as to the process of mixing loess into cover beds. All previously discussed syn- or postdepositional mechanisms able to homogeneously mix substrates would have destroyed existing structures or prevented

FIGURE 2.16 Involution bag at the base of the intermediate layer with the larger clasts following the shape of the cryoturbation structure, Kolbeterberg, Vienna Forest (*photo by B. Terhorst*). (For color version of this figure, the reader is referred to the online version of this chapter.)

sedimentary structures from forming, which is not the case, at least for the intermediate layer. The consequence would be that layers incorporated loess, which had been deposited before they began forming (cf. Sauer, 2002). Since thick deposits of intermediate layers are usually stratified into more than one particular sublayer, there is no need for syn-depositional processes to explain a loess admixture. All loess may have deposited before the formation of each particular layer. However, there is a process other than cryoturbation, as proposed by Bertran and Texier (1999), which could be responsible for syngenetic admixture: if ice lensing interferes with the accumulation through gelifluction, silt may become incorporated into the entire matrix; ice lensing is indicated by a lenticular-platy fabric (Harris, 1998; Van Vliet-Lanoë et al., 2004), which is also frequently found in intermediate layers.

Whatever may be most important for the formation of intermediate and upper layers, in this section we focus on the reworking of older into younger materials, as to say on inheritance, which is often overlooked in current work on cover beds. Inheritance likely plays a decisive role for layer properties as statistical calculations indicate, which commonly exhibit significant correlation between the properties of layers overlying one another (Section 2.5.5). In principle, there are two major types of layers created by gelifluction with regard to their relationship with underlying materials: overriding and inheritance. (1) "Overriding": Gelifluction may form lobes slowly migrating over older ground. In this case, the lobe substrates form an entirely new deposit, overlying an older one in a classical geological sense. In modern analogs, this may be identified by overridden soils just beneath the lobe material (Veit, 1988; see also Figs. 6.12 and 6.13). It may be doubted whether soils could have been preserved in Pleistocene layers, which anyway had very weakly developed soils at most. Thus, there is no simple identification of this process. However, such mechanism needs to be inferred for layers with extremely different properties directly overlying one another, for example, basal layers immediately above intermediate layers (Semmel, 1990) or above loess (Veit, 1991; Zöller and Nehring, 2002) without any incorporation of the underlying materials into the younger ones. (2) "Inheritance": The second type comprises new layers entirely derived from previous ones (Fig. 2.17). Both types may occur side by side (Jaesche, 1999). The latter type is not absolutely concordant with the "Law of Superposition" because substrates, more exactly numerical dates derived from them, may be essentially the same and only the time at which they reached their final position may differ.

Both types are able to explain a phenomenon described in Section 2.5.4: in areas with relatively gentle slopes, clasts in upper layers were found to correlate with clasts in intermediate layers. In the area with steeper slopes included in the study, however, there was correlation between upper and basal layers. Explaining this appears simple at first sight (Fig. 2.18). This is, however, not a clear indicator of the "overriding" or the "inheritance" type of dislocation, because both may result in a similar constellation. Anyway, the results indicate that

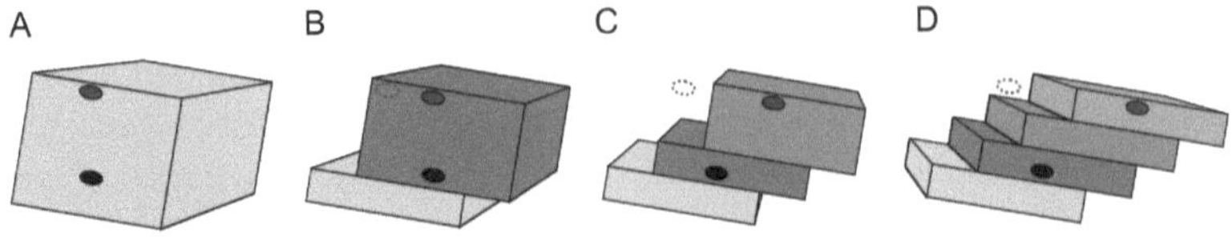

FIGURE 2.17 Inheritance of layer properties. Four stages are shown from left to right with movement depicted as discrete blocks. (A) Initial stage with two rock fragments, the fate of which is to be observed, in their initial positions (ovals); materials are derived from bedrock through weathering; (B) stage reached after a first episode of dislocation when the fragments migrated down-slope (dotted: original position of the upper rock fragment) but maintained the same relative position to one another and to other materials within their block; (C) subsequent stage with the upper rock fragment migrating on and on, whereas the lower one is no more involved in dislocation and remains in place; (D) the upper fragment has been dislocated down-slope but has remained unchanged in its context. The final uppermost block comprises displaced but identical material compared to the initial stage. However, relative to underlying blocks, including the lower rock fragment, which were not involved in the last dislocation episode(s), proportions have changed; disconformities may have come into being where the blocks adjoin each other. Note this schematic view of inheritance neglects admixture of eolian matter. (For color version of this figure, the reader is referred to the online version of this chapter.)

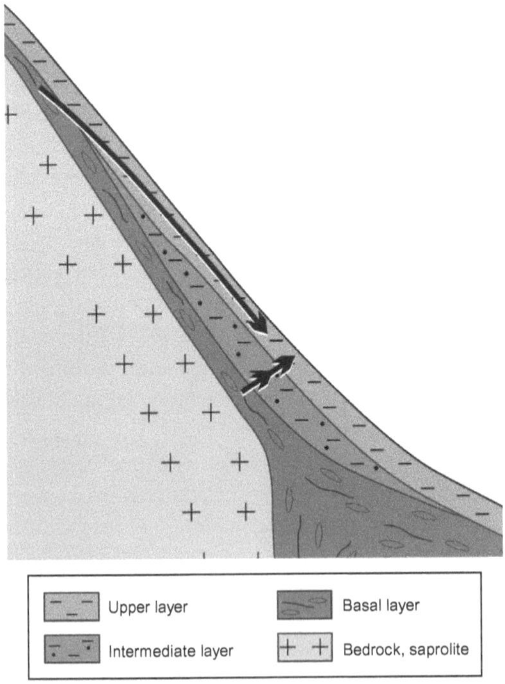

FIGURE 2.18 Provenance of basal-layer material in an upper layer overlying an intermediate layer. The arrows depict dislocation paths as overriding essentially parallel to the slope, as inheritance more or less perpendicular to the slope. (For color version of this figure, the reader is referred to the online version of this chapter.)

the likeliness of preferred incorporation of material from a penultimate deposit increases with slope inclination (cf. Thalheim and Fiedler, 1990; Scholten, 2003). This is not necessarily due to larger dislocation distances in steeper terrain but may also be due to the fact that intermediate-layer distribution is more restricted in such terrain, which is less prone for eolian deposition than flatter, especially concave, slope positions. One may assume that relief steepness is important for the probability of one of these two types of incorporation to prevail: in steep relief overriding likely is the acting process, on gentler slopes inheritance.

^{10}Be measurements (Schaller et al., 2002) from two basal layers allow to hypothesize that they have never been very much closer to the surface than they are now. This would indicate that the overlying deposits existed already at the time of the deposition of these basal layers, which later were either truncated or, more likely, incorporated into overlying layers.

2.6.4.2 Case Study I: Sandreuth, Northeastern Bavaria

Figure 2.19 is a sketch of a slope extending from a relict gravel layer deposited on a local hill-top in northeastern Bavaria (Fig. 2.1). Rock types in this gravel differ markedly from the underlying Triassic bedrock, simplifying the identification of dislocated materials. The left edge of the figure is portrayed in Fig. 2.4. In the right part of the sketch, a bipartite layer succession is visible, consisting of a basal and an upper layer. This changes into a four-partite succession within a short-distance down-slope. This comprises a basal layer overlain by two different intermediate layers and an upper layer. The deeper intermediate layer has a

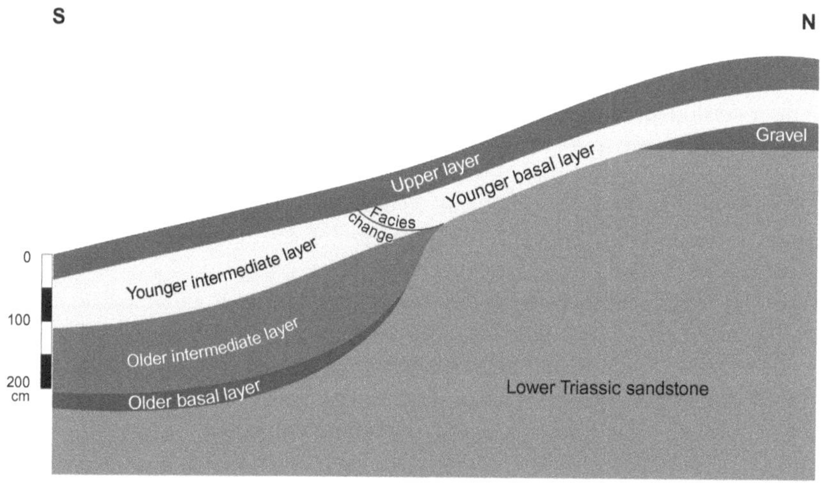

FIGURE 2.19 Cover beds along a slope exposed in a sand quarry near Sandreuth, northeastern Bavaria. Horizontally not to scale (*modified from Kleber, 1992a; Kleber and Stingl, 2000*). (For color version of this figure, the reader is referred to the online version of this chapter.)

strong argillic[3] horizon with an equally strong imprint by stagnant soil water leading to thick, white-bleached former root tracks and a coarse soil structure. This is overlain by another intermediate layer with a less mature soil horizon in it. It is argillic, too, but cutans are remarkably thinner, the stagnic[4] color pattern comprises smaller stains and is yellowish gray rather than white, and peds do not reach the same size as in the underlying horizon. The boundary between the two is wavy but abrupt. The upper layer carries the A and E horizons of the soil profile. This implies that there is a more maturely developed paleosol in the older intermediate layer, which indicates a pedogenic imprint at the time when the overlying materials were not yet the same as they are now. Moreover, the maturity of this soil is barely explained through an interstadial but rather an interglacial type of pedogenesis.

In the transitional zone between the two successions, the basal layer cannot be traced as the respective deepest layer in both sections of the profile wall. Rather, basal layers in both section are unconnected. The up-slope basal layer undergoes a facies change within less than a meter, starting exactly where it meets the underlying intermediate layer with the mature paleosol. Down-slope of this facies change, the basal layer contains a significant amount of loess. Therefore, it has to be designated an intermediate layer; its facies changes from a basal to an intermediate layer. Such a short-distance difference in loess enrichment cannot be explained through a local diversity of loess deposition during layer formation. Rather, the loess was incorporated during dislocation from previous substrates, the remnants of which are the underlying substrates now. The first conclusion drawn from this is that intermediate layers do not necessarily incorporate loess that has been deposited during or at least in a timely relationship to the layer's activity. The second inference is that the upper intermediate layer in the down-slope succession owns its entire substrate from the older layer, thus being an impressive example of the "inheritance" type of gelifluction. The facies change may be compared to the edges of the discrete blocks in Fig. 2.17. As long as no "overriding" takes place, basal layers cannot be present where loess-bearing materials already exist. As soon as loess has been deposited at a certain site and is not entirely removed, no classified basal layer can ever come into being through "inheritance." As long as there is no reliable discrimination between the two types, it cannot be ascertained whether intermediate layers did develop from noticeably older loess-rich material or whether the admixed loess is approximately as old as the layer dislocation. As a last conclusion, intermediate layers do not necessarily carry signals of the paleoclimate during their dislocation.

In principle, the same may hold true for the upper layer. There also is chance of inheritance, so that upper layers may contain inherited material from intermediate layers, most of which may have entirely been incorporated into upper layers.

---

3. A subsurface soil horizon formed by clay illuviation (SSS, 2010).
4. A soil horizon formed by stagnant soil water (WRB, 2006).

This leads to the conclusion that the formation of upper layers was not necessarily connected with a phase of loess deposition (see Section 2.7.1).

2.6.4.3 Case Study II: Spessart Mountains, Southeastern Hesse

Müller and Thiemeyer (2009, 2010) and Müller (2011) applied micromorphological analyses combined with heavy-mineral analyses to periglacial slope deposits in soil pits in the Spessart Mountains (Fig. 2.1). The bedrock, dominated by micaceous schist in the northwestern part of the Spessart Mountains, is covered by a two-layer succession of upper and basal layers on the upper slopes and in very steep slope sections. Lower and less steep slopes are usually covered by three-layer successions comprising upper, intermediate, and basal layers.

Figure 2.20 depicts a typical three-layer profile: the basal layer consists of weathered and dislocated micaceous schist, whereas both the intermediate and the upper layer contain a considerable amount of decalcified loess. Furthermore, minerals from the Allerød eruption of Lake Laach were discovered in the upper layer (Table 2.4). An Epiaqualf (SSS, 2010) has developed within this succession.

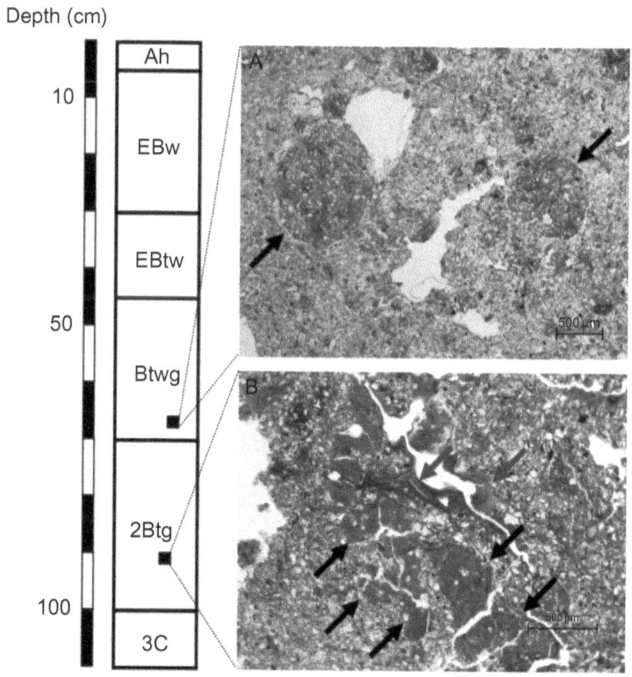

FIGURE 2.20 Sketch of the profile in the Spessart Mountains with microphotographs from thin sections: (A) reworked, rounded aggregates (intermediate-layer material, black arrows) in the lower part of the upper layer; (B) clayey aggregates in the silty groundmass of the intermediate layer (reworked paleosol material, black arrows) and modern clay coatings (red arrows). (For interpretation of the references to color in this figure legend, the reader is referred to the online version of this chapter.)

The upper layer carries four of its soil horizons (A-EBw-EBt-Btg1). An even more clayey illuvial horizon (2Btg2) has formed in the intermediate layer. The strongly developed argillic properties of the 2Btg2 horizon underlying less mature argillic horizons (EBt and Btg1) within another, younger deposit suggest that both belong to different phases of soil formation. Undisturbed soil samples were taken from each horizon for thin-section preparation, description, and interpretation.

The micromorphological analyses confirm the above suggestion: the lower part of the upper layer (Btg1) shows clay coatings, which are attributed to modern illuviation processes. Moreover, it contains aggregates of clayey material embedded in a silty groundmass rounded by transport processes (Fig. 2.20A). This material is interpreted as being reworked from the underlying intermediate layer in the course of the formation of the upper layer. The 2Btg2 horizon in the intermediate layer contains separated, clay-rich aggregates in a silty groundmass (Fig. 2.20B), typical for most of the loess in the slope deposit. Besides these, fractured clay coatings are frequent. Because of this fracturing we conclude that these aggregates and fragments are reworked matter derived from a mature paleosol rather than an *in situ* product of pedogenesis in the intermediate layer. This inferred paleosol must have come into being between the development of the basal and the preserved intermediate layer, when stable conditions must have allowed soil formation. The age of this paleosol has not yet been determined. According to Felix-Henningsen (1991), who found reworked paleosol material in the intermediate layer of a soil profile in the Rhenish Massif, this might represent the last interglacial soil. This would date the basal layer into an earlier glacial epoch and render pre-Pleniglacial denudation just marginally effective. In our (S. Müller and H. Thiemeyer) opinion, this is hardly conceivable. Therefore, we favor reworking of an intra-Würm/Weichselian soil of yet unknown age. Since interglacial conditions are not necessarily required for soil formation (Kühn, 2003), an interstadial genesis of this paleosol might be possible. However, considering the very scattered age suggestions that exist for basal layers and the preliminary numerical datings of basal layers, some of which also cover the penultimate glaciation (see below), such an old basal-layer age cannot be excluded. This issue remains to be clarified by further research on the micromorphological properties of cover beds and by their numerical dating.

The profile (Table 2.8) displays two types of incorporation: (1) inherited material from the underlying stratum and (2) inherited soil formation. The rounded aggregates in the lower part of the upper layer represent reworked material from the underlying stratum (intermediate layer) or, in the case of the aggregates within the intermediate layer, entirely incorporated residues of an older pedogenic episode. Moreover, gelifluction and cryoturbation as main formation processes of the slope deposits cannot be clearly identified, because there is no evidence of features assumed to be indicative of frost action (e.g., Harris, 1998; Kühn et al., 2006) such as a lenticular-platy microstructure in connection with vesicles or fine-material cappings on peds in the thin sections of the soil. Nevertheless, aggregates are fragile structures. Their persistence

TABLE 2.8 Heavy Minerals in a Soil Profile, Spessart Mountains, Hesse (*Reproduced from Müller and Thiemeyer, 2009*)

	EBw	EBt	Btg1	2Btg2	3C
Pyroxene	5	4	4	+	–
Brown hornblende	23	16	14	2	–
Sphene (titanite)	6	7	6	+	–
Anatase	3	3	2	5	19
Andalusite	–	+	–	–	–
Apatite	–	1	–	–	–
Brookite	–	–	+	–	–
Kyanite	1	–	1	1	–
Epidote	19	26	32	36	–
Mica	1	1	1	2	–
Garnet	2	1	2	2	–
Green hornblende	9	8	8	20	+
Rutile	1	1	1	+	–
Sillimanite	1	–	–	–	–
Staurolite	3	2	6	4	+
Tourmaline	12	16	12	20	39
Zircon	13	14	11	8	41

Numbers are percentages; +: contained with less than 1%; –: not contained. Especially, the minerals diagnostic for Lake Laach tephra (printed in italics) allow to differentiate between the upper layer (EBw to Btg1 horizons) and the intermediate layer (2Btg2 horizon), in which they are remarkably decreased in percentage; the basal layer (3C horizon), derived from underlying micaceous schist, is easily kept apart as well in the field as through its meager mineral content.

suggests transport in a frozen state, otherwise, they probably would have been destroyed. However, currently the material is undergoing homogenization processes by bioturbation as also indicated by micromorphological analyses.

The soil shows a difference in clay content between top- and subsoil associated with visible clay cutans in the latter. The main result of the study with regard to the issue of inheritance is that the modern clay illuviation, which is responsible for the cutans, only masks the major difference in clay content, which is not the result of *in situ* pedogenesis but caused initially by stratification (see Section 3.3). Cover beds may contain substrates derived from underlying layers, which may be reconstructed though they have not even persisted to our times as distinct deposits.

2.7 CHRONOLOGY OF PERIGLACIAL COVER BEDS

D. Hülle, A. Kleber

2.7.1 Relative-Age Criteria

2.7.1.1 Introduction

Whether the "Law of Superposition" applies to cover beds may be disputed. It appears obvious that, if not the substrates (Section 2.6.4), then at least their respective dislocation follows a sequence of the deepest layer being the oldest, the uppermost being the youngest in a succession. However, there is speculation of variable depths of dislocation at around the same time so that more than one cover bed might have been dislocated simultaneously, but being disproportionated with time according to different dislocation velocities at depth (Frühauf, 1996; Stahr, 1979). The idea behind this is that active layers in permafrost areas may encounter different thaw depths at different times of the year, leading to faster moving upper sections and slower moving lower ones (Bibus et al., 1976; Semmel and Terhorst, 2010). Since we do not consider permafrost to be a *sine qua non* for the formation of upper layers, we have no need to argue for such a disproportion and we will consequently treat layers in the following as if they were formed at different times.

There has always been a dispute whether cover beds are chronostratigraphic units, that is, whether a distinct type of layer may represent a certain time slot. This was taken for granted by some early workers (Altermann et al., 1977; Jäger and Altermann, 1991; Schilling and Wiefel, 1962). However, this soon was doubted (Fiedler and Hunger, 1970; Fried, 1984; Semmel, 1968; Stahr, 1979). The latter authors pointed out that cover beds are lithological units, that is, they are defined based upon facies rather than age, which led them to the conclusion that basal and intermediate layers have no defined ages but may vary in age regionally or even locally, but, on the other hand, they believed that the upper layer has been formed during the Younger Dryas (Frühauf, 1990, 1991; Kleber, 1992a, 1999b; Scholten, 2003; Semmel and Terhorst, 2010; Völkel, 1994, 1995). Even this last assumption has come to doubt recently (Völkel and Mahr, 1997; Völkel et al., 2002a; see Section 2.7.1.4).

2.7.1.2 Age of the Basal Layer

The bandwidth of age suggestions for particular basal layers reaches from Early Quaternary (Frühauf, 1996), through third-to-last glaciation (Kleber and Stingl, 2000), penultimate glaciation (Felix-Henningsen et al., 1991; Semmel, 1968; Veit, 1991), older than Pleniglacial (Schäfer et al., 1991; Zöller and Nehring, 2002), and younger than the 30 ka Lohne Soil (Semmel, 1968) to younger than the Allerød Lake Laach tephra (Fried, 1984). The last-mentioned author based his assessment on tephra-derived mineral grains found within basal layers, which finding is not very compelling, because tephra may have been dislocated to depth later. The other age estimates rest on under- or overlying paleosols or successions of paleosols.

The proposed older-than Last Glacial ages mostly concern very particular cases where basal layers bifurcate and intercalate with, mostly, numerically and/or pedostratigraphically dated loess successions. Especially, Zöller and Nehring (2002) demonstrated profiles where basal layers form single, undivided layers on the slope but split up into several sublayers in the basin, where they interfinger with loesses. Their findings let one conclude that the layers on the slope have been dislocated one last time to reach their final positions, but that there were several earlier episodes of dislocation, affecting always the same matter again and again; this did not result in distinguishable layers except where fresh material, for example, loess, became deposited between the particular episodes (see Kleber, 1998, 2004a and Sauer, 2002 for this line of argumentation) or where deposited matter became trapped and jammed with time mainly in slope depressions and on foot slopes. The authors also concluded from the relationship between layers and loess paleosols that gelifluction was especially active just after periods of relative stability when the soils had developed. Their explanation for this is environmental inertia, with the stabilizing elements of the environment adjusting to climate deterioration with a remarkable time lag (see Hoek, 2001), similar to the concept of vegetation inertia proposed by Kleber (1994a) as a geomorphic agent.

Zöller and Nehring (2002) described also vertical and lateral facies changes between basal and intermediate layers within one layer or layer series and interpreted them as an effect of loess that, once mixed with a slope deposit, does not necessarily place that deposit into a certain type of paleoenvironment (see Section 2.6.4). This casts doubt to the idea that basal and intermediate layers always formed during cold/moist phases and during cold/dry episodes, respectively (Kleber, 1992a; Völkel, 1995; Völkel et al., 2002a), as also expressed by Sauer (2002).

None of the cited authors attempted to transfer their findings of old basal layers to all or a larger number of occurrences of basal layers. Rather, they agreed that the majority of basal layers be placed in the last, Weichselian (Würm) glaciation, mainly in its upper part (summarized by Frühauf, 1996), which is in agreement with evidences of extensive erosion at the onset of the Weichselian (Rohdenburg, 1965; Schönhals et al., 1964; Semmel, 1968). Especially in cases of very thick, multilayered basal layers, deeper sections may be remarkably older than higher ones (Völkel et al., 2002a), though there usually is no supporting evidence other than sheer thickness for this. Thus, the theoretical maximum age of basal layers is the onset of glaciations at the end of the Neogene.

However, there is some indication allowing to constrain the minimum age of basal layers. Several independent works from formerly glaciated areas in Central Europe yielded a coherent picture of basal layers being absent on moraines and tills of the Last Glacial Maximum. This holds true for the higher parts of subdued mountains with local glaciers such as in the Bavarian Forest (Raab, 1999; Raab and Völkel, 2002, 2003; see Section 2.8.1), in the German Alpine foreland (Bussemer, 2003; Kösel, 1996, 1997), and in the domain of the Scandinavian ice sheet in northern Central Europe (Bussemer, 1994, 2002). Often these studies had difficulties to keep basal layers and tills apart, because tills in their upper parts typically have a very high bulk density, similar to basal layers. Differentiating sediment types was achieved by, for instance, clast-orientation measurement (see Fig. 2.28 later in this chapter). The astonishing compactness of till in the relative position of a basal layer is, by the way, another indication that this consolidation is not an original property of the basal layer but due to secondary overprint, probably by later permafrost (see Section 2.6.1). Raab (1999) described a thin, less consolidated transitional layer between a till and an upper layer, the genesis of which he could not finally clarify through physical, mineralogical and macrofabric analyses. Other studies just mention the entire lack of basal layers or equivalent deposits. In conclusion, typical basal layers have not formed after the Last Glacial Maximum in three rather different and distant areas, so that there is a chance that this result is transferable.

However, there comes a caveat with this conclusion: some authors studying soil parent material on till of earlier glaciations also report missing basal layers (Lorz, 2008; Mailänder and Veit, 2001). This might be due to confusion of basal layers with till, as Mailänder and Veit (2001) concede, but might also mean that till and basal layers are mutually exclusive for some yet unknown reason (on the other hand, see Section 6.2.1 for evidence of a basal layer overlying older till).

2.7.1.3 Age of the Intermediate Layer

Age handles on intermediate layers are even scantier. In a few cases, there are paleosol relics preserved in intermediate layers which may serve as age markers (Kleber and Stingl, 2000; also see Fig. 2.19). Thereafter, intermediate layers may occasionally have formed as early as the penultimate glaciation. Cold and dry climate conditions are commonly attributed to the formation of intermediate layers (Kleber, 1992a; Völkel, 1995; Völkel et al., 2002b) or to the deposition of the loess contained in them prior to their formation (Sauer, 2002). Evidence for incorporation of older material into intermediate layers demonstrates that intermediate layers may develop independent of synchronous loess admixture, owing their allochthonous components to earlier episodes and, thus, paleoenvironmental conditions. This may hold true as long as gelifluction acts as an "inheritance" rather than an "overriding" type of dislocation, the latter presumably being much less frequent (see Section 2.6.4).

Just one study provides some clues on the minimum age of intermediate layers. In the Alpine foreland of Switzerland, intermediate layers are absent on moraines (Mailänder, 2002; Mailänder and Veit, 2001) that are older than about 14 ka (^{14}C uncal.: Hadorn et al., 2002), probably older than 15 ka (^{10}Be: Ivy-Ochs et al., 2008). How far this result from a single area quite peripheral to Central Europe may be transferred to the entire region is questionable.

2.7.1.4 Age of the Upper Layer

Since admixed minerals from the Allerød (12.9 ka: Schmincke et al., 1999) eruption of Lake Laach (Bogaard and Schmincke, 1985) had been reported in the upper layer (Semmel, 1968), its timing has appeared unquestionable for a long time: This finding was interpreted to the effect that the tephra shortly preceded the formation of the upper layer and were incorporated into the latter during its dislocation. As a consequence, the upper layer was believed to have formed during the Younger Dryas (e.g., Frühauf, 1996; Kleber, 1991, 1994b; Semmel, 1968, 1998a; Semmel and Terhorst, 2010; Völkel, 1994; Zöller, 1980). Rohdenburg (1968) assigned a partially older age to an upper layer, when he described bipartite upper layers intercalated by Lake Laach Tephra (see Kösel, 1996, 1997, for a comparable observation, though without intercalated tephra).

In the recent past, arguments were raised challenging this common sense regarding the correlation between tephra and upper layer, that is, of the previously assumed timing of upper-layer formation:

- Mixing of tephra laid upon a sediment into this sediment through pedo- and/or bioturbation would be an equally valid explanation of incorporated tephra particles. Unintentionally, Fried (1984) demonstrated that later incorporation occurs, when he substantiated components of the tephra in intermediate and even in basal layers, which are quite certainly older than the Younger Dryas. On the other hand, the frequency of diagnostic tephra minerals in deeper layers decreases with depth (Fried, 1984), contrary to the finding of the highest mineral abundances near the bottom of upper layers (Stöhr, 1963).
- Bogs and fens in some subdued mountains were dated palynologically and by C-14 as Allerød or even Bölling, that is, older than the Younger Dryas. They were found to overlay upper layers, which could uninterruptedly be traced from surrounding slopes to beneath the peat (Völkel and Leopold, 2001; Völkel and Mahr, 1997; Völkel et al., 2001, 2002a,b; Fig. 2.21). Accordingly, these authors dated the upper layer as Older or, even more likely, Oldest Dryas, that is, to approx. 13.6 varve-ka before present (Brauer et al., 1999a).

A simple explanation for these seemingly incompatible results would be regionally different ages of the upper layer. However, calling for paleoclimatic gradients that were significantly steeper than, and geographically very different from, today's, this is unsatisfactory. It should be discussed, therefore, whether both lines of argumentation may in fact be in agreement and do not rule out one another. The hypothesis proposed to unify the contrasting views is that the upper layer was dislocated several times, viz., during each of the Late Glacial climate oscillations (Oldest, Older, and Younger Dryas), with the same material involved in most cases (Kleber, 2004a; Sauer, 2002). Stacked layers usually did not form. Whenever dislocation reached a bog, it was ceased because gelifluction does not operate in open water (e.g., Washburn, 1979), because this

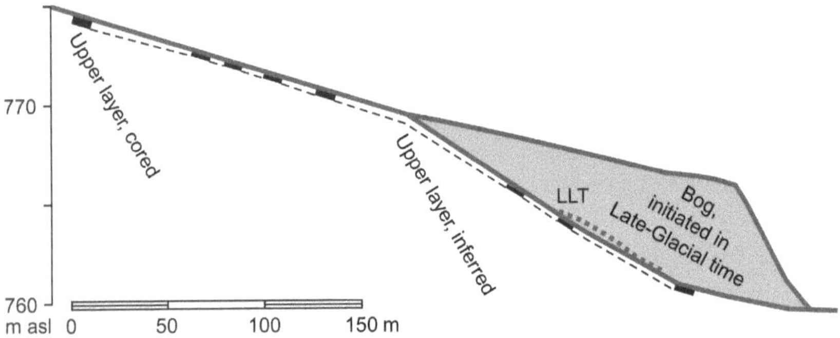

FIGURE 2.21 Relationship between a bog, Lake Laach Tephra (LLT), and the upper layer, Black Fen, Rhön Mountains, Hesse, Germany. The distribution of the upper layer was derived from a coring motor auger and geophysical sounding using ground-penetrating radar (*redrawn and simplified from Völkel et al., 2002a*). (For color version of this figure, the reader is referred to the online version of this chapter.)

would render pore-water pressure gradients impossible. Thus, where there was a lake or a bog, layers could have developed only before the advent of the open water body, except the peat body was frozen (Kleber, 2004a).

Stacked gelifluction layers (Section 2.3.1) may form under three conditions: (1) in the case of an "overriding" type of dislocation (Section 2.6.4), (2) in the case of a sufficiently long duration of formation to produce jam effects in a concave terrain, provided that the dislocation is retarded on slope flattening at all, or (3) by addition of fresh, allochthonous matter to the surface, which becomes affected by dislocation (Kleber, 2004a). If none of these conditions applies, dislocation would involve always the same material so that individual dislocation episodes would not leave traces in the sediment.

The absence from the peat bogs of a Younger Dryas mineral layer (Völkel and Leopold, 2001) does not preclude gelifluction having acted on the slopes surrounding the bogs at just that time, provided there was open water in the bog to abruptly terminate gelifluctional dislocation. A prerequisite for this termination, thus, is an open, unfrozen water body during the season of slope activity. Obviously, this is incompatible with synchronous, continuous permafrost, which is an often supposed requirement of upper-layer dislocation; however, permafrost as a necessary driver of upper-layer genesis has already been rejected for other reasons (Section 2.6.3).

It is important to discuss the relative importance of the aforementioned three conditions for the formation of stacked layers with regard to upper layers: (1) The formation of overriding lobes may be regarded as failure of gelifluted materials. Failure of such materials is mainly triggered by grain size (especially clay content), stress history, and slope inclination (Harris et al., 2008c). If these thresholds were not reached, upper layers would not have formed lobes. (2) According to measurements of active gelifluction in seasonally frozen ground (Jaesche, 1999), slope inclination plays a minor role in this process, so that stacked layers due to jam effects are unlikely to occur. (3) The most difficult issue comes from the lack of external additions, because loess influx is usually believed to be synchronous to layer dislocation. However, there is a general lack of loess during the Younger Dryas in Central Europe (Semmel, 2002a). If so, loess in the upper layer may be derived from older layers, explaining the virtual absence of Younger Dryas loess in peat bogs (Völkel and Leopold, 2001). Koster (2005), in a literature review, dated the last loess deposition episode in West (including Central) Europe as Oldest Dryas. In northern Central Europe, eolian sand has been deposited during the Younger Dryas, but not loess (Enters et al., 2010; Isarin and Renssen, 1999; Latałowa and Borówka, 2006; cf. Litt and Stebich, 1999; and Brauer et al., 1999b for maar sediments of the Eifel Mountains); in the southern part, there is even no evidence for eolian sands, let alone loess (Niessen et al., 1992, for Lake Constance). Activity of gelifluction during the Younger Dryas has been rated insignificant over all, because considerable material input into fluvial systems at the time has not been found (Andres et al., 2001).

2.7.2 Numerical Dating of Periglacial Cover Beds

2.7.2.1 Introduction

As previously discussed, the age of periglacial cover beds has been a matter of debate for decades. This is promoted by the fact that handles for numerical age determination of these deposits are limited. They are often devoid of organic material, which could be used for radiocarbon or dendrochronological dating, or of pedogenic carbonate, which could be used for the uranium series method (Bateman, 2008). Therefore, optically stimulated luminescence may be one of the few valuable tools to derive numerical ages on cover beds. This radiometric dating method enables to assess the time that has elapsed since sediments were last exposed to sunlight. It is frequently used for age determination of Quaternary deposits of, for example, eolian, fluvial, and glacial origin. Luminescence dating was carried out in a recent study to derive the ages of periglacial slope deposits at three sites in the Taunus Mountains (Hohe Wurzel and southeastern Taunus Mountains, Fig. 2.1, see Hülle et al., 2009).

Völkel and Mahr (2001) demonstrated considerable problems of luminescence dating of periglacial slope deposits. Based on multiple aliquot infrared stimulated luminescence dating of polymineral fine-grain samples, they found a broad age range for both upper (3.0 ± 0.7 to 18.3 ± 3.9 ka) and intermediate layers (6.0 ± 1.3 to 25.5 ± 2.6 ka). They interpreted the overestimates in terms of inhomogeneous bleaching related to gelifluction, which does not necessarily expose all involved grains to sunlight, and the underestimates by postdepositional admixture of younger grains by bioturbation, pedoturbation, and uprooting.

Given that Völkel and Mahr (2001) demonstrated the incapability of multiple aliquot techniques to yield precise, accurate age estimates, Hülle et al.

(2009) used single-aliquot regenerative-dose (SAR) protocols for equivalent-dose determination. These protocols allow for a more detailed investigation of equivalent-dose distributions. This is advantageous in particular for sediments which possibly have not been fully bleached during transport and deposition such as, presumably, cover beds.

SAR protocols were applied to sand (quartz and K-feldspar, separately) as well as silt (polymineral fine-grain fraction) to take advantage of the specific characteristics of the different fractions. Quartz bleaches faster than feldspar (Godfrey-Smith et al., 1988), which is important in case of unknown bleaching conditions. The reduction of insecurities in dose-rate determination for K-feldspars resulting from the internal dose from the decay of ^{40}K in the crystal lattice is advantageous (Preusser, 2003). However, K-feldspar contents in sediments are often low due to faster weathering, and feldspars often are affected by severe age underestimation due to anomalous fading (Huntley and Lamothe, 2001). The latter also applies to the feldspar-dominated luminescence signal of polymineral fine-grained material (Zink, 2008). However, this fraction yields high signal intensities and consistent dating results (Rees-Jones, 1995). Presumably, this also is the grain-size fraction with the highest percentage of allochthonous eolian loess components in cover beds; loess is one of the most appropriate materials for this dating technique. Furthermore, various techniques for dose-rate estimation were applied to elaborate which approach is best suited to obtain a realistic assessment of the natural radiation field in these heterogeneous deposits.

Luminescence dating was carried out at three sites in the Taunus Mountains (Hülle et al., 2009). Site 1 was located in a slope hollow on a slope exposed to the west and site 2 on a straight slope exposed to the east (see Section 2.6.2 for the influence of landform and aspect especially on intermediate layers). Samples from all three layers were taken from these profiles, each including two or three subsamples to test for reproducibility of the results. The third site was "Hohe Wurzel" (see Figs. 2.22 and 2.23 for the relationship between cover beds and tephra), which yielded independent age control via Lake Laach Tephra.

Techniques and experimental details used for analyses, especially with regard to equivalent-dose estimation, the potentials of different mineral fractions (coarse grains of quartz and feldspar, polymineral fine grain) for dating cover beds, and dose-rate determination were elaborated by Hülle et al. (2009).

2.7.2.2 *Methodical Difficulties in Dating Cover Beds*

Insufficient bleaching due to the mode of transport and, consequently, age overestimation was expected to be a major problem for luminescence dating of periglacial slope deposits (Völkel and Mahr, 2001). Contrary to this, tight dose distributions obtained for sand-sized K-feldspars from intermediate layers and scarcely skewed distributions of the quartz results are interpreted as evidence for sufficient bleaching. Measurements of large and small aliquots at

FIGURE 2.22 Lake Laach Tephra underlying an upper layer, Amöneburg Basin, Hesse (*photo by A. Kleber*). (For color version of this figure, the reader is referred to the online version of this chapter.)

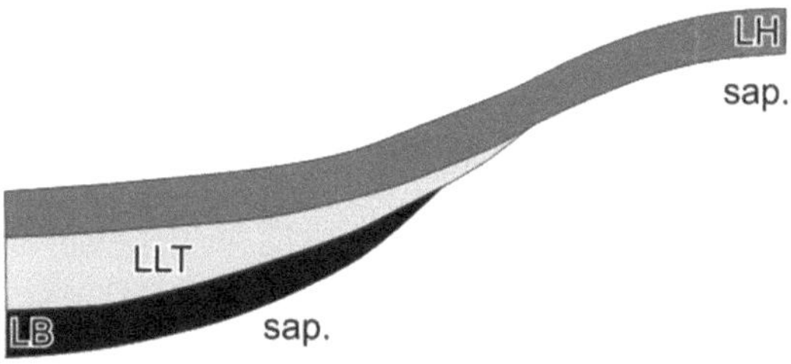

FIGURE 2.23 Schematic cross-section of cover beds under- and overlying Lake Laach Tephra at Hohe Wurzel, Taunus Mountains, Hesse, Germany (*simplified from Semmel, 2003*). LLT = Lake Laach Tephra, LH = upper layer, LB = basal layer, sap. = saprolite.

two of the sites yielded consistent results (Hülle et al., 2009). This would also argue for sufficient resetting of OSL signals in the majority of grains (bleaching conditions of basal layers will be discussed below).

Nevertheless, concerning the dating of cover beds, there are other substantial problems related to dose-rate determination, equivalent-dose determination, and postdepositional history of the sites. The main problems of equivalent-dose determination are poor luminescence properties of quartz. Figure 2.24 gives a brief

		Site 1 Fine grain	Site 1 Quartz	Site 1 Feldspar	Site 2 Fine grain	Site 2 Quartz	Site 2 Feldspar
LH	Signal	(+)	(−)	(−)	(+)	(?)	(−)
	LM-OSL		(?)			(?)	
	fading	(−)		(−)	(−)		(−)
	scatter	(+)	(−)		(+)	(−)	
	dev. DRT	(+)	(+)		(−)	(−)	(+)
LM	Signal	(+)		(+)	(+)		
	LM-OSL		(?)			(+)	
	fading	(−)		(−)	(−)		(−)
	scatter	(+)	(−)	(+)	(+)	(−)	(+)
	dev. DRT	(+)	(+)	(+)	(−)	(−)	(+)
LB	Signal	(−)	(o)	(−)	(−)	(o)	(−)
	LM-OSL		(−)			(−)	
	scatter		(−)			(−)	

FIGURE 2.24 Summary of properties relevant for dating of the different mineral fractions for two sites. LM-OSL = linearly modulated OSL; dev. DRT = deviation from the given dose in dose-recovery tests; scatter = standard deviation observed for n individual equivalent-dose values; (+) = the attribute yielded plausible results; (o) = the attribute yielded plausible results, which may be due to incomplete bleaching, and may be regarded as maximum ages only; (−) = the attribute yielded no or no plausible results; (?) = no conclusive assessment possible; missing symbol = either there was no conspicuity concerning the examined feature or the feature was not examined because all previous tests had failed already. LH = upper layer; LM = intermediate layer; LB = basal layer (*modified from Hülle et al., 2009*).

summary of the potentials and pitfalls for equivalent-dose determination of the different mineral fractions for the two sites with all three periglacial layers (upper, intermediate, and basal layers) examined. Some characteristics seem to be layer-dependent (dating of sand-sized K-feldspars), whereas others seem to be related to the sites (dating of the sand-sized quartz fraction). However, to check for an underlying regularity, additional sites need to be investigated.

The average relative error in age for the presented OSL dates is 11%. However, as shown by comparing different approaches for dose-rate determination (Hülle et al., 2009), the inhomogeneous radiation field introduces the major uncertainty into age calculation, which is not even included in this analytical error. Figure 2.25 demonstrates how strikingly the derived ages depend on the method of dose-rate estimation: Ages may even be almost doubled, that is, shifted from the Last Glacial Maximum to the Younger Dryas (sample a from upper layer, age derived from quartz, method I vs. V). Thus, it is difficult to evaluate the true age of deposition of the eolian matter, let alone the age of dislocation (Section 2.6.4), without any independent age control.

Other than luminescence-related issues influence the results. There is post-sedimentary mixing within the upper layer (Section 2.6.3), which complicates establishing a reliable luminescence age determination (Bateman et al., 2003, 2007). Heimsath et al. (2002) demonstrated by single-grain OSL measurements

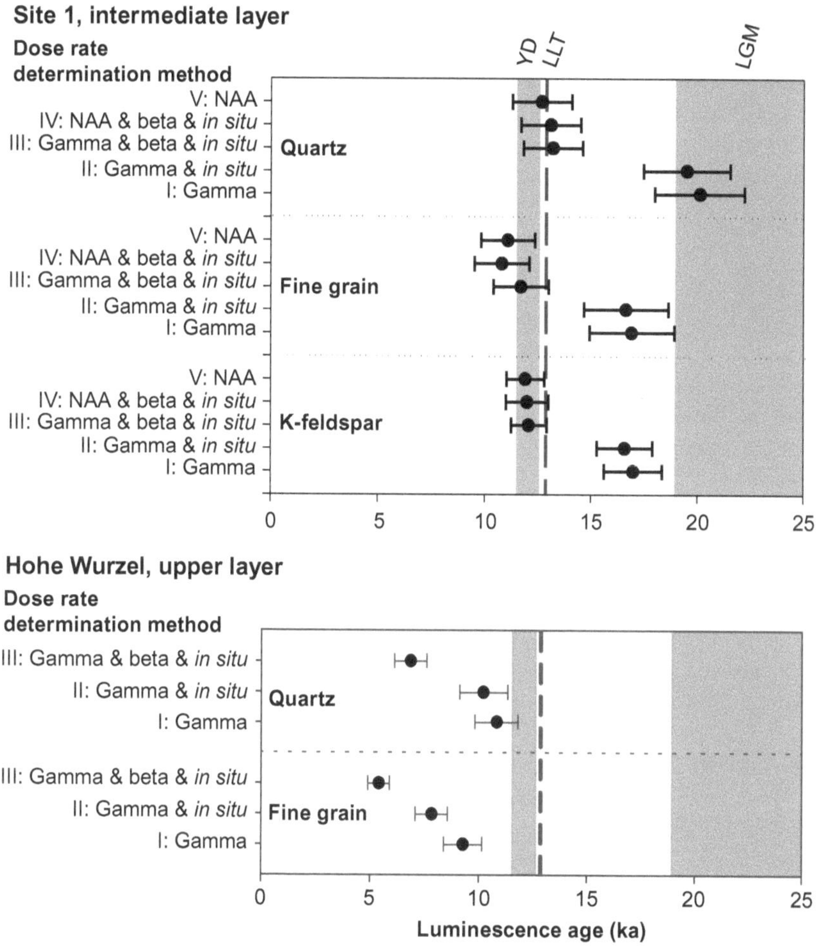

FIGURE 2.25 Comparison of age results for different dose-rate determination methods (*modified from Hülle et al., 2009*). The ages were calculated for a sample from an upper layer and from an intermediate layer using different dose rates: I = dose rate was calculated with U/Th and K contents determined by laboratory γ-spectrometry; II = α- and β-dose were determined by laboratory γ-spectrometry, γ-dose determined by *in situ* γ-scintillation; III = α-dose was determined by laboratory γ-spectrometry, β-dose determined by β-counting, γ-dose determined by *in situ* γ-scintillation; IV = α-dose was calculated with U/Th and K contents determined by neutron activation analysis, β-dose determined by β-counting, γ-dose determined by *in situ* γ-scintillation; and V = dose rate was calculated with U/Th and K contents determined by neutron activation analysis. YD = Younger Dryas (ages *modified from Litt et al., 2003*); LLT = Lake Laach Tephra (as dated by Schmincke et al., 1999); LGM = Last Glacial Maximum (ages *modified from Clark et al., 2009*).

and Monte Carlo simulations that repeated grain displacement may result in mixing of an entire soil column. In the profiles under study, micromorphological examination indicated mixture through rooting and burrowing by animals, which was even able to rotate rock fragments (Hülle et al., 2009) and which may have dislocated sand-sized grains from surface to depth.

The uppermost parts of the cover beds may have been disturbed by human impact, as Roman and Medieval settlements are known from this region (RGZM, 1972). Furthermore, remobilization of the sediment or rinsing out of fine material, especially in the case of catastrophic events and at times of extensive forest clearing, was enhanced at least on steeper slopes. It is barely possible to assess to which degree the various external factors actually affected the cover beds at the different sites throughout the past.

2.7.2.3 Luminescence-Dating Results from Basal Layers

Samples collected from basal layers at two sites were examined in addition to the loess-rich layers. OSL dating is even more complicated for basal layers than for other layers: all problems that affect upper and intermediate layers also apply to basal layers. Furthermore, there is evidence of incomplete bleaching, and the luminescence properties of the minerals are even worse than in the other layers.

Coarse-grain K-rich feldspars and the polymineral fine-grain fraction showed no luminescence signals at all. Only the measurements of coarse-grained quartz appeared promising. At least from large aliquots (>1000 grains), equivalent-dose values could be calculated. As the basal layer has a high content of autochthonous material (Section 2.2.1), poor bleaching is very likely, and evidence for this is obvious (Hülle et al., 2009). In basal layers, there was also indication of a slow component of the luminescence signal, requiring substantially longer times of exposure to light for resetting than the fast component usually utilized for OSL dating (Singarayer et al., 2005).

In summary, the resulting ages may only be interpreted as maximum ages: the results are <59 and <106 ka for site 1 and <83 and <97 ka for site 2, respectively. The variation of ages calculated for the lowest and highest equivalent dose of the distributions ranges between 43 and 195 ka. These tentative estimates would support previous age assumptions that basal layers date from before the Last Glacial Maximum.

2.7.2.4 Luminescence-Dating Results from Intermediate Layers

All dating results from loess-enriched layers are presented in Fig. 2.26. In the light of the described problems, interpretation of the derived ages is arguable. Even though consistent results for the same fraction from one layer could be achieved in some cases, ages of different mineral fractions obtained from the same sample often differ considerably—especially if ages were corrected for anomalous fading. These discrepancies cannot be explained by different bleaching characteristics. Nevertheless, a tentative attempt in interpreting the results is given in this and the subsequent section.

Two sites in the southeastern Taunus Mountains where samples from intermediate layers were obtained yielded remarkably different results. At site 1, with the intermediate layer located in a slope depression on a west-facing slope, the quartz ages indicate deposition around the last glacial maximum (19–23 ka).

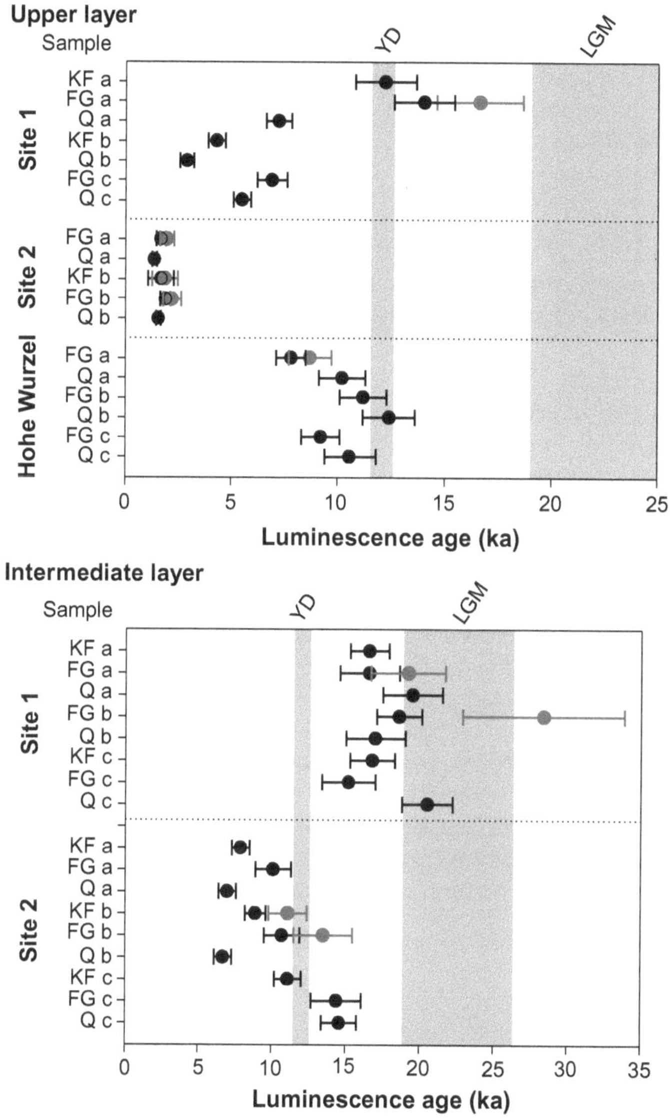

FIGURE 2.26 Luminescence dating results of upper and intermediate layers. FG = polymineral fine grain, Q = coarse-grain quartz, KF = coarse-grain potassium-rich feldspars, a, b, c = repetitive samples. Additionally, fading-corrected ages are given in gray. YD and LGM as in Fig. 2.23 (*reproduced from Hülle et al., 2009*).

Both the polymineral fine-grain fraction and the K-feldspar sand fraction show a severe signal loss due to anomalous fading and, thus, an age underestimation of 15% and 53%, respectively. In contrast, most of the dates from the intermediate layer of site 2 (whether fading-corrected or not) on an east-facing slope are in an age range younger than the Pleniglacial but not younger than the Younger Dryas.

This discrepancy might be an effect of the many pitfalls inherent to the method as applied in this study. However, if they are correct, the results may be interpreted to imply that intermediate layers have been formed at different times; maybe the different location (slope depression in the luff vs. straight slope in the lee) plays a role in this. In the Vienna Forest, the intermediate layer was OSL dated as younger than 25.4 ± 4.8 ka (Frank et al., 2011).

2.7.2.5 Luminescence-Dating Results from Upper Layers

Upper-layer ages were derived at all three sites. The OSL results show that the best conditions for application of luminescence dating were provided at "Hohe Wurzel." Because of a smaller amount of rock fragments in the upper layer, the radiation field appears to be more homogeneous than at the other sites. This is documented by good agreement of the γ-spectrometry results for different subsamples (Hülle et al., 2009). The consistency in age between the subsamples a, b, and c gives an additional hint on the reliability of the OSL ages. After fading correction, the ages for the fine-grained material (8.7 ± 1.0, 12.5 ± 1.4, 10.2 ± 1.1 ka; uncorrected: 7.8 ± 0.7, 11.2 ± 1.1, 9.2 ± 0.9 ka) and the quartz (10.2 ± 1.1, 12.4 ± 1.2, 10.6 ± 1.2 ka) still are consistent within the error limits. Finally, the independent age control by underlying Lake Laach Tephra (Semmel, 2001a, 2003) corroborates the hypothesis that the upper layer at this site was deposited during the Younger Dryas, when a cold reversal of the climate may have re-enabled periglacial processes to some extent.

Regardless of the mineral fraction or dose-rate values used, the model ages of upper-layer samples at the other two sections are almost all Holocene in age and thus apparently far more influenced by the described problems. Whether the young ages may be explained by re-deposition and/or postsedimentary mixing remains speculative. However, it is noticeable that the only Late Glacial age was obtained for a sample from a less mixed upper layer on a less inclined slope.

2.7.2.6 Conclusions

The potential for reconstructing depositional ages of cover beds by optical luminescence dating is mainly limited by the determination of the true dose rate, due to the heterogeneous nature of these deposits. As this problem can neither be sufficiently solved by laboratory analyses nor by computer simulation, the best approach to assess the true dose rate in future studies of gelifluction layers would be a long-time survey with *in situ* dosimeters inserted into the sediment as close as possible to the later sampling position.

Because of the observed fading of feldspars and the feldspar-dominated fine-grained fraction, quartz turned out to be preferable for equivalent-dose determination. However, as all quartz samples dated in this study were characterized by rather dim OSL emissions, coarse-grained populations are needed to improve the reproducibility of the OSL measurements. However, such large subsamples do not provide detailed insight into the bleaching conditions.

Therefore, additional single-grain or small subsample analyses of K-rich feldspars may be the best approach toward investigating the influence of postdepositional mixing and the degree of signal resetting.

In addition to the difficulties in precise dose-rate and equivalent-dose estimation, optical dating of periglacial slope deposits is further complicated by other factors such as pedoturbation, anthropogenic disturbance, and/or topography.

Despite all these problems, the results at the site "Hohe Wurzel" show at least some potential for optical luminescence dating. The luminescence ages correspond reasonably well with the expected ages after the deposition of the well-dated Lake Laach Tephra (12.9 ka, e.g., Schmincke et al., 1999) and before the beginning of the Holocene (Semmel, 2003).

Altogether, the database of this study does not allow a profound interpretation of the succession of geomorphic processes at the study sites. Even if the OSL dates appear to be precise, insufficient accuracy limits their significance for paleolandscape reconstruction especially in the Late Glacial with its rather short-lived climate oscillations. Further systematic studies with emphasis on high-resolution downhill transects are desirable. Thereby, the variable impact of the aforementioned external factors may be elucidated in more detail and perhaps better accounted for.

2.8 REGIONAL DIFFERENCES IN COVER-BED PROPERTIES AND DISTRIBUTION

T. Raab, D. Sauer, H. Bullmann, P. Felix-Henningsen, T. Scholten, E.-D. Spies

2.8.1 Highest Altitudes of Subdued Mountains

T. Raab

During Pleistocene cold stages, higher elevations of subdued mountains were affected not only by periglacial but also by glacial processes. Depression of the snowline led to plateau and valley glaciation in, for example, the Black Forest (equilibrium line altitude (ELA) 850–1100 m asl), with glaciers up to 25 km long (Metz, 1997), and to spatially more restricted valley and cirque glaciers with maximum lengths of 7 km in the Bavarian Forest (ELA 1000–1150 m asl; Raab and Völkel, 2003). Quaternary sediments in these areas typically include two major components: (1) glacigenic till, directly deposited by glacial ice and (2) nonglacigenic sediments in the vicinity of the glacial environments which may be ascribed to mass movement or re-sedimentation processes. Stratigraphic concepts and genetic classifications of glacigenic sediments (Brodzikowski and van Loon, 1987, 1991; Dreimanis, 1989) that have been developed for decidedly glacial environments may successfully be applied to these settings. In combination with detailed surveying and geomorphological mapping, Raab (1999) classified periglacial and glacial environments in the

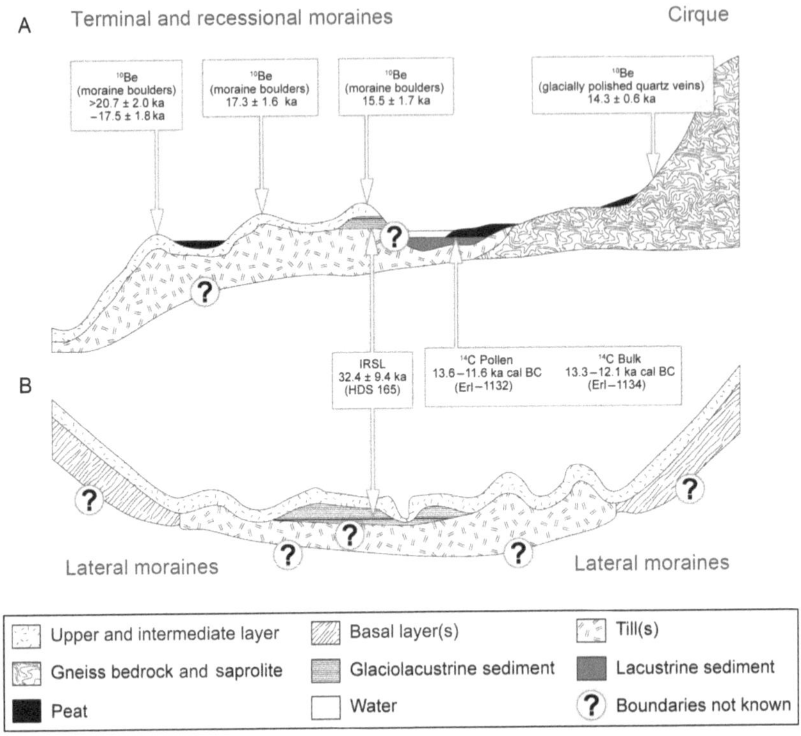

FIGURE 2.27 Stratigraphic and age relationships between periglacial, glacigenic, glaciolacustrine, lacustrine, and organic sediments in the formerly glaciated landscape of the Kleiner Arbersee Valley, Bavarian Forest, Germany (*redrawn from Raab et al., 2007*).

Arber area, Bavarian Forest (Fig. 2.27). The genetic imprints of gelifluction and cryoturbation especially may be derived from clast shape, clast fabric, and occasionally from the grain-size distribution of the sediments (see Section 2.4 and Fig. 2.28).

The glacial history in subdued mountains has major impact on the distribution not only of glacigenic but also of periglacial sediments. The existence of well-developed soils such as Inceptisols[5] and Alfisols[6] is restricted to areas that were not affected by glacial erosion. In cirques and other areas of glacial abrasion, Holocene pedogenesis was not able to proceed beyond the Entisol[7] stage. Semmel (1985) already emphasized the relationship between the maturity of

5. Soil type characterized by a cambic subsurface horizon with evident oxidization of iron-bearing minerals.
6. Soil type characterized by an argillic subsurface horizon, enriched by translocated clay.
7. Immature soil type characterized by slight to absent weathering and thin to discontinuous humus accumulation.

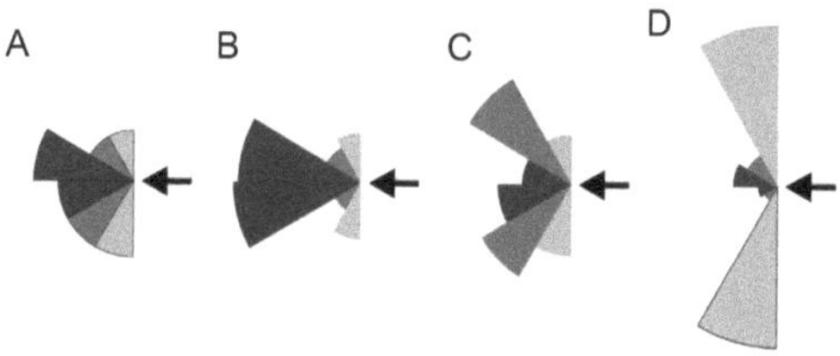

FIGURE 2.28 Clast orientation of cover beds and tills. The dip direction of clasts' longitudinal axes was measured in six classes of 30° width each, as described in Raab and Völkel (2003). Tilted clasts were omitted. The radius of the sector denotes the relative abundance in the respective class, while the arrow expresses the direction of transport (based on data of Raab & Völkel, 2003). (A) Clasts' longitudinal axes of the upper layer are random, (B) those of the basal layer are well oriented parallel to slope inclination, and (C) those of till are either random or (D) in the case of lodgment till perpendicular to the flow direction.

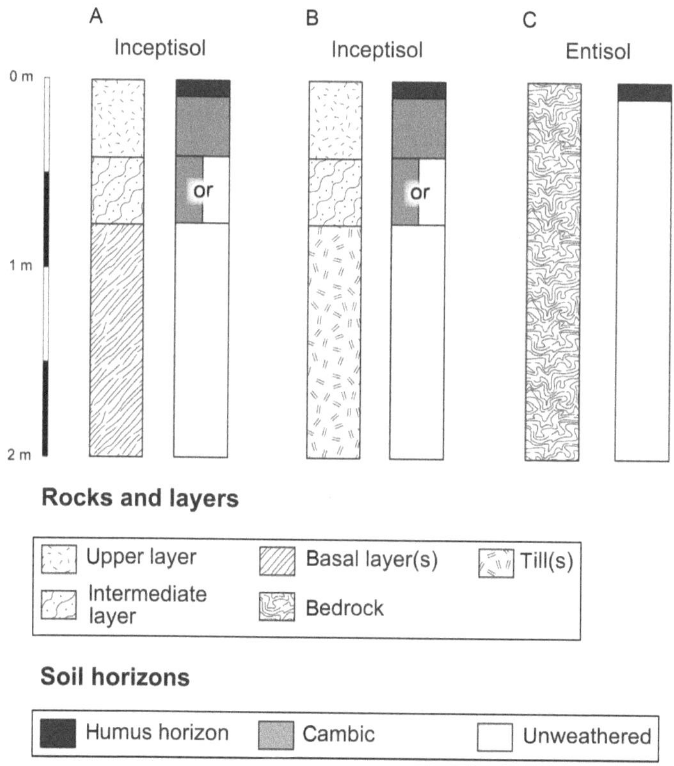

FIGURE 2.29 Schematic soil profiles from the formerly glaciated landscape of the Kleiner Arbersee Valley, Bavarian Forest, Germany, showing different intensities of pedogenesis in (A) periglacial cover beds, (B) periglacial cover beds overlying glacigenic sediments, and (C) stripped bedrock (*modified from Raab et al., 2007*).

soils and the cover-bed distribution and concluded that the depth of the solum and the intensity of pedogenesis are more influenced by sediment characteristics than by the duration of Holocene weathering processes. By small-scale mapping of soils and sediments, Raab et al. (2007) could provide support for this view (Fig. 2.29).

As a consequence of the differing physical properties of glacigenic and periglacial sediments, distribution, stratigraphy, and intercalation have a major impact on the drainage patterns and dynamics of mountainous landscapes (e.g., Kleber et al., 1998b; Rezzoug et al., 2005). In the Bavarian Forest, for example, slopes which consist exclusively of cover beds show less surface runoff but a dominating interflow component, mainly in the basal layer (Völkel, 1995, also see Chapter 4). In Pleistocene glacial landscapes, the interflow may be blocked by overconsolidated glacigenic sediments like lodgment tills. Therefore, especially at the borderline between landscapes dominated by periglacial and glacigenic sediments, changes in surface and subsurface slope-water dynamics are common (Fig. 2.30).

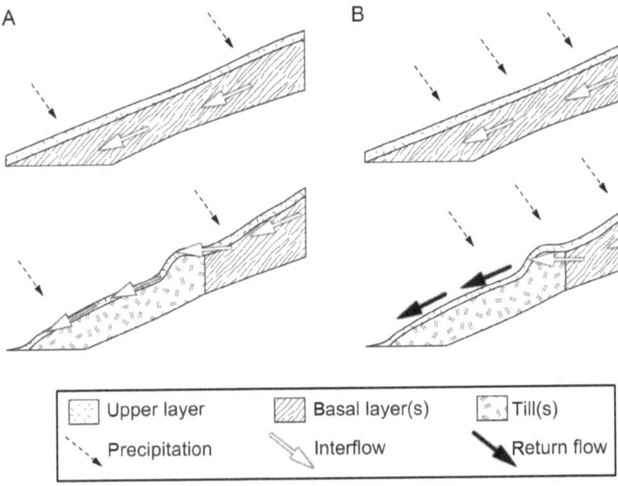

FIGURE 2.30 Interrelations of surface runoff, interflow, and Quaternary sediments at Kleiner Arbersee, Bavarian Forest, Germany, under different meteorological conditions: (A) normal precipitation and (B) intensive precipitation. During rainfall of moderate intensity, water first infiltrates the poorly consolidated upper layer, which covers the entire slope, and then percolates vertically into the underlying substrates. In the basal layer, water runs parallel to the surface, that is, down-slope. On intermediate slopes where most Pleistocene valley glaciers were located as indicated by lateral moraines, basal layers are not present down-slope from the moraines (Fig. 2.27). Instead, the upper layer is directly underlain by overconsolidated lodgment tills, which serve as aquitards. Consequently, interflow through the upper layer persists and, as rainfall increases, the interflow water table rises toward the surface to eventually cause return flow (lower row), whereas on slopes without till, the precipitation intensity does not make much difference (upper row) (*modified from Raab et al., 2007*).

2.8.2 Rhenish Massif

D. Sauer, T. Scholten, P. Felix-Henningsen, E.-D. Spies

The Rhenish Massif is mainly located in the western part of Central Germany, but its very western part, the Ardennes, is located in Belgium, Luxembourg, and France. The river Rhine flows through the Rhenish Massif from SSE to NNW. The Rhenish Massif was formed during the Variscan orogeny, which compressed predominantly Devonian marine sedimentary rocks into folded slates with folds running SW–NE. Petrologically, the Rhenish Massif mainly consists of slates and sandstones overtopped by quarzite ridges. Furthermore, in the Westerwald area, Devonian submarine volcanism led to widespread occurrences of dolerite.

Periglacial cover beds were studied in three subunits of the Rhenish Massif in western Germany: in the Hunsrück (W of the Rhine, S of the Mosel), the Eifel (W of the Rhine, N of the Mosel), and in the eastern Westerwald Mountains (E of the Rhine) (Fig. 2.1).

The central part of the Rhenish Massif is characterized by extensive Mesozoic to Tertiary planation surfaces at elevations between around 300 and 600 m asl, which were deeply dissected by rivers during the Pleistocene (Felix-Henningsen, 1990). The most peculiar phenomenon in the Rhenish Massif is deep kaolinitic saprolitization, which is widespread, particularly in the Hunsrück area. The saprolites show undisturbed structures of the slates and consist of kaolinite, muscovite, and quartz. They display mostly light grayish colors, however, irregularly distributed dark gray, brown, reddish, yellow, and white colors occur as well (Fig. 2.31, left and middle). Clayey and loamy material reworked from kaolinitic saprolite comprises an important component in periglacial slope deposits of the area (Felix-Henningsen et al., 1991) (Fig. 2.32).

Basal layers consisting of reworked saprolite are commonly very dense because of their fine texture. The substrate has completely lost its original structure during gelifluction. Hence, most soils are characterized by a perched water table (Albaquults[8]) if derived from successions comprising an upper layer, a basal layer, and saprolite (Sauer and Felix-Henningsen, 2006).

Intermediate layers are more frequent on the plateaus of the Eifel and Hunsrück areas, compared to the Westerwald area, and also compared to many other subdued mountain areas of Central Europe. This is assumed to be due to differences in topographic properties, that is, that the Eifel and Hunsrück comprise much gentler slopes and plateaus than most subdued mountain areas of Central Europe. In successions comprising upper, intermediate, and basal layers, as well as saprolite, clay coatings are frequent on ped surfaces of the intermediate layer (see Section 2.4.1). The water permeability of intermediate layers is higher than

8. Deeply weathered soil material with indication of soil water stagnation and an abrupt textural change between upper and lower horizons.

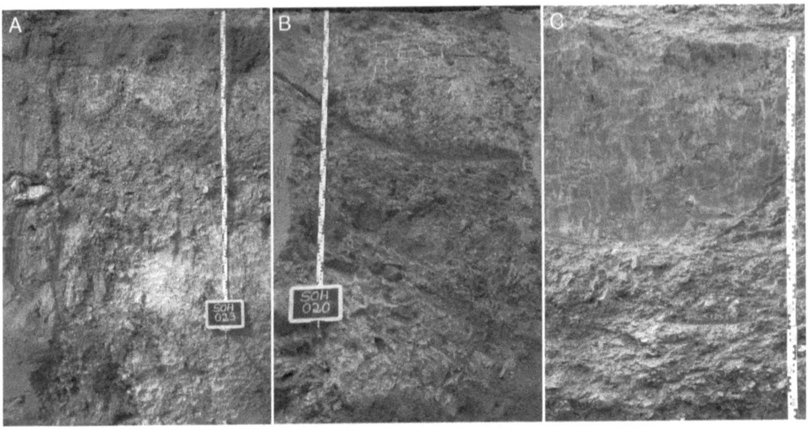

FIGURE 2.31 Brown, yellow, white, and red saprolites (A) as well as relicts of red Tertiary soils (B) from the material underlying the periglacial slope deposits have been incorporated into the slope deposits in many places in the Rhenish Massif, particularly in the Hunsrück area. Where intermediate layers occur (C), their silty texture and light-brown color reflect a high loess contents. Frost cracks, bleached due to impeded drainage, are very common in intermediate layers of the Rhenish Massif. (For interpretation of the references to color in this figure legend, the reader is referred to the online version of this chapter.)

FIGURE 2.32 Gas pipeline ditch, approx. 2.5 m deep, exposing saprolite derived from slate, covered by Pleistocene periglacial slope deposits in the Hunsrück area (western Rhenish Massif, S of the Mosel river). A: the saprolite frequently displays grayish colors. Weathered quartz veins in the saprolite are accompanied by red to orange iron oxides, the so-called Hunsrück ore, because the quartz veins were better aerated than the dense, loamy saprolite around them. The tilted upper part of the rusty quartz veins indicates down-slope movement due to creep and thus the transition from autochthonous rock into a basal layer. B: the different colors in the saprolite allow for tracing back the source area of the overlying slope deposits. (For interpretation of the references to color in this figure legend, the reader is referred to the online version of this chapter.)

that of saprolite-derived basal layers, but lower than that of upper layers. This is indicated by bleached tongues along former frost cracks and root channels penetrating into the argillic horizon in the intermediate layer (Fig. 2.31, right). Most of these soils may be classified as Albaqualfs.[9]

9. Same as Albaquult but less weathered.

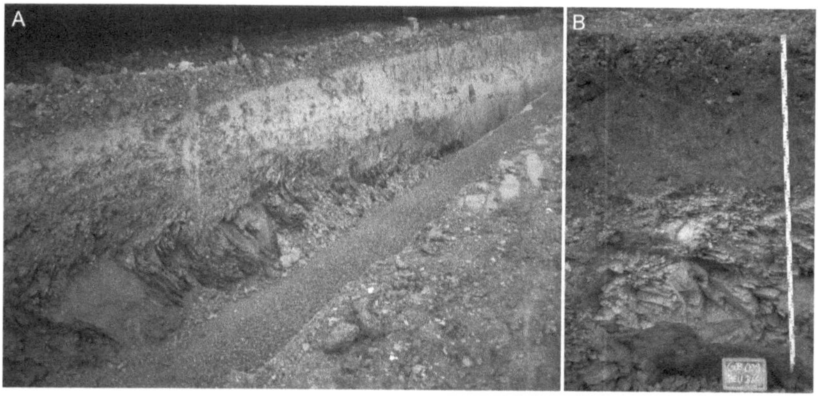

FIGURE 2.33 Left: the underlying rock in the Eifel area (western Rhenish Massif, N of the Mosel river) is dominated by alternating layers of kaolinitic saprolite and weakly or unweathered slates (A: fresh rock left, saprolite right). On top of the slate, a gray basal layer follows, composed of saprolitic material, overlain by a light-brown intermediate layer with significant amount of loess, and a deposit artificially disturbed by heavy machinery. B: the unweathered rock at this site must previously have been covered by periglacial slope deposits, which were eroded and later replaced by a Holocene anthropogenic sediment. As typical for upper layers and anthropogenic sediment in the area, the material is rich in moderately weathered glass shards from the Lake Laach eruption, which lead to its characteristic intense brown color. (For interpretation of the references to color in this figure legend, the reader is referred to the online version of this chapter.)

In addition to the slates and sandstones, Tertiary and Quaternary volcanic rocks cover considerable parts of the Eifel area. Saprolites are widespread there as well, but due to strong erosion in the course of Quaternary tectonical uplift, the basal parts of the weathering mantle had become exposed to the surface when periglacial processes were initiated. Therefore, layers of saprolite alternate with layers of less- or unweathered rock more frequently than in the Hunsrück area (Fig. 2.33, left). Soils on periglacial cover beds overlying and partially derived from unweathered or only weakly weathered rock mostly are Inceptisols (in upper-layer–basal-layer successions) or Alfisols (with the argillic horizon in intercalated intermediate layers). The Eifel is where Lake Laach is located, which is the remnant of the Allerød volcanic eruption of Lake Laach. That eruption released and distributed vast amounts of tephra over hundreds of square kilometers, which serve as a valuable chronological marker in sediments (see Section 2.7.1.4). Upper layers and Holocene anthropogenic sediments in the Eifel area, as in other parts of the Rhenish Massif, contain large proportions of Lake Laach Tephra (Fig. 2.33, right).

The eastern Westerwald area is composed of slates and sandstones, with some quartzite ridges, as well. In addition, dolerite covers a considerable area. Saprolitization is much less common in the eastern Westerwald than in the other two areas, due to tectonically induced erosive removal of the thick Mesozoic–-Tertiary weathering mantle, whereas the Hunsrück and Eifel Mountains are located in more central parts of the Rhenish Massif, which were less affected by this erosive impact. Distribution pattern and properties of cover beds were investigated along toposequences on the three rock types (Fig. 2.34). The toposequences were established perpendicular to the folds of the Rhenish Massif (on NW and SE slopes), and parallel to the folds (on SW and NE slopes). Altogether twelve toposequences were studied, four on each of the three rock types (see Sauer, 2002). The total thickness of the periglacial slope deposits generally increases down-slope. It is greatest on quartzite (increasing from 80 cm at the upper slope to 180 cm at the foot slope), followed by slate (increasing from 60 to 110 cm), whereas the slope deposits on dolerite are least thick (increasing from 39 to 86 cm) (Fig. 2.35). The toposequences were established along straight (neither convex nor concave) slope sections as far as possible. In addition, a concave slope section was investigated by coring motor auger along

FIGURE 2.34 Periglacial cover beds on slate (A), quartzite (B), and dolerite (C). (For color version of this figure, the reader is referred to the online version of this chapter.)

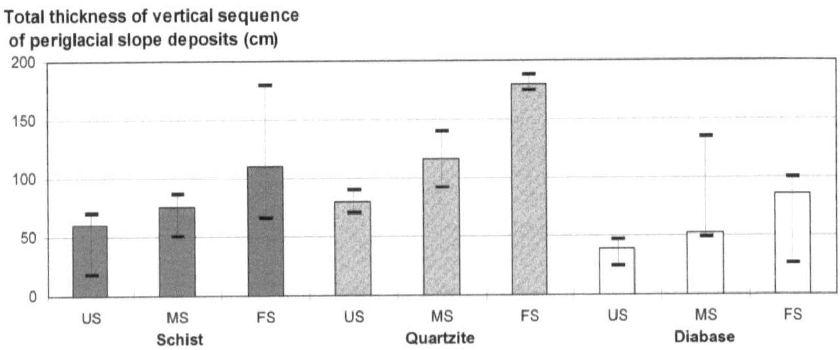

FIGURE 2.35 Total thickness of the vertical succession of Pleistocene periglacial slope deposits on slates, quartzite, and dolerite in different slope positions (US = upper slope, MS = middle slope, FS = foot slope). The figure shows medians, minimum and maximum values, obtained from 12 profiles on slates, 8 profiles on quartzite, and 13 profiles on dolerite.

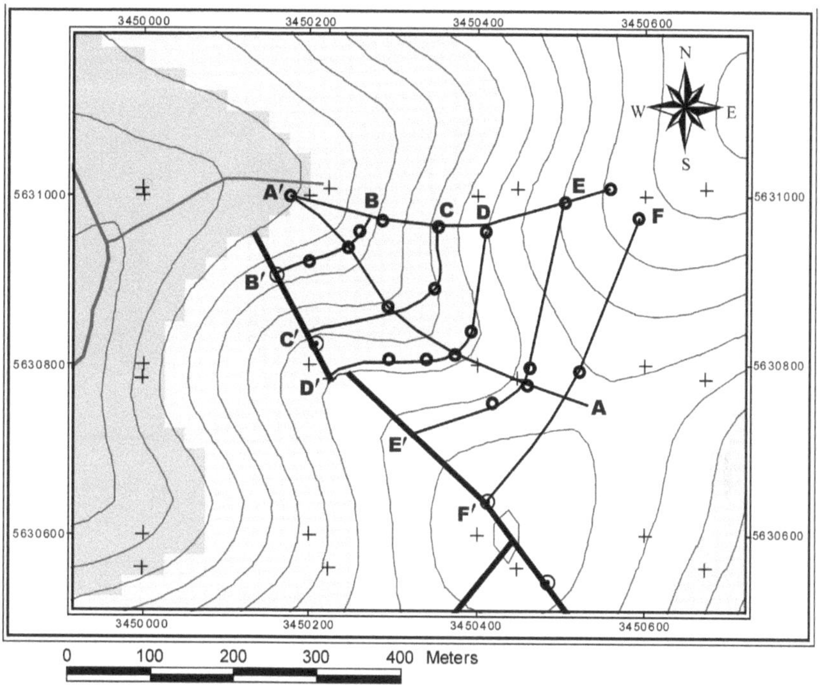

FIGURE 2.36 Concave section of a NW-exposed slope in the Westerwald area (eastern Rhenish Massif). This slope section was investigated in three soil profiles (HÜN-NW 1, 2, and 3) by coring with a motor-driven auger along a toposequence (A–A′), and along five transects running parallel to the slope (B–B′ to E–E′). Coordinates according to the Gauss–Krüger system.

transects, running vertically and horizontally through the concave slope section (Fig. 2.36). This study showed that periglacial slope deposits may be more than 5 m thick in concave slope sections, and that intermediate layers may contribute considerably to this thickness, even in the eastern Westerwald area where the distribution of intermediate layers is commonly limited (Fig. 2.37). The basal layer in this concave area is in most transects very thin or missing, pointing to an erosional phase before the depression was filled with younger slope deposits. An intermediate layer sets in already in the second transect from the top (E–E′). Two intermediate layers are present in the next lower transect (D–D′), and four intermediate layers with down-slope increasing thickness occur in the two lower transects (C–C′ and B–B′). The upper layer shows a rather constant thickness, with one exception in transect D–D′, where it has been partly eroded. A Holocene anthropogenic layer is widespread on top of the periglacial cover beds. The slope deposits on the three rock types in the Westerwald area show some differences in their contents of rock fragments (Fig. 2.38), which is, despite of considerable variability, typically in the range of 40–60 mass% in the upper layer, whereas it is lower in anthropogenic layers, that is, between 10%

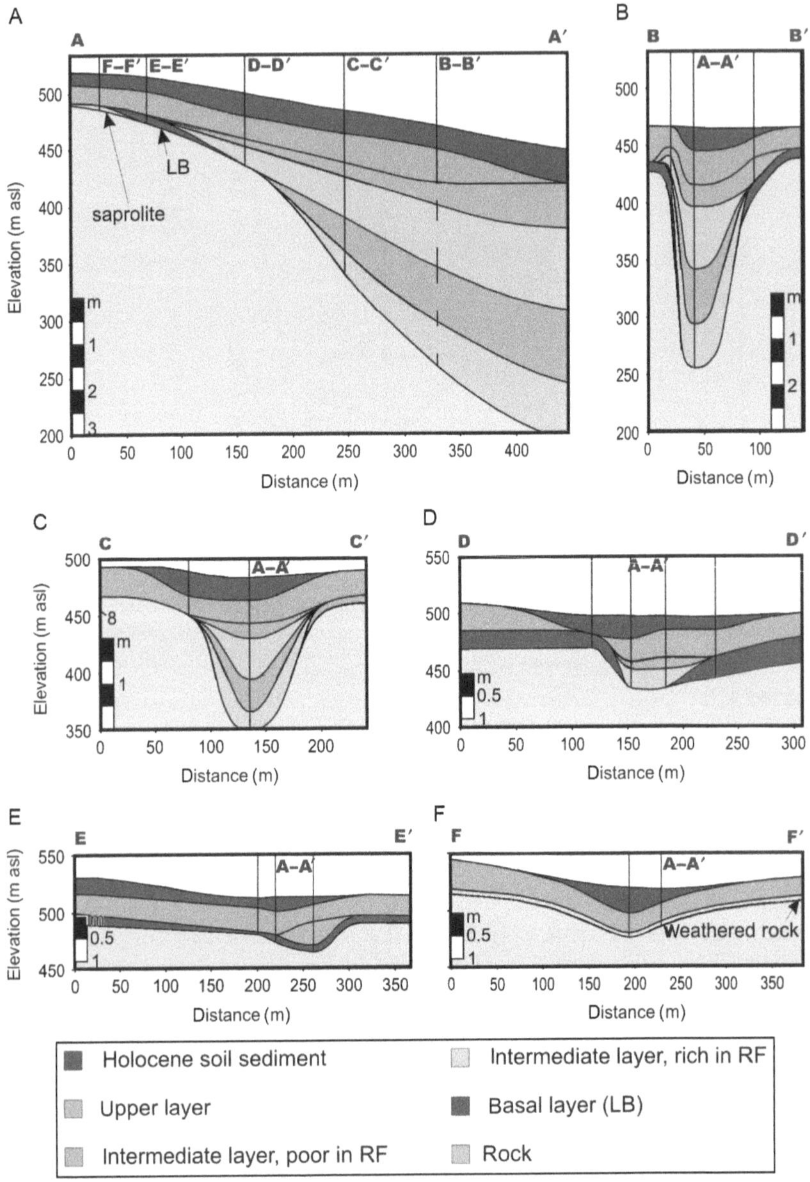

FIGURE 2.37 Vertical successions and thicknesses of periglacial slope deposits along the transects shown in Fig. 2.36. RF = rock fragments. (For color version of this figure, the reader is referred to the online version of this chapter.)

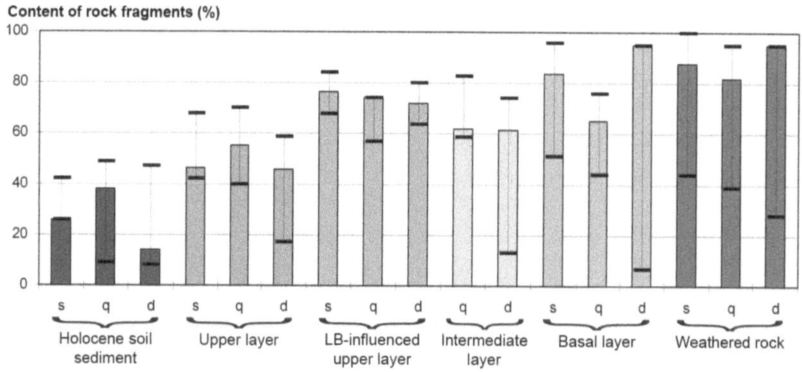

FIGURE 2.38 Contents of rock fragments in periglacial cover beds, anthropogenic layers, and underlying, partly weathered rock in the Westerwald area (s = slate, q = quartzite, d = dolerite). The figure shows medians, minimum and maximum values, obtained from 12 profiles on slate, 8 profiles on quartzite, and 13 profiles on dolerite. (For color version of this figure, the reader is referred to the online version of this chapter.)

and 40%. However, if the upper layer directly overlies a basal layer, its lower part often shows an increased content of rock fragments, which typically varies between 60% and 80% (termed "LB-influenced upper layer" in Fig. 2.38). Intermediate layers were found in five profiles on quartzite and four profiles on dolerite only, exhibiting rock-fragment contents of 60–80% rock fragments on quartzite and a very wide range between ~15% and 75% on dolerite. The basal layers have 80–85% rock fragments on slate, 65% on quartzite, and 95% on dolerite (median values). The degree of weathering and saprolitization and, hence, rock-fragment content of the basal layers is highly variable, nonetheless.

2.8.3 Carbonate Rocks of the Eastern Thuringian Basin

H. Bullmann, J. Heinrich

2.8.3.1 Introduction

Landscapes developed on calcareous bedrock have rarely been in the focus of research on periglacial cover beds, compared to numerous studies on siliceous bedrock (see Mahr, 1998; Sauer, 2002). Nevertheless, some particular dissimilarities have been pointed out in those previous overview studies:

(1) Periglacial cover beds on calcareous rock of the Middle Triassic Muschelkalk are markedly rich in rock debris and lack fines. The cover beds are often thin or even absent (Bartels, 1967; Hempel, 1955; Mensching, 1960; Rohdenburg and Meyer, 1963; Semmel, 1968; Steinmüller, 1996). Several explanations have been proposed for this: (i) calcareous rocks are relatively resistant to frost weathering under periglacial conditions. Therefore, the overall

amount of frost-debris produced was low (Braukämper, 1990; Hempel, 1955); (ii) dissolution of limestone reduced the thickness of the debris cover during the Holocene (Rohdenburg and Meyer, 1963); (iii) soil erosion processes largely shortened the original cover-bed thicknesses. Therefore, profiles on limestone are assumed to be degraded to some extent (Semmel, 1968; Terhorst, 1997). (2) Gelifluction upon limestone is inhibited compared to adjacent siliceous rocks, because frost-shattered debris is highly permeable due to the lack of fines. Together with jointed underlying bedrock, this results in lower soil moisture which further limits the gelifluction process (Washburn, 1973; Weise, 1983). As a result, decreased transport distances and less effective geomorphologic denudation may occur (Artmann and Völkel, 1999; Fritz, 1976; Hempel, 1955). (3) Furthermore, calcareous landscapes ubiquitously feature fine-grained, clayey sediments generated by limestone solution processes. They are distinguished into "red loam" (in German "Kalksteinrotlehm") and "brown loam" ("Kalksteinbraunlehm") according to Kubiena (1953), with the brown loam being much more common in the mid-latitudes. Here, the term "brown loam" is used for brown tinted, clayey-loamy substrates that are residues of limestone dissolution. The loamy material may be dislocated or mixed with other sediments like limestone rock debris or, in minor quantities, with loess. There are conflicting conceptions regarding the age of Central European brown loams, including the Tertiary, interglacials of the Pleistocene, and interstadials of the last glaciation (e.g., Altermann and Rabitzsch, 1976; Mückenhausen, 1979; Pfeffer, 2004; Rehfuess, 1990; Rohdenburg and Meyer, 1968; Šály et al., 1976; Seidel and Steinmüller, 1993; Smolikova and Lozek, 1962). Nevertheless, middle and eastern European brown loams of subdued mountain areas were not formed solely during the Holocene. This is indicated by mass-balance estimates and dissolution rates calculated for various calcareous rocks of the Muschelkalk (Rohdenburg and Meyer, 1963; Scheffer et al., 1960) and Jurassic (Häusler and Niederbudde, 1992; Werner, 1958). Notably, the dissolution of pure limestones (with less than 5% insoluble residue (ISR)) could not provide the currently present amounts. Nevertheless, (the thickness of) brown loams may further be modified by eolian admixtures (AGB, 2005; Kubiena, 1986; Kuntze et al., 1994; Rohdenburg and Meyer, 1968). This process has been demonstrated for analogous situations elsewhere (Durn, 2003; Muhs and Budahn, 2009; Muhs et al., 2007; Yaalon, 1997).

Sedimentary rocks of the Muschelkalk mainly consist of limestones, which vary in their genesis (e.g., bedding intervals) and in absolute carbonate content. Therefore, these calcareous rocks are suitable to study the influence of rock properties on the distribution and properties of periglacial cover beds.

The calcareous rocks of the Muschelkalk have been grouped into three main subunits. The regional division of the Muschelkalk in the area of the Thuringian Basin (e.g., Seidel, 1995) is as follows. (1) Rocks of the Lower Muschelkalk are fine-grained marly limestones, showing parallel, horizontal, or undulose stratification with thin interbeds of marl or crystalline calcite (locally designated

"Wellenkalk"). The amount of ISR, which represents the siliceous rock components, typically ranges between 7% and 15% (maximum: 25%: Langbein and Stepansky, 1996). Three thick-bedded limestone units occur between these thin-bedded limestone and have a fine-crystalline and dense (Terebratel Limestone) or an oolitic, bioclastic, and porous (locally designated "Schaumkalk") structure. The ISR of these intersecting units is less than 7%, occasionally even less than 3% ("Schaumkalk"). (2) The Middle Muschelkalk is formed of evaporites (carbonates, gypsum, rock salt) with interbeds of calcareous marl and clay. The ISR is typically high, between 10% and 25%, occasionally reaching up to 35% within interbedded mudstones and clayey layers. These interbedded strata are subdivided by thin layers of dolomite with low contents of ISRs. (3) The Upper Muschelkalk consists of thick-bedded, middle to coarse-grained, and partly porous limestone that may contain fossils (crinoid ossicles) in rock-forming abundance (locally designated "Trochitenkalk"). The second unit of the Upper Muschelkalk is an alternation of calcareous mudstone and fine- to middle-grained limestones with rare ammonites (locally designated "Ceratiten-Schichten"). Due to the muddy beds, the ISR of the "Ceratiten-Schichten" is high, with a maximum around 30% and typical values between 5% and 17%. The thick-bedded "Trochitenkalk" contains less than 6% insoluble siliceous residue.

These calcareous rocks are variably resistant to both physical (fracture) and chemical weathering processes ("specific solution velocity," cf. Scheffer et al., 1962) based on their lithologic properties (Bullmann and Heinrich, 2012; Prick, 1997).

The study area is located between 300 and 470 m asl at the Ilm-Saale-Muschelkalk Plateau of the eastern margin of the Thuringian Basin (Fig. 2.1).

2.8.3.2 *Properties and Distribution of Cover Beds*

Periglacial cover beds overlying calcareous bedrock of the Muschelkalk are derived from loess sediments, from frost-weathered debris, as well as from residues of limestone solution. They may be distinguished in the following sediment types: (a) limestone debris (rock fragments with a small proportion of calcareous fine matter); (b) brown-loam debris (rock fragments with a lot of fine matter from limestone dissolution); (c) debris-free brown loam; (d) debris-rich or debris-free, clay-rich, and dense loess loam; and (e) debris-free, silt-rich, or clay-rich loess loam. These sediments form layered cover beds of varying vertical structure and thickness. Furthermore, limestone rubble covers scarp slopes as well as some plain relief positions, showing no or only weak stratification. This is important because scarp slopes are frequent in the study areas, which appears characteristic for limestone landscapes.

The following soil-profile descriptions illustrate layering and cover-bed properties, as they typically occur in the study areas of the Muschelkalk landscape of the eastern Thuringian Basin. The cover-bed distribution is exemplified by an asymmetric valley cross-section (Fig. 2.39), a plain on "Wellenkalk"

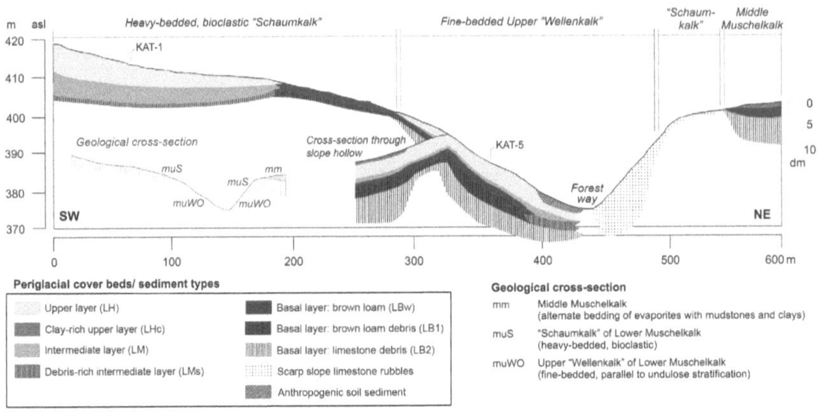

FIGURE 2.39 Distribution and vertical structure of periglacial cover beds in an asymmetric valley cross-section, eastern Thuringian Basin. The thickness and vertical composition of the cover beds depend on bedrock properties and relief position. A limestone rubble cover is developed on the SW-facing scarp slope, the thickness and clast properties of which are also influenced by the bedrock. Vertical exaggeration is approx. 3×; sediments are not drawn on the same scale.

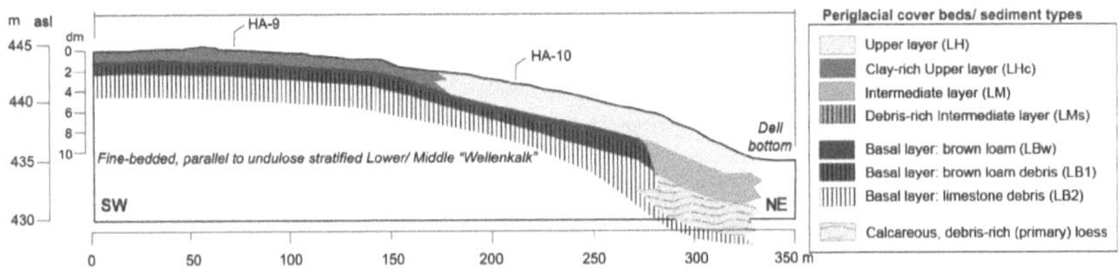

FIGURE 2.40 Facies change from a clay-rich upper layer (LHc) to a silt-rich upper layer (LH) probably due to increased loess admixture. The section is located on a plain which is inclined about 2° to the NE (note exaggeration of drawing). An intermediate layer overlying a calcareous loess wedge at the slightly steeper (5°) slope to a slope depression, whereas limestone debris, brown-loam debris, and brown loam (basal-layer sediments) wedge out. The bottom of the slope depression is filled with fluvial Pleistocene limestone gravel with a chaotic fabric. Vertical exaggeration is approx. 5×; sediments are not drawn on the same scale. "Dell" is a synonym for "slope depression."

(Fig. 2.40), and a cross-section including "Trochitenkalk" and Middle Muschelkalk (Fig. 2.41).

Basal layers on the "Wellenkalk" of the Lower Muschelkalk are generally stratified and composed of three sediment types, as exemplified by profile KAT-5 (Fig. 2.42, left): (1) Bedrock is overlain by limestone debris (LB2) that often merges into the bulk bedrock. Clasts derived from the "Wellenkalk" are angular, sharp-edged, predominantly platy, and of intermediate size (2–20 cm). The clasts generally make up more than 75 vol.% (by volume) of the sediment. The fine material is silty and strongly calcareous. It also has a relatively high sand content, which may be traced back to calcareous sand. Most platy clasts are oriented parallel to the slope inclination. (2) Brown-loam debris (LB1) is found on top of limestone debris. The disconformity between both debris types

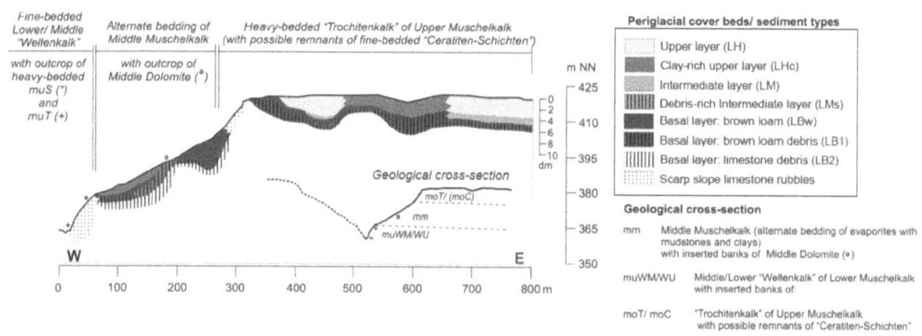

FIGURE 2.41 Distribution and vertical structure of periglacial cover beds in the eastern part of a valley cross-section in the Middle Muschelkalk. Within the "Trochitenkalk" plain, areas with upper and intermediate layers are interconnected with areas with clay-rich upper layers as well as with loess-free areas. The rocks of the Middle Muschelkalk have been changed into silt-rich, calcareous fines, which are relatively poor in debris. Larger amounts of debris in the basal layers originate from dolomite (Middle Dolomite) or from displaced "Trochitenkalk." The thickness of the brown loam is enhanced in the concave slope section directly beneath the small scarp in the "Trochitenkalk." Due to high soil moisture, caused by overflowing percolating water from the "Trochitenkalk," chemical weathering and accumulation of the brown loam is accelerated. Vertical exaggeration is approx. 4×; sediments are not drawn on the same scale.

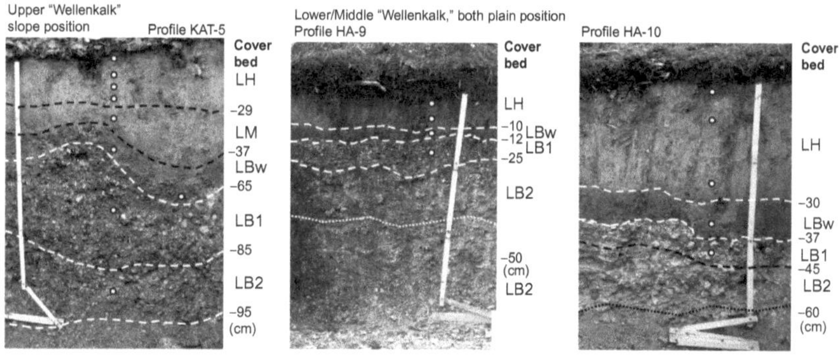

FIGURE 2.42 Vertical structure and properties of cover beds on slopes developed in the "Wellenkalk" of the Lower Muschelkalk (left, profile KAT-5) and in plain position (middle and right, profiles HA-9 and HA-10). For location of the profiles, see Figs. 2.39 and 2.40. White dots indicate the relative position of soil samples taken for analyses (see Table 2.9). *Reproduced from Bullmann and Heinrich (2012).*

is clear. The brown-loam debris contains clay-rich, noncalcareous fines, which amount to 50% in the profile KAT-5. Clasts are predominantly intermediate to fine in size (<2 cm), rounded or very well rounded (comparable to "grèzes litées," De Wolf, 1988). Clasts derived from the "Wellenkalk" often feature a loose weathering rind of calcareous silt particles, which adhere to the surrounding clay-rich matter. This feature has been named "mazeration crust" by Rohdenburg and Meyer (1963) and is often found where limestone clasts are embedded in clay-rich sediments or brown loam (cf. Thiemeyer, 2007).

Moreover, some small clasts also are completely brittle and crumble. Both features are assumed to be the result of periglacial permeation weathering ("Durchdringungsverwitterung," Scheffer et al., 1962). (3) Noncalcareous brown loam (LBw) covers the brown-loam debris and completes the basal-layer succession. The brown loam is free of clasts and rich in clay, with a clay content of typically about 60% (Table 2.9), ranging from 50% to >70% throughout all studied profiles. Generally, the properties of the brown loam (clay content, color) are similar to those of the clayey fines found in the brown-loam debris. Fines in profile KAT-5 are dark brown tinted. The brown loam in this profile fills a small slope depression carved into brown-loam debris and obtains a thickness of 18 cm.

In profile KAT-5, intermediate (LM) and upper layer (LH) follow on top of the basal-layer succession. They contain admixed loess as indicated by a high coarse-silt content, a high coarse-silt to fine-silt ratio, and by the soil color (Table 2.9). Both upper and intermediate layers are free of debris and entirely decalcified. The intermediate layer has a higher bulk density and clay content than the upper layer. The intermediate layer's structure is homogeneous in this particular profile; however, some intermediate layers are characterized by interfingering of silt-rich and clayey parts, suggesting participation of cryoturbation in their genesis. The clay content of such intermediate layers is assumed to be partly derived from admixed underlying brown loam, suggesting that inheritance (see Section 2.6.4) may have contributed to the layer formation.

Profiles HA-9 and HA-10 are examples of vertical structures of cover beds on a plain relief in the "Wellenkalk" of the Lower Muschelkalk (Fig. 2.42, middle and right). The basal layer's structure is similar to the one on slopes, that is, limestone debris, brown-loam debris, and brown loam in succession. Notably, the thickness of each particular layer is distinctly less than on slopes.

In profile HA-10, an intermediate layer is missing; the upper layer is about 25 cm thick and directly overlies the brown loam. Profile HA-9 is located 200 m from profile HA-10. Distinct loess-rich layers are entirely absent, whereas the basal-layer structure remains similar to that in profile HA-10. The brown loam is covered by a substrate with a high silt and clay content (Table 2.9). This type of upper layer is labeled "clay-rich upper layer" (LHc). We assume that such clay-rich upper layers formed by mixing of brown loam and only minor quantities of loess; as the loess influence was low, and because of the textural dependence of the gelifluction depth (see Section 2.6.3), these layers have limited thicknesses only, reaching a median of just 10 cm in all investigated clay-rich upper layers (Bullmann, 2010). Due to this overall marginal loess deposition, intermediate layers never were sufficiently thick to remain preserved after the formation of upper layers, if they had existed at all. Little eolian admixture is supported by the small coarse-silt to fine-silt ratio of ULc (example HA-9, Table 2.9).

Despite of their similarity in texture, clay-rich upper layers may be distinguished from brown loam (LBw) by their soil fabric and soil color: the LHc is characterized by a subangular blocky fabric (with edges of less than 3 mm) and by a brownish yellow (10YR 7/6–6/6) color. In contrast, brown loam has a

TABLE 2.9 Physical and Chemical Properties of Some Soil Profiles from Periglacial Cover Beds Present on Different Bedrock Types of the Muschelkalk

			Fine earth										
			% of fine earth										
				Silt									
Depth (cm)	Cover bed	Soil horizon	Clay	Fine	Middle	Coarse	Sand	cSi/fSi	Lime %	pH ($CaCl_2$)	Color (dry)	Clast %	Clast roundness
KAT-1 (on thick-bedded "Schaumkalk," plain relief, not inclined)													
3		A	16.6	14.1	32.1	32.8	4.4	2.32	0.0	3.8	10YR 7/1	0	
23		E	16.2	10.1	36.0	36.7	1.0	3.63	0.0	3.6	10YR 8/4	0	
28		EBt	20.1	12.5	31.0	35.2	1.0	2.81	0.0	3.6	10YR 7/4	0	
32	LH	Bt	26.2	10.3	30.0	32.5	0.9	3.16	0.0	3.6	7.5YR 7/6–6/6	0	
55	LM-a	2Bt	36.3	9.7	25.6	27.3	1.0	2.80	0.0	3.7	7.5YR 5/6	0	
74	LM-b	3Bw	38.5	7.7	23.5	29.5	0.7	3.81	0.0	4.5	10YR 6/ 6+5YR 4/4	0	
90	LMs	4CBw	37.5	9.4	23.5	28.1	1.5	3.00	1.8	7.0	10YR 6/ 6+5YR 3/3	78	+
HA-7 (on thick-bedded "Terebratel-Kalkstein," plain relief, 1–2°, ENE aspect)													
2		A	n.d.	n.d.	n.d.	n.d.	n.d.	n.d.	0.0	3.6	10YR 7/1	0	
25		E	15.4	10.1	34.6	36.3	3.6	3.59	0.0	3.7	10YR 8/4	0	
38	LH	EBt	23.4	10.7	30.2	32.6	3.1	3.05	0.0	3.8	10YR 7/4	0	
55	LM	2Bt	45.3	7.6	20.0	23.8	3.2	3.12	0.0	4.1	10YR 5/6	0	

Continued

TABLE 2.9 Physical and Chemical Properties of Some Soil Profiles from Periglacial Cover Beds Present on Different Bedrock Types of the Muschelkalk—Cont'd

			Fine earth										
			% of fine earth										
				Silt									
Depth (cm)	Cover bed	Soil horizon	Clay	Fine	Middle	Coarse	Sand	cSi/fSi	Lime %	pH ($CaCl_2$)	Color (dry)	Clast %	Clast roundness
65	LBw	3Bw	52.1	8.3	14.9	14.7	9.9	1.77	0.3	6.0	10YR 5/6+4/4	1	+++
85	(LB1)	3CBw	n.d.	n.d.	n.d.	n.d.	n.d.	n.d.	n.d.	n.d.	10YR 5/6+4/4	75*	+
HA-9 (on thin-bedded "Wellenkalk," plain relief, 1°, NNE aspect)													
6		A	47.5	12.8	20.7	17.1	1.9	1.33	0.0	3.7	10YR 4/2	0	
10	LHc	Bw	44.6	14.6	20.2	17.7	2.9	1.21	0.0	3.8	10YR 7/4–7/5	1	
12	LBw	2Bw	59.8	10.4	15.9	12.2	1.7	1.17	0.4	6.3	7.5YR 4/6	3	++++
25	LB1	2CBw	56.7	10.8	12.6	16.7	3.2	1.55	9.6	6.9	10YR 4/4–4/5	52	++++
50	LB2	2C	n.d.	n.d.	n.d.	n.d.	n.d.	n.d.	n.d.	n.d.	n.a.	>75*	—
HA-10 (on thin-bedded "Wellenkalk," plain relief, 2°, NNE aspect)													
4		A	n.d.	n.d.	n.d.	n.d.	n.d.	n.d.	0.0	3.8	10YR 7/1	0	
8		E1	24.9	10.4	28.1	32.2	4.5	3.11	0.0	3.5	10YR 7/4	0	
25		E2	26.8	11.8	26.2	31.8	3.5	2.70	0.0	3.7	10 YR 7/4	0	
30	LH	Bt	33.2	9.0	27.0	27.7	3.1	3.07	0.0	4.1	10YR 6/4	<2	

Continued

TABLE 2.9 Physical and Chemical Properties of Some Soil Profiles from Periglacial Cover Beds Present on Different Bedrock Types of the Muschelkalk—Cont'd

Depth (cm)	Cover bed	Soil horizon	Fine earth: % of fine earth: Clay	Silt: Fine	Silt: Middle	Silt: Coarse	Sand	cSi/fSi	Lime %	pH ($CaCl_2$)	Color (dry)	Clast %	Clast roundness
37	LBw	2Bt	70.7	8.5	11.2	8.6	1.0	1.01	0.1	6.3	7.5YR 4/6	<2	++++
45	LB1	2CBw	n.d.	n.d.	n.d.	n.d.	n.d.	n.d.	n.d.	n.d.	7.5YR 4/6	75	++++
60	LB2	2C	n.d.	n.d.	n.d.	n.d.	n.d.	n.d.	n.d.	n.d.	n.a.	>75*	—
KAT-5 (on thin-bedded Upper "Wellenkalk," backslope, 8–10°, NE aspect)													
4		A	n.d.	n.d.	n.d.	n.d.	n.d.	n.d.	0.0	3.9	10YR 6/2	0	
11		E	18.5	13.8	32.2	34.5	1.0	2.51	0.0	3.7	10YR 8/3	0	
20		E	20.5	13.3	34.5	30.8	0.9	2.32	0.0	3.8	10YR 8/3	0	
29	LH	Bt	30.2	12.1	27.0	29.9	0.8	2.47	0.0	4.0	10YR 7/4+6/6	0	
37	LM	2Bt	46.5	10.8	22.1	19.7	0.9	1.82	0.0	4.4	10YR 6/ 6+7.5YR 5/4	0	
50	LBw	3Bw1	54.8	9.9	15.4	19.0	0.8	1.91	0.1	6.0	7.5YR 3/3+5/6	0	
65	LBw	3Bw2	60.2	8.3	14.2	16.7	0.6	2.02	0.3	6.5	7.5YR 4/4+5/6	1	++++
85	LB1	4CBw	64.0	8.9	15.3	9.0	2.8	1.01	11.6	7.2	7.5YR 4/4	73	+++
95	LB2	5C	37.6	12.7	31.8	10.0	7.8	0.79	46.8	7.4	10YR 7/3	86	—

n.a., not applicable; n.d., not determined; LH, typical silt-rich upper layer; LHc, clay-rich upper layer; LM, typical intermediate layer; LM-a, intermediate layer with uniform structure; LM-b, intermediate layer with interfingering structure; LMs, debris-rich intermediate layer; LBw, basal layer constituted by limestone-weathering residue; LB1, brown-loam debris; LB2, limestone debris. Grain-size distribution of fines is given in %, clay: <2 μm; fine silt: 6.3–2 μm; middle silt: 20–6.3 μm; coarse silt: 63–20 μm; sand: 0.63–2 mm. Lime = carbonate content given in mass%; clasts: content is given in mass%, except for □, where the given value is in field-estimated% by volume. Clast roundness: ++++ very well rounded with distinct weathering rinds; +++ very well rounded; ++ well rounded; + rounded; — angular.

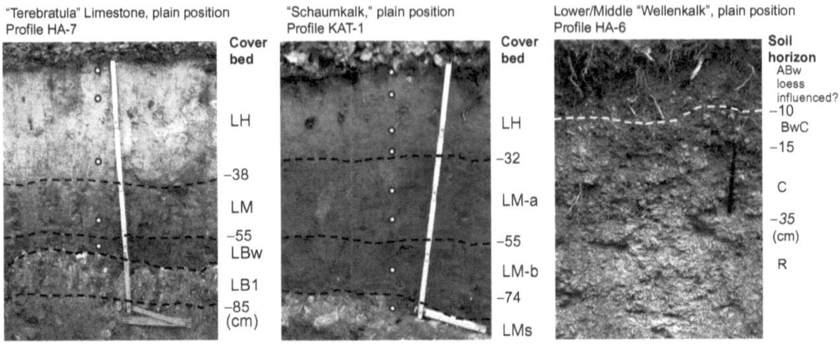

FIGURE 2.43 Soil profiles depicting the vertical structure and properties of cover beds on thick-bedded limestone in plain relief: "Terebratel-Kalkstein" (left), "Schaumkalk" (middle), and "Wellenkalk" (right; note that soil horizons rather than layers are indicated for this profile). For location of profile KAT-1, see Fig. 2.39. HA-6 and -7 are located closely spaced ca. 400 m northeast of the toposequence depicted in Fig. 2.40. White dots in KAT-1 and HA-6 indicate the relative position of soil samples taken for analyses (see Table 2.9).

distinct angular soil fabric (with edges of 5–10 mm) and a strong brown, yellowish, or reddish brown soil color. These properties may still be identified in profiles in which brown loam has been brought to the surface by soil erosion, rendering an alternative interpretation of the LHc as brown loam modified by bioturbation and surficial pedogenesis unlikely.

Basal layers overlying thick-bedded limestone differ in their vertical structures from those found on "Wellenkalk." Profile HA-7 is an example from the "Terebratel-Kalkstein" (Fig. 2.43, left). The bedrock is disintegrated into coarse debris. This loosened bedrock is covered by brown loam that also penetrates into the fissures of the debris beneath. Clasts rapidly increase in size from 2 to 6.3 cm at the upper boundary of the brown loam to >20 cm at its lower boundary. In the same vertical direction, the shape of the clasts changes from well to weakly rounded. Because in these cases the fines consist of brown loam, we refer to this mixed debris as "brown-loam debris" (LB1). Limestone debris is absent, which is found to be typical for thick-bedded limestones. This bipartite basal layer is covered in profile HA-7 by an intermediate and an upper layer (Table 2.9).

Profile KAT-1 on a "Schaumkalk" plain provides another example for the weak rock disintegration of thick-bedded limestone types (Fig. 2.43, middle). The overlying debris layer is dominated by edge-rounded cuboidal rock fragments (5 to >20 cm in diameter). It is covered by a succession of two intermediate layers and an upper layer. The clay content, the coarse-silt content, and the coarse-silt to fine-silt ratio (cSi/fSi) are the same for the lower intermediate layer and the fines in the debris layer (Table 2.9). Coarse silt is assumed to be mainly derived from eolian sources, whereas limestone weathering (brown loam) yields coarse silt to a considerably lesser extent (see Alaily, 1984). Both the high coarse-silt content and the high cSi/fSi ratio indicate that the debris' fines contain eolian material (compare with the remarkably lower cSi/fSi ratios of brown loams in Table 2.9). The admixture was probably favored by the rough surface of the coarse debris that may have been incompletely covered by brown loam at the beginning of loess deposition (see McFadden et al., 1998 or Dietze et al., 2010 for mechanisms). Therefore, the lithofacies of this layer is described as a debris-rich intermediate layer (LMs). Both brown loam and debris-rich basal layers are absent.

The intermediate layers in profile KAT-1 (Table 2.9) are not distinguished by their texture, but by their structure, which is uniform in the upper (LM-a) part and interfingering in the lower. Furthermore, clay-illuviation coatings occur predominantly in the lower intermediate layer (IL-b), which are also evident in the fines of the underlying debris-rich intermediate layer. Thus, a certain amount of clay in the intermediate layers may be traced back to clay illuviation. The entire profile is decalcified to a depth of 74 cm, including the LM-b (see pH in Table 2.9), facilitating dispersion and translocation of clay. Clay-illuviation coatings in the debris-rich intermediate layer (LMs) were found in the center of fissures filled with fines. However, fines adjacent to limestone rock fragments hold a certain amount of carbonate because they are affected by modern limestone dissolution. Due to this closely spaced geochemical change, both components were included in mixed soil samples, resulting in a "mixed" carbonate content of 1.8% and a "mixed" pH value of 7.0.

Limestone rubble covers scarp slopes with inclinations of more than 20°, but partly covers less inclined plains as well. Scarp-slope limestone rubble derived from the "Wellenkalk" may achieve thicknesses of more than 1 m. Clasts are angular with sharp edges, except for the uppermost part, where the clasts are rounded, smaller and less frequent than in the lower part, probably due to more intensive weathering. The clasts in the lowest part of the rubble cover are oriented parallel to the slope inclination, suggesting gelifluction movement. These clasts were later successively buried by gravity-transported clasts, as indicated by the clasts' chaotic structure in the middle and upper part of the rubble. Organic material and clay-rich fines are enriched in the upper part of the rubble. Within the outcrop of thick-bedded limestone types on scarp slopes (e.g., see "Schaumkalk"), the bedrock is partly exposed or just covered by blocky, humous rubble.

Limestone rubbles with similar properties to those found on scarp slopes also occur on plains derived solely from the "Wellenkalk" (Fig. 2.43, right). They are up to 1 m thick; with increasing depth, the clast size increases, whereas the clast rounding decreases. Organic material and clay-rich fines are enriched in their upper. Internal stratification is absent. The rubble passes gradually into loose bedrock. The limestone rubble profiles on plains are closely interconnected with layered, loess-rich cover beds, a feature already described from other Central European Muschelkalk landscapes (e.g., Hempel, 1955; Rohdenburg and Meyer, 1963).

Thick-bedded limestone bedrock on plain relief is only weakly disintegrated, as shown in profile HA-7. Areas not affected by loess deposition have a vertical structure with a thin brown-loam cover under brown-loam debris of loose bedrock. The total thicknesses do not exceed 30 cm.

2.8.3.3 Conclusions

Periglacial cover beds on calcareous rock exhibit some properties not found in cover beds on other bedrock. Basal layers, especially those on thin-bedded "Wellenkalk," show a three-partite succession with three sediments. The various properties of the stacked sediments are not attributed to alternating bedrock properties up-slope. Rather, they are due to the transformation of limestone debris. The sharp edges of clasts in limestone debris (LB2) indicate that the clasts are derived from frost shattering. Calcareous fines of limestone debris emphasize an unweathered state that coincides with the unweathered state of the clasts. The small-sized, rounded, partly brittle clasts of brown-loam debris (LB1) indicate alteration by chemical weathering, which is confirmed by the noncalcareous, clay-rich fines of these deposits. The basal-layer succession, formed by limestone debris—brown-loam debris—brown loam, suggests that the limestone debris partly underwent chemical weathering, which resulted in rounding of clasts and reduction of the clasts' size as well as in accumulation of dissolution residues (cf. Pawelec, 2006). Thus, the occurrence of brown loam, a product of dissolution weathering, is responsible for the complex structure of the basal layers (cf. Altermann and Rabitzsch, 1976; Apel and Fiedler, 1996), rendering landscapes in calcareous rocks unique.

Thin-bedded limestone (e.g., "Wellenkalk") seems to be more susceptible to frost weathering than thick-bedded limestone (e.g., "Schaumkalk") (see Bullmann and Heinrich, 2012), which is reflected in varying amounts of clasts and in different thicknesses of the basal layer.

"Wellenkalk," which partly consists of marl sections and intercalations, may be described as a permeable, porous rock that may become water-saturated under periglacial, cold-humid climate conditions. Disintegration of rock due to frost action is, therefore, favored (Matsuoka and Murton, 2008). The thin-bedded intervals facilitate horizontal fracturing of the bedrock, producing preferably platy and middle-sized debris (6.3–20 cm). Frost weathering of thick-bedded limestone, for example, the "Schaumkalk," almost exclusively takes place along joints, resulting in blocky debris (>20 cm). The alternating bedding of the Middle Muschelkalk is destroyed by frost action, resulting in calcareous, mainly silty fines with a moderate debris content within a depth of 1–2 m. The limestone debris of the Middle Muschelkalk has a higher debris content only down-slope of dolomite outcrops, which is attributed to displaced dolomite clasts (Bullmann, 2010).

Thus, permeability and porosity (nontrapped pore volume) as well as noncalcareous intercalations effectively control the frost weathering of limestones (see Bellanger et al., 1993; Prick, 1997; Scheffer et al., 1960) and are reflected in basal-layer and clast properties.

The thickness of basal layers is furthermore affected by the relief. Stacked limestone debris (LB2) occurs in slope depressions. Regarding the "Wellenkalk," debris is more extensive on slopes than in plain relief positions. In plain relief, there is a thin limestone-debris cover (LB2), which is not completely altered to brown-loam debris (LB1) by dissolution weathering. Both features suggest that, in plain relief position, both rock disintegration and chemical weathering were inhibited. In contrast, both processes are favored on slopes, probably due to gelifluction, and result in thicker accumulations of limestone debris (LB2) and of brown-loam debris (LB1) (Bullmann, 2010; Bullmann and Heinrich, 2012).

Dissolution is stronger in limestone debris derived from thin-bedded limestone than in debris from thick-bedded limestone. This is due to both a higher content of ISRs and an increased dissolution tendency, as indicated by better rounding and by weathering rinds on clasts. Moreover, the fines content and the smaller clasts sizes may have favored chemical dissolution.

Due to the sedimentary origin of the Muschelkalk rocks, petrologic and bedding properties change quickly in a vertical succession within a single sedimentological subunit (e.g., in the "Wellenkalk," cf. Langbein and Stepansky, 1996). These features are assumed to be responsible for the small-scale heterogeneity of debris thicknesses and properties as well as for the distribution of brown loam (Bullmann, 2010). Whereas small-scale variations are rather compensated by gelifluction on slopes, varying debris properties may remain on plain relief positions and hence result in varying sediment patterns (cf. Bullmann and Heinrich, 2012).

The distribution of loess sediments in the Thuringian Basin is effectively controlled by elevation, aspect (Steinmüller, 1996), and slope morphology (Bullmann, 2010). Intermediate layers (LM) occur on concave and lower slope sections as well as in slightly incised slope depressions, probably attributed to karstic origin. Moreover, loess deposition might have been sensitive to microscale changes in soil moisture and vegetation cover, which were controlled by the (karstic) microrelief and the variable distribution of brown loam and brown-loam debris during the onset of loess deposition (Bullmann and Heinrich, 2012).

Chapter 3

Influence of Cover Beds on Soils

C. Lorz, M. Frühauf, R. Mailänder, J.D. Phillips and A. Kleber

3.1 INTRODUCTION

The fact that soil and upper regolith form a functional and genetic unit is well accepted by soil geomorphologists and pedologists. Pursuing this idea, the soil profile must not be seen exclusively as a weathering zone or a pedogenic zone, but as a four-dimensional body formed by a complex set of near-surface processes (Birkeland, 1999; Kleber, 1997; Lorz, 2008; Lorz and Phillips, 2006; Ollier and Pain, 1996; Phillips and Lorz, 2008; Schaetzl and Anderson, 2005; Semmel, 1993b). To simplify this approach, it is useful to consider the soil column as split into a geologic profile and a pedologic profile (Fig. 3.1). The geologic profile is subdivided into two sections: (1) the mobile upper zone (allochthonous regolith) and (2) the *in situ* weathered lower zone (autochthonous regolith). Only for the latter, the constancy of the relationship (see Brewer, 1972) with the bedrock is maintained, as is characteristic for saprolite or frost-shattered debris.

The upper regolith in Central European subdued mountains is usually made up of cover-bed successions, which have formed from material displaced by near-surface processes (the "mobile zone" of Ollier and Pain, 1996). Their composition is mostly a mixture of substrates from up-slope, from beneath (bedrock, regolith), and/or of eolian origin (see Fig. 3.1A and Section 2.1). According to their different derivations, cover beds may vary in composition and properties. Hence, cover beds are separated from one another by disconformities. The major disconformity is located between the allochthonous and the autochthonous regolith, where the lack of a close relationship between bedrock and regolith is most obvious. Surface soils formed in cover-bed successions always have layered parent materials with disconformities (e.g., Arnold, 1968; Lorz, 2008; WRB, 2006). Horizons have developed either independent from geogenic layers (i.e., primary layers, formed before pedogenic alteration, as opposed to "soil layers" sometimes found in the literature) or coincide with them (Fig. 3.1B). A phenomenon common at the disconformity between upper layers and underlying materials is that boundaries of pedogenic horizons coincide with geogenic layers (Kleber, 1992a, 1997; Semmel, 1968). The term we use for the

Developments in Sedimentology, Vol. 66. http://dx.doi.org/10.1016/B978-0-444-53118-6.00003-9

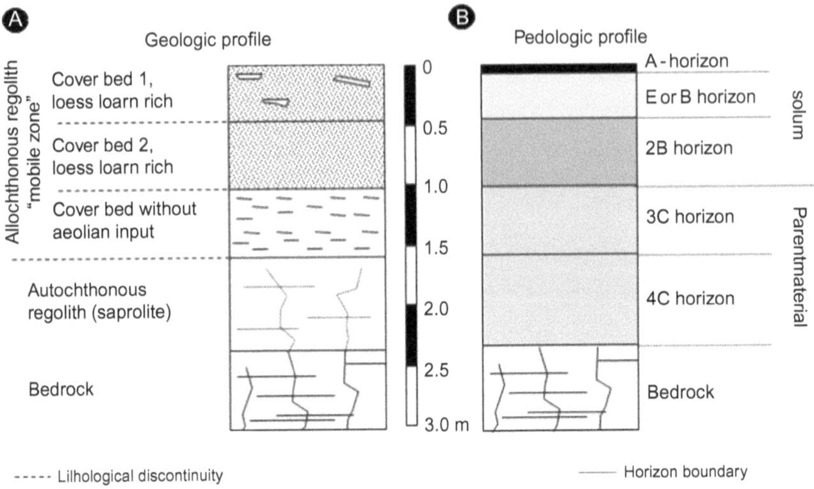

FIGURE 3.1 The geologic profile and the pedologic profile together form the soil profile. Scheme for Central European subdued mountains, following the approaches of Schwanecke (1970) and Kopp (1970).

phenomenon of soils developed in stratified sedimentary successions is "lithologically discontinuous soil" because the interface between two layers forms a disconformity (see Section 1.3.2).

Soils with lithologic discontinuities are also called "duplex soils" (in Australia), "texture-contrast soils," and "fabric-contrast soils," though texture and fabric contrasts are not always related to disconformities. For example, duplex soils have a texture contrast of more than 1.5 texture groups and the boundary is at most 0.1 m thick (Northcote, 1979). However, mostly there is no distinction made as to whether the discontinuity is of geogenic, pedogenic, or mixed origin (Chittleborough, 1992; Daniels and Hammer, 1992; Paton et al., 1995; see Phillips, 2001b, 2004).

The idea that the upper regolith is essentially decoupled from bedrock does not gain much attention in the literature for explaining the vertical organization of surface soils. However, the fact that most parent material is not derived from underlying bedrock has far-reaching implications for the assessment of soil evolution and development of soil properties. The formation of soil profiles is not the exclusive result of vertical top-down processes (Almond and Tonkin, 1999; Schaetzl and Anderson, 2005)—as is assumed by traditional soil-evolution models—but of a number of complex processes involving near-surface processes such as material addition and removal. The consideration of addition and loss of material has fundamental implications for many soil-genetic, soil-ecologic, and environmental topics, because (1) geologic layers and disconformities affect soil properties and, as a consequence, properties of sites and landscapes and (2) soils with lithologic discontinuities are archives of earth-surface processes, which embrace information on the evolution of landscapes. In other words, the evolution and the properties of soil profiles and (soil) landscapes cannot be fully understood if soil is considered separately from regolith.

The identification of horizons and the implicit distinction from geologic features are fundamental for a scientific approach in pedology. At the same time, the identification of stratification might be the key to pedogenesis in primarily layered parent material (Birkeland, 1999). However, the distinction between a horizon and a geologic layer is a fundamental problem in the interpretation of profiles with disconformities. There are only a few efforts to distinguish geogenic, relic pedogenic, and current pedogenic processes (e.g., Phillips, 2001a; Tandarich et al., 2002). However, the formal distinction between (geologic) layer and horizon may be kept simple. According to Bridges (1997), layers are not created by pedogenic processes, whereas soil horizons are. However, the distinction of horizon and layer remains difficult because of (1) a lack of general criteria, (2) the concurrent operation of erosion/sedimentation and pedogenic processes, (3) the modification of geogenic layering by pedogenic processes and vice versa, and (4), consequently, the coincidence of horizon and layer (Birkeland, 1974, 1999; Breemen van and Buurman, 2002; Jenny, 1980; Nikiforoff, 1949).

Section 3.2 provides the basic framework for pedogenesis in stratified material. Section 3.3 outlines the specific lines of soil evolution taken in cover beds in varying constellations. Section 3.4 details the consequences of soil formation in cover beds for soil properties, with subheadings on a general theory of these consequences, and on physical and chemical soil properties mainly depicting case studies.

Soil terminology in this chapter follows Soil Taxonomy (SSS, 2010) with nomenclature according to World Reference Base for Soil Resources (WRB, 2006) in parentheses. The study sites of this chapter are depicted in Fig. 2.1.

3.2 AN INTEGRATED SOIL-EVOLUTION MODEL FOR LITHOLOGICALLY DISCONTINUOUS SOIL

The traditional A–B–C approach, that is, the idea of soil evolution from top down into a uniform parent material under downward formation of soil horizons approximately parallel to one another, is not appropriate for soils with multiple substrates, for example, soils developed in (complex) cover-bed successions or floodplain sediments (Gerrard, 1995). As a consequence, the A–B–C model (Tandarich et al., 2002) for the soil formation during the Holocene has to be questioned as well. The formation of the parent material may be explained only by considering allochthonous input of material, such as loess, solifluction layers, alluvial sediments, etc. Pedologists and geomorphologists, depending on their training, predisposition, and other factors, have tended to uncritically accept conceptual models such as the top-down horizonation concept (A–B–C model, Fig. 3.2A) or basic sedimentological stratigraphic principles, missing

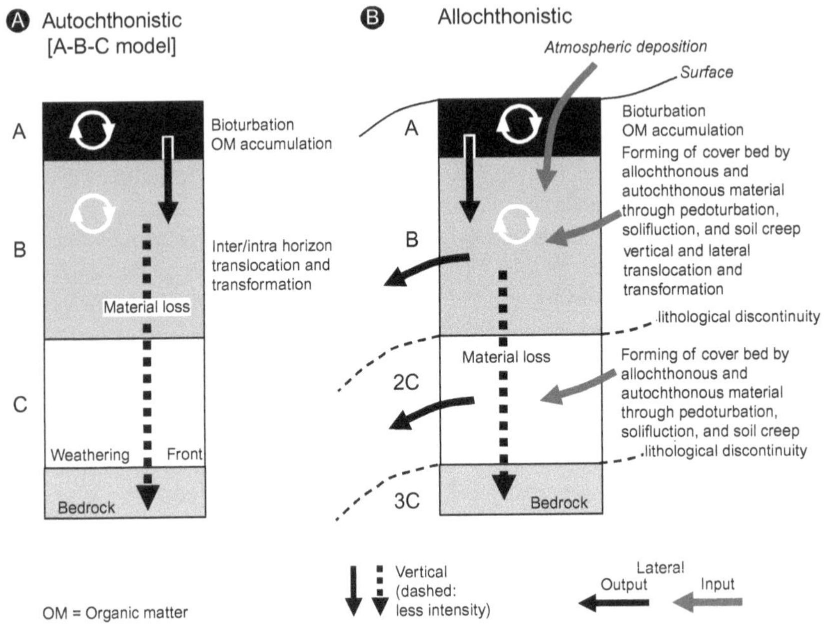

FIGURE 3.2 The autochthonistic (A) and the allochthonistic (B) approach to soil formation (*modified from Lorz and Phillips, 2006*).

the opportunity to connect these approaches with regolith-based stratigraphic principles (i.e., whole-regolith pedology of Ollier and Pain, 1996).

As a consequence, general genetic models which are capable of extending the dominant current approach of autochthonistic soil formation (term by Lorz and Phillips, 2006) toward a dynamic regolith-oriented model are scarce (allochthonistic approach, see Fig. 3.2B). This is of even greater importance because assessing pedogenic and ecologic processes using soil properties (soil morphology, soil chemistry, and soil physics) without taking stratification into consideration may result in considerable errors (Arnold, 1968; Ehwald, 1970).

Our conceptual model (Fig. 3.2B) combines soil-forming processes *sensu stricto* and geomorphic–geological processes. The model was inspired by Simonson's (1959, 1978) generalized theory of soil genesis. Processes of horizon formation (pedogenesis) and layer building (substrate genesis) are responsible for the vertical organization of the regolith (Johnson et al., 1987; Phillips, 1993). This idea leads to an allochthonistic approach complementing the autochthonistic idea of soil formation (Fig. 3.2B). The allochthonistic concept of pedogenesis recognizes that regolith formation is a complex, nonlinear function of time (Birkeland, 1999). The simultaneous operation of regressive and progressive pedogenic pathways (Johnson and Watson-Stegner, 1987) raises the possibility that soil evolution may exhibit patterns far more complex than increasing development over time, as it is assumed in traditional soil-evolution concepts. The allochthonistic approach considers the early knowledge of pedologists that soil is a spatial body, the genesis of which is closely related to landscape changes (i.e., Butler, 1982). Thus, we understand soils as long-term and short-term open systems, with vertical and horizontal dynamics.

In an autochthonistic approach (A–B–C model: Fig. 3.2A), soil forms through pedogenesis exclusively from the underlying geological material (bedrock). Taking t_0 as the time of deposition or exposure of the underlying material and t_n as the present time, the age of the soil is $t_n - t_0$.

In an allochthonistic approach, layering indicates disturbance or modification of the weathering profile, so that the consistent relationship (Brewer, 1972) with the bedrock is not maintained. Therefore, the properties of the whole regolith depend very much on the composition of all layers. The soil morphology of the regolith profile is determined by pre- and syn-sedimentary pedogenesis or weathering and *in situ* pedogenesis or weathering after deposition. Within the allochthonistic approach, sedimentary processes (layer formation) and pedogenic processes (horizon formation) may act by turns or even simultaneously.

First, stratification may have existed at t_0, for example, in stratified alluvium or sedimentary rocks as a part of a geological process. This geologic layering may persist as weathering proceeds and other geomorphologic and pedologic processes occur and may either be sharpened and enhanced or obscured by the latter.

Second, layering may have resulted from relatively abrupt episodic creation (e.g., climate dependent) at time t_i, with $0 < i < n$. For instance, a flood, landslide, or storm event may deposit a lithologically distinct layer on an existing surface, thereby creating a second layer at least—more if the deposited material is stratified itself. This type of layering is more likely to be associated with surficial processes, but episodic layer creation could be associated with subsurface processes as well, perhaps triggered by abrupt changes in water tables or by subrosion (e.g., piping), for example.

Third, layering may be created by gradual (though probably variable in rate) processes operating over some time period $t_i - t_j$, with $0 < j < i < n$. Vertical differentiation by creation of a surface biomantle (Johnson, 1990) is one example. A distinction between active ($t_i = t_n$) and relic ($t_i < t_n$) properties may be made. Again, the processes may operate primarily at or below the surface. Further, they may be primarily vertical (top-down or bottom-up) or horizontal, that is, any lateral transport of solids (Lorz and Phillips, 2006) or solutes (Birkeland, 1999).

3.3 PEDOGENESIS IN COVER BEDS

Despite the variability of cover beds, several (though simplified) conclusions may be drawn as to their influence on pedogenic processes and resulting soil types. In two-layer successions without an intermediate layer, having an upper layer poor in loess, the cover beds influence the formation of soil as they provide loose, weathered material. In addition, their disconformities determine where soil-horizon boundaries typically form and may occasionally cause soil-water stagnation. This coincidence of boundaries between soil horizons and layers is in general one of the most conspicuous phenomena observed in cover-bed profiles. It is supposedly due to the fact that layer boundaries usually mark changes in soil-parent materials, leading to changes in pedogenic processes, that is, their types or intensities, at exactly that depth. Properties such as texture, fabric, bulk density, mineral composition, and—influenced by these—permeability may change at boundaries. Accordingly, soil-horizon boundaries develop just at the disconformities and may be enhanced (or occasionally blurred) during soil formation. The disconformity at the base of the upper layer typically is a rather distinct horizon boundary. For instance, the boundaries between the Bw- and 2C horizons,[1,2] of Inceptisols[3] (Cambisols) or between the E- and 2Bt horizons of Alfisols[4] (Luvisols) usually coincide with the lower boundary of the upper layer.

It has been argued that changes in color between the Bw horizon developed in an upper layer and an underlying 2C horizon of a basal layer are not pedogenic in nature but reflect the original color of the loess-rich upper layer (Bargon et al., 1971). However, there is no correlation between the extent of color change and the loess content (Fried, 1984). Thus, this change in color is assumed to be primarily due to pedogenic alteration.

Spodosols[5] (Podzols) may develop where the upper layer contains very small amounts of eolian silt but is dominated by derivatives of felsic rock. Just like Entisols[6] (Leptosols), they are usually restricted to elevated sites (Stahr, 1979) and to locations where the upper layer is missing due to soil erosion or where it is overlain by anthropogenic soil sediment which may be depleted in nutrients due to former cultivation. Furthermore, Spodosols frequently have formed from cover beds of the North German lowlands with the eolian contribution being quartz-rich sands rather than silty loess.

With increasing silt content, some pedogenic processes become promoted (depletion of clay) or hindered (podzolization) by loess admixed to upper layers, and these influences correspond to loess content and, especially, to the total thickness of loess-enriched layers. On felsic rock, underlying bedrock alone would support the evolution of acidic soil, that is, of Spodosols, but in fact the contribution of the loess content to the buffering system of the soil is often able to prevent this (Kleber, 1997). Rather, Inceptisols largely predominate

1. Leading numbers indicate major disconformities in the soil.
2. "C" refers to material not remarkably altered by pedogenesis.
3. Soil type characterized by a cambic (Bw) subsurface horizon with evident oxidization of iron-bearing minerals.
4. Soil type characterized by an argillic (Bt) subsurface horizon, enriched by translocated clay, usually overlain by a depleted E horizon.
5. Soil type characterized by spodic (Bh, Bs) subsurface horizons, enriched by translocated iron, aluminum, and humus, usually overlain by a depleted E horizon.
6. Immature soil type characterized by slight to absent weathering and thin to discontinuous humus accumulation.

two-layer successions comprising loess-rich upper layers. If the upper layer overlies a basal layer that is permeable and that has lower soil acidity, there is often weak evidence of translocated clay below the upper layer, reflected in clay bridges or in clay lamellae and bendings.

Intermediate layers within tripartite successions, which are more than two or three decimeters thick, seem to largely decouple pedogenesis from bedrock influence and usually lead to the development of Alfisols with its argillic (argic) horizon within the intermediate layer. In aging argillic horizons, clay enrichment often causes stagnation of soil water, especially on insignificantly inclined slopes, preventing fast lateral percolation of slope water.

It is not the sheer silt content of underlying substrates that triggers clay dislocation: from some bedrock, such as shale, slate, or phyllite, weathering results in higher amounts of silt. Then, the basal layer may contain silt that is not of eolian origin. Inceptisols tend to form in such cases with Bw horizons developing in both the upper and the basal layers and with boundaries between the two layers that are less distinct than usual. Because of the acidic chemical character most of these rocks provide to their soils, there is no clay displacement. This usually allows one to keep these soils apart from soils having an intermediate layer. Basalt, on the other hand, is chemically almost neutral, in cases enabling the formation of Alfisols (Kleber, 1991).

The upper layer is usually not very thick on carbonate bedrock (see Section 2.8.3). Since Ca ions dispersed throughout the soil—mainly derived from carbonate clasts—hinder clay translocation, Rendolls (Rendzic Leptosols) are the typical soils. Their humus-rich A horizons[7] are developed in, and usually restricted to, the upper layer. Many limestone plateaus have a very long history of weathering. On these, clay that is mainly derived from carbonates as a residue of weathering may be incorporated into the cover beds, particularly into the basal layer, pretending clay translocation through a strong texture contrast. However, on rock with an intermediate carbonate content (some alluvial gravels, marl), the layers may be deeply decalcified (depending on elevation) so that Inceptisols could form. Marl and basal layers derived from marl may even have provided suitable conditions to form argillic horizons (Semmel, 1977).

Along slopes, there often is a transition between a bipartite succession with Inceptisols (A–Bw–2C) and a tripartite succession bearing Alfisols (A–E–2Bt–3C). The Bw/2C and the E/2Bt boundaries coincide with the disconformity below the upper layer. Texture (especially clay content), color, and other soil properties of the soil horizon in the upper layer are identical, regardless whether it overlies a basal or an intermediate layer, that is, whether the soil is an Inceptisol or an Alfisol. Identical materials and horizons are thus classified as a Bw- or an E horizon, depending on their position relative to other horizons deeper in the profile, in preference over process-related considerations.

7. "A" refers to soil horizons dominated by their humus content.

Therefore, Bargon et al. (1971) reached the conclusion that there is no striking pedogenetic relationship between the E- and the 2Bt horizons in the tripartite case, because the E horizon should differ significantly from the neighboring Bw horizon, if there acted different pedogenic processes. They called this type of soil-like feature "pseudo-Alfisol." Recently, Müller (2011) could strongly support this interpretation when she demonstrated that argillic horizons own their high clay contents mainly from reworked and inherited older soil materials rather than pedogenesis from top down. The consequences of this idea for landscape history are serious.

- Either these argillic horizons are remnants of soils that formed during lateglacial interstadials, prior to accumulation of the upper layer (Harrach, 1974; Kühn, 2003; Plass, 1968; Poetsch, 1975). However, there are several objections: even interstadials were cold, and a cold and dry climate does not tend to form argillic horizons (e.g., Clayton et al., 1977; Tedrow, 1977). Furthermore, if a thick layer of Allerød Lake Laach tephra separates the upper layer from the intermediate layer, the evidence of clay displacement is restricted to where (Holocene) pedogenesis was able to penetrate this layer along cracks (Rohdenburg, 1978; Rohdenburg and Meyer, 1968).
- Or, argillic horizons can form without a proprietary E horizon. Indeed, Goss et al. (1973) seemed to demonstrate this as possible under certain circumstances. However, during their experiment, dispersed clay was washed into large pores that were kept open artificially. Nature probably did not provide such conditions for a longer time. Therefore, a typical Alfisol, unless it is truncated, is supposed to possess an upper part where the clay becomes dispersed—the E horizon—and a lower part where the clay flocculates.
- However, clay dispersed in the upper layer with a Bw horizon does not necessarily have to flocculate directly beneath but may be dislocated along the slope with interflowing slope water; thus, the eluvial horizon does not necessarily have to overlay the related illuvial horizon in a vertical profile.
- Local redistribution of clay within argillic horizons may play an important role in producing clay films (Bronger, 1978; Stephan, 2000). Thus, the intermediate layer may itself be the source of the clay, which may have been dislocated over very small distance and become enriched within the same horizon which had set it free.

The occurrence and vertical composition of the slope deposits in Central Europe are not random. Therefore, knowledge of their properties allows a much better prediction of the soils at a distinct location than bedrock, climate, slope-angle, etc. alone can provide. This is of special importance for soil surveying (Semmel, 1966). Explaining the evolution of soils—particularly the nature of horizon boundaries or the presence of clay skins—is greatly supported by taking their stratified and partially allochthonous parent materials into account. For balancing pedogenic processes, the problem arises that it is difficult to satisfy the need for homogeneous soil-parent material (Maus and Stahr, 1977; Wang and Arnold, 1973). For instance, pollution with heavy metals in soils may not be compared with deeper parts of the profile, as the primary contents may differ remarkably (and predictably) between the various layers (Kleber et al., 1998a; Sabel, 1989; Spies, 1991). Or, the calculation of clay formation yields unrealistic results as long as the primary clay content of the various layers is unknown (see also Colman, 1982).

3.4 CONSEQUENCES FOR SOIL PROPERTIES

3.4.1 Theory of Soil Properties in Lithologically Discontinuous Soil

As outlined in the previous sections, properties of soil profiles depend on the properties of their parent material and on pedogenic alteration. If parent material is stratified, soil properties in their vertical sequence are likely to be affected by this (see Section 3.3). As a result, the distribution of vertical properties in soils with lithological discontinuities may deviate from expected depth functions that are based upon traditional top-down soil-formation models.

In soil science, depth gradients are often used to make inferences about processes, typically based on the notion that differentiation is associated primarily with vertical translocation by water. This interpretation, which is based on the autochthonistic approach (Fig. 3.3A), is plausible, if smooth pedogenic gradients (e.g., for oxalate-extractable iron) occur and a homogenous parent material is assumed (Fig. 3.3A, I). A pure pedogenic interpretation would be less plausible for case II of Fig. 3.3, if a gradient exists that obviously cannot be explained by pedogenesis alone. If the same depth function is seen in the context of an allochthonistic approach, the interpretation is more complex (Fig. 3.3B). It becomes plausible

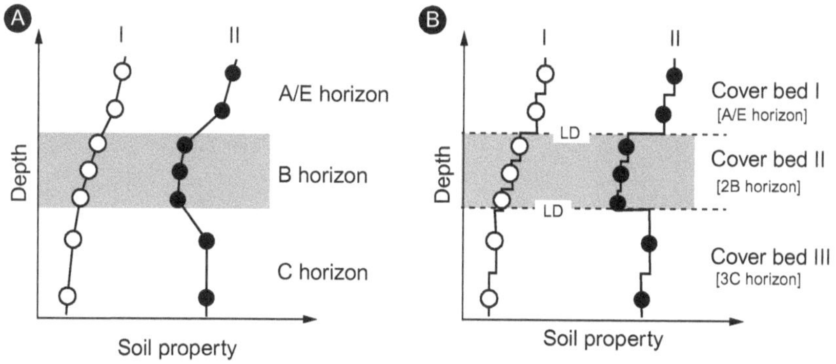

FIGURE 3.3 Interpretation of two principal depth functions (I + II) in lithologically discontinuous soil. The soil property is oxalate-extractable iron or similar (every open dot represents a soil sample with a hypothetical property; (A) interpretation of samples following the A–B–C model; (B) interpretation following the allochthonistic model).

only, if stratification is taken into account. In contrast, the interpretation of profiles with a depth function of type I becomes more complex, if the stratification is taken into account, because the stratification is blurred by pedogenesis.

The depth function of specific soil properties in lithologically discontinuous soil may be depicted in four simplified patterns (Fig. 3.4). A simple contrast (Fig. 3.4A) occurs, if a soil property is caused by two layers differing in primary properties. The difference between the upper and lower section of the depth gradient cannot be explained by pedogenesis (e.g., differences in silt or sand content). In more complex lithologically discontinuous soils, a multiple contrast may be found (Fig. 3.4B). Examples would be complex cover-bed successions, alluvial soils with multiple fluvial sediment successions, or anthropogenic layers at foot slopes.

From our observations, we conclude that pedogenesis in lithologically discontinuous soil may lead to intensification of the original contrast (Fig. 3.4C). This is often the case with texture contrasts in soils with denser (e.g., clay-rich) subsoils. A higher bulk density in the subsoil enhances accumulation of illuviated clay in the subsoil (a reinforced feedback). Contrasting layer properties may affect other parameters as well. In lithologically discontinuous soil with

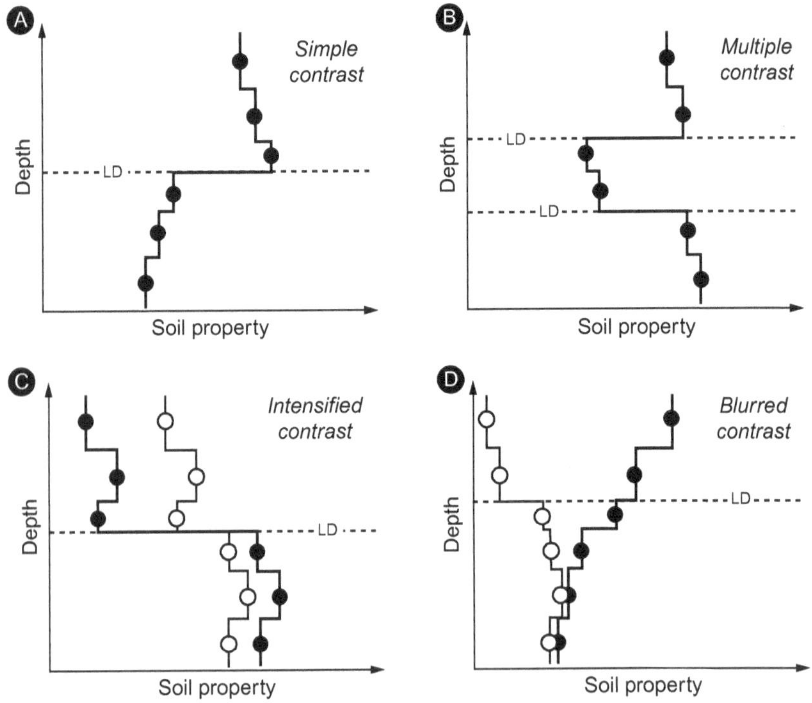

FIGURE 3.4 Four idealized types of depth function in lithologically discontinuous soil (every dot represents a soil sample with a soil property, open dots indicate the assumed geogenic properties before pedogenesis) (*modified from Lorz et al., 2011*).

a denser subsoil, water-logging is common, as infiltration is hindered (see Section 3.3). In the upper water-logged layer, hydromorphic properties[8] will be frequent, often in concert with clay illuviation. The restriction of roots to the upper layer because of the denser subsoil (especially of basal layers) may lead to intensified rooting (Frühauf, 1990; for Australian duplex soils, see Pankhurst et al., 2002) and preferred uptake of nutrients. Consequently, the upper layer becomes depleted of nutrients. Then again, stratification may be blurred by mobile elements that are either set free by pedogenesis or supplied by soil-external sources (blurred contrast, Fig. 3.4D). This is especially the case for cover-bed successions with shallow upper and missing intermediate layers. In the case of thicker upper and intermediate layers and a dense subsoil, for example, a basal layer, dislocation of mobile elements into the subsoil may be hindered and accumulation may be restricted to the upper layer.

These patterns are idealized and will typically occur as combinations of two or more types rather than as a pure type. In case, there are, in a lithologically discontinuous soil, no differences for one or more properties between samples below and above a disconformity, this could be because the layers are too similar and pedogenesis was too weak to alter this. This pattern may be found mainly if the intensity of pedogenesis is low.

3.4.2 Physical Soil Properties

3.4.2.1 Texture in Lithologically Discontinuous Soil

The most significant effect in lithologically discontinuous soil is deviation of grain-size distributions from expected pedogenic depth functions. A texture contrast within a profile may indicate stratification and/or pedogenic alteration of the parent material (regolith).

A good example for a simple texture contrast caused by stratification is a silt mantle covering a skeletic[9] or sandy-clayey subsoil, as is characteristic for shallow (<1 m) loess or sandy loess covers. In most of the profiles in Fig. 3.5, a silt mantle is the most prominent feature. The profiles Dahlener Heide and Wermsdorf 1 and 2 are textbook-like examples for texture-contrast soils with at least two contrasting geologic layers, an upper sandy-loess layer and an underlying glaciofluvial sand or till, respectively. The silt and sand fractions may be used to distinguish the upper cover bed with eolian influence (silt ~40–80%) from the underlying material with large (coarse) sand (~40–80%) and small silt contents (~5–35%). The contrast factor for silt between the upper and the lowest cover bed is >2. In the profile Dorndorf, the silt mantle is nearly 1 m thick and covers limestone debris very rich in rock fragments. Based upon pH (see Fig. 3.11), color, and field-estimated bulk density, the silt mantle itself may be split into

8. Soil properties induced by reducing (sometimes alternating with oxidizing) conditions by permanent or seasonal water logging.

9. Soil material rich in coarse fragments.

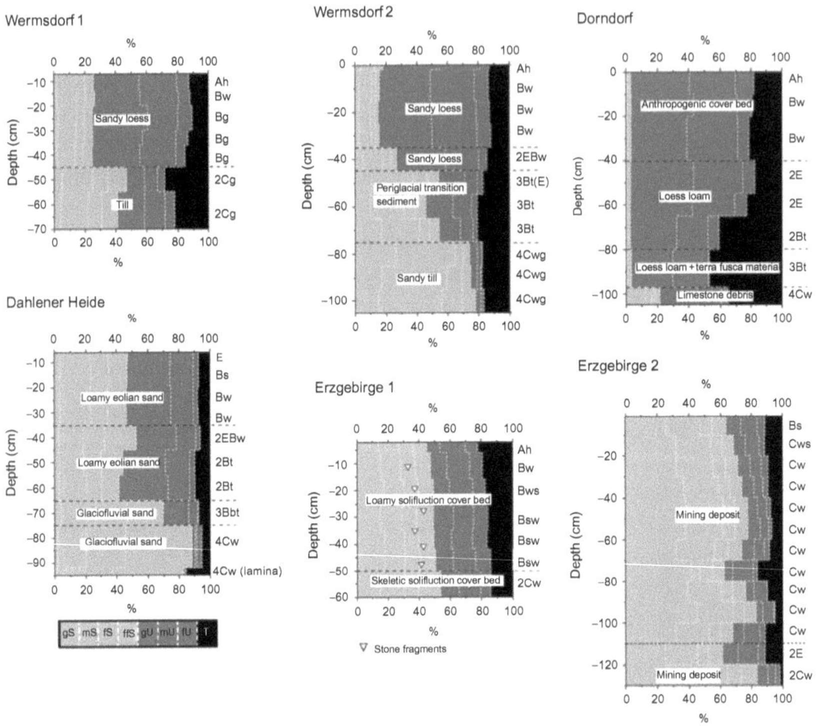

FIGURE 3.5 Texture of example profiles (*reproduced from Lorz, 2008*). Wermsdorf 1, NW-Saxony, Albaquult (Planosol); Wermsdorf 2, NW-Saxony, Hapludalf (Luvisol); Dorndorf, E-Thuringia, Paleudalf (Luvisol) overlain by anthropic epipedon—see the text for the term "terra fusca material"; Dahlener Heide, NW-Saxony, Hapludalf (Luvisol); Erzgebirge 1, S-Saxony, Dystrudept (Cambisol); Erzgebirge 2, S-Saxony, Haplanthrept (Regosol).

an anthropogenic (see Section 2.4.2) and underlying periglacial cover beds. In the profile Erzgebirge 1, the cover beds show a generally lower silt content (~30%) due to decreased loess input assumed to be caused by the higher altitude (705 m asl) compared with the other profiles (~150–200 m asl). The silt content is not suitable for the identification of disconformities in the profile Erzgebirge 1, as the layers are very similar in fine-soil texture. Layers may be identified by field-estimated bulk density, color, and content of rock fragments. In the profile Erzgebirge 2, the parent materials are mining wastes, probably deposited when the exploration on tin in this region began (thirteen to fifteenth century). Within this deposit, no stratification may be recognized from field or laboratory analysis, with the exception of a disconformity at a depth of 110 cm indicated by buried soil horizons.

Texture contrasts are often caused by differing clay contents, which may arise solely from pedogenic processes, if there is no disconformity between eluvial and illuvial horizons. However, in cases where the boundary coincides with a disconformity, it is doubtful whether the clay content is purely pedogenic. Paton et al. (1995) and Phillips (2004) concluded for texture-contrast soils that the clay contents of subsoils are often much too high to be caused solely by clay illuviation. Therefore, the texture contrast (clay) between two horizons is not decisive as to whether the two horizons are pedogenically linked by translocation (Brewer, 1972). In this context, the clay content of argillic horizons in Central Europe is often seen as at least partially geogenic (Semmel, 1993b, 2001b; also see the "pseudo-Alfisols" in Section 3.3). Recently, Jacobs and Mason (2007) reported loess profiles of the Great Plains in the United States to have a disconformity between the A and B horizons. They proposed *in situ* clay formation in the B horizon and, later, eolian sedimentation of the A-horizon material.

In the profile Erzgebirge 1, the clay content decreases with depth from around 20% in the A horizon to approximately 15% in the Bg horizon[10] as an effect of decreasing clay production. In the profile Erzgebirge 2, there is a similar decrease of clay in the upper section of the profile. In the profile Wermsdorf 2, we interpret the increase of clay below a depth of 35 cm as primarily attributable to clay illuviation (Lorz, 2008). A mixture of geologic properties and translocation effects may be assumed for the profile Wermsdorf 1. While in the sandy loess an increase of clay with depth as a result of inner-horizon translocation is observed, the stark contrast to the till must be seen as a mixed effect of a higher geogenic clay content and additional clay illuviation (intensified contrast). A complex situation is also found in the profile Dorndorf. In the anthropogenic layer, first signs of clay dislocation may be assumed from the slight increase of clay with depth. Below 40 cm, in the buried soil profile, the grain-size distribution characteristic of Alfisols (Luvisols) may be found. An exception from this is the 3Bt horizon with a clay content of about 45%. This high clay content is not the effect of clay illuviation but presumably the effect of long-lasting limestone dissolution setting free residual clay from the rocks (simple to blurred contrast, see below); this clay-rich matter is referred to as "terra fusca" material (see Section 2.8.3.2; also see Muhs and Budahn, 2009 for similar soils). In the profile Dahlener Heide, the total clay content is remarkably low. Clay accumulation by illuviation may be assumed for the weakly developed 2Bt horizon. Its difference in clay content with the eluvial horizon (2EBw) is small (2–3%). There are clay laminae in the subsoil (4Bt), interpreted as a second zone of clay accumulation, formed during an older phase.

3.4.2.2 Soil Water in Lithologically Discontinuous Soil

Even though the potential effect of stratification on available water capacity and hydraulic conductivity at the site scale is obvious, research on the effect of stratification on water flux and water storage in lithologically discontinuous soil is scarce (see Chapter 4). Furthermore, extensive work has been done on water dynamics of Australian duplex soils (Chittleborough, 1992; Northcote, 1979).

10. Soil horizon influenced by hydromorphosis.

Available water capacity (AWC) depends to a large extent on the bulk density and the silt content. Therefore, silty mantles are crucial for water storage and water supply of sites with subsoils rich in clay, sand, or a skeleton with typically remarkably lower field capacities. This applies to areas adjacent to loess regions where the silty cover beds are shallow and the subsoils have a high bulk density or sandy texture. The importance of silt mantles for the AWC is obvious (Table 3.1). In combination with lower bulk densities, the importance of the upper layer for AWC is assumed to be even higher. Its removal by soil erosion would leave sites with considerably lower AWC and worse site quality.

TABLE 3.1 Texture, Estimated Available Water Capacities, and Estimated Hydraulic Conductivity for Example Profiles (*reproduced from Lorz, 2008*)

Layer (no.)	Thickness (cm)	Texture class[a]	Silt (%)	AWC (%)[b]	SHC (cm d^{-1})[c]
Profile Wermsdorf 1					
Sandy loess (1)	45	Silt loam	62	24	43
Glacial till (2)	10	Sandy clay loam	24	17	17
Glacial till (2)	15	Loam	37	16	27
Profile Dorndorf					
Anthropogenic layer (1)	40	Silt loam	76	28	35
Loess loam (2)	40	Silt loam	76	23	51
Loess loam + terra fusca (3)	15	Silty clay	48	17	14
Limestone debris (4)	10	Clay loam	44	14	18
Profile Dahlener Heide					
Loamy eolian sand (1)	35	Loam	45	27	60
Loamy eolian sand (2)	30	Loam	44	23	60
Glaciofluvial sand (3)	10	Sandy loam	22	17	77
Glaciofluvial sand (3)	20	Loamy sand	7	16	94

[a]Texture class (*modified from Brady and Weil, 2008*).
[b]Available water capacity estimated (*modified from AGB, 2005: table 70*).
[c]Saturated hydraulic conductivity, estimated (*modified from AGB, 2005: table 76*).

A second aspect of stratification is the impact on vertical and horizontal flow of water through the soil. Percolation is impeded if denser subsoils have a water-logging effect for the topsoil; this is often found in soils with consolidated, dense subsoils. Frühauf (1990) measured saturated hydraulic conductivities in upper cover beds between 138 and 3 cm d^{-1} and in basal layers ranging from 27 to <1 cm d^{-1}. This pattern of less permeable subsoil and less dense topsoil also holds true in two of our profiles (Table 3.1: Wermsdorf 1 and Dorndorf). However, in the profile Dahlener Heide, the pattern is inverse, as the subsoil has a higher sand content than the upper cover bed.

In another context (plant yield), extensive literature exists on water-logging in Australian duplex soils with dense subsoils. Typically, a shallow seasonal aquifer perched on the B horizon of texture-contrast soils is formed (McFarlane and Williamson, 2002; Silberstein et al., 2002). The susceptibility factors for water-logging in lithologically discontinuous soil are (1) a sufficient difference in hydraulic conductivity, (2) a high bulk density of the subsoil, which causes fast saturation and water-logging, (3) a shallow topsoil (<4 dm), and (4) a relief position favoring water-logging over lateral flowage (Cox and McFarlane, 1995; Frühauf, 1990).

3.4.3 Chemical Soil Properties

3.4.3.1 Geochemical Composition of Cover Beds

The impact of cover beds on the geochemical composition of soil-parent materials is illustrated by a schematic example depicting a typical situation in the subdued mountains of Central Europe (Fig. 3.6). In the first stage (Fig. 3.6a), one or more basal layers form, consisting to a large extent of up-slope material. Their composition depends on the mixing members, which could be frost-shattered rock debris, saprolite, or a combination thereof. Varying compositions and substantially higher trace-element concentrations may be caused by metal-bearing veins mixed into the basal layer (also see Stückrad et al., 2008). The composition of the basal layer (allochthonous regolith) is a mixture of up-slope and underlying materials, hence, it may differ decisively from the lower (par) autochthonous regolith or saprolite.

Intermediate layers are typically found at specific relief positions (Section 2.3.2). Their composition is usually remarkably different from the autochthonous regolith (Fig. 3.6b), given that the accumulated eolian material is derived from distant sources. For these profiles, decoupling of parent material from bedrock (Kleber, 1992a) is most evident. The composition of the upper layer (Fig. 3.6c) depends on the pre-existing, already heterogeneous layers on one hand and on eolian input on the other. The upper layer is comparatively uniform due to loess admixture. With increasing altitude, the influence of loess decreases, whereas local material and, therefore, the local geochemical composition gain importance. A further complication arises from widespread anthropogenic erosion and the

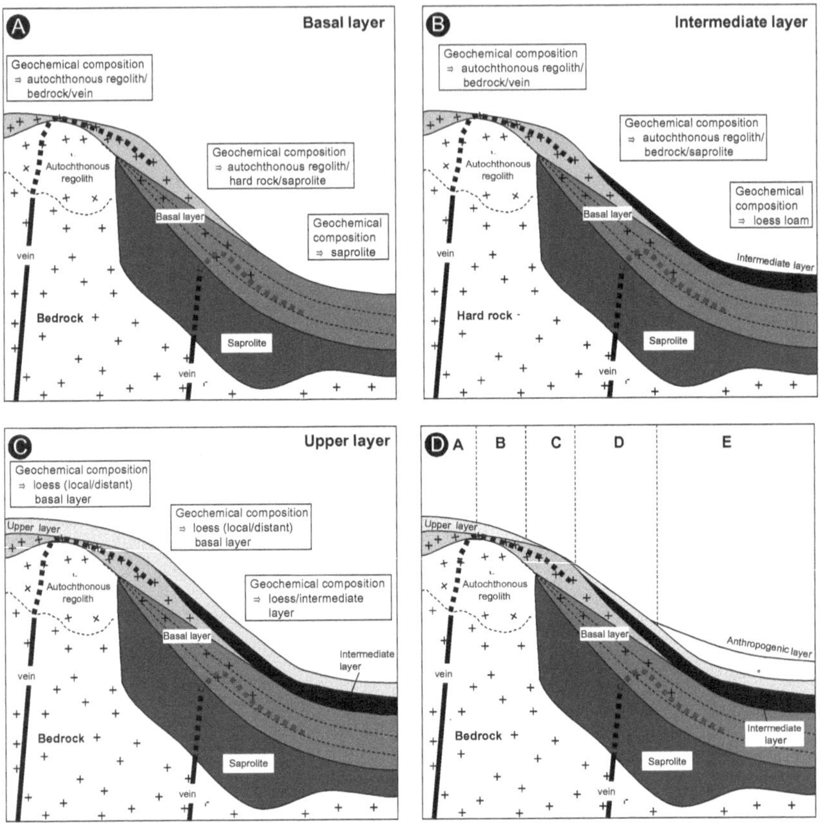

FIGURE 3.6 Evolution scheme of the soil cover (*modified from Lorz, 2008*). This example represents an idealized slope section in four stages of the formation of cover beds. The possible geochemical composition is shown for different slope positions.

formation of anthropogenic layers (Fig. 3.6d; see Section 2.4.2). These are direct and indirect mixtures of topsoil material, mostly from the upper layer, often rich in organic matter. However, at sites with severe soil erosion material from underlying cover beds may also be involved. Younger anthropogenic cover beds often have higher concentrations of contaminants (Clemens and Stahr, 1994).

Comparatively simple cover-bed successions are two-layer profiles consisting of an upper over a basal layer (Fig. 3.6d, A). Their mineralogical and geochemical composition is determined by the autochthonous regolith and admixed loess. A specific situation with higher concentrations of metals exists if materials originating from veins with metal enrichment are mixed in (Fig. 3.6b; see also Stückrad et al., 2008). In steep and/or convex slope sections, the upper layer is often removed by anthropogenic soil erosion, and the basal layer is exposed again (Fig. 3.6d, C). As a consequence, truncated profiles may be found. However, the standard situation for the foot slope and concave relief forms are multilayer profiles (Fig. 3.6d, D and E) with geochemical properties only marginally controlled by bedrock.

3.4.3.2 Heavy Metals

Traditional approaches to explain the distribution of heavy metals in soil profiles are based on the idea of progressive top-down processes (A–B–C concept, Fig. 3.2A), assuming that soil is pedogenically altered bedrock. Depth/concentration functions of heavy metals are thought to be characteristic for each soil type (Alloway, 1990; Blume, 1992). Several authors criticized this concept (e.g., Beck, 2002; Eiberweiser, 1995; Hindel et al., 2004; Kleber and Mailänder, 2000; Kleber et al., 1998a; Lorz, 1996, 2008; Lorz and Phillips, 2006; Rosenberg, 1988; Sabel, 1989). They concluded that the interpretation of metal distribution in stratified soils is more complex than assumed by a solely pedogenic approach. In another context—distribution of cosmogenic nuclides—Schaller et al. (2002) supported this conclusion with regard to lithologically discontinuous soil.

Studying the dispersal of geochemical parameters, for example, heavy metals, in soils often reveals a complex situation. As elements are relocated within the profile by pedogenesis, pre-existing differences in concentrations may change during soil evolution and, even more challenging, this may be modified by modern human activities. However, we conclude for cover-bed successions that each cover bed is a geogenic layer with its own primary composition leading to different spectra and amounts of heavy metals. Each cover bed has to be taken as a particular unit in terms of its metal contents. Rosenberg (1988) and Ruppert and Schmidt (1987) emphasized the following aspects that directly influence the distribution of heavy metals:

- Turbation and translocation processes have formed mixed materials. Erosion, sedimentation, and, thus, landform properties are important parameters that influence the distribution of heavy metals in soils (Opp, 1998; Ruppert, 1994).
- The addition of allochthonous material (e.g., Brimhall et al., 1988) may cause interferences with the "constancy of relationship" (Brewer, 1972). As a consequence of loess input or solifluidal overriding, high metal contents of underlying substrates may be obscured. However, cover beds may also be richer in heavy metals above rocks or pedorelicts with low metal contents. Altogether, eolian admixture tends to attenuate extremes if provided by underlying materials.

To assess the effects of disconformities on the vertical distribution of heavy metals, depth functions of metal fractions of example profiles are used, using a sequential extraction of three fractions (I–III) of increasing binding strengths (Lorz and Phillips, 2006).

An obvious pattern of most depth functions are decreasing Pb concentrations (fractions I and II) with depth within the upper soil, reflecting (past) atmospheric deposition (Fig. 3.7). Disconformities do not show effects with regard to the least

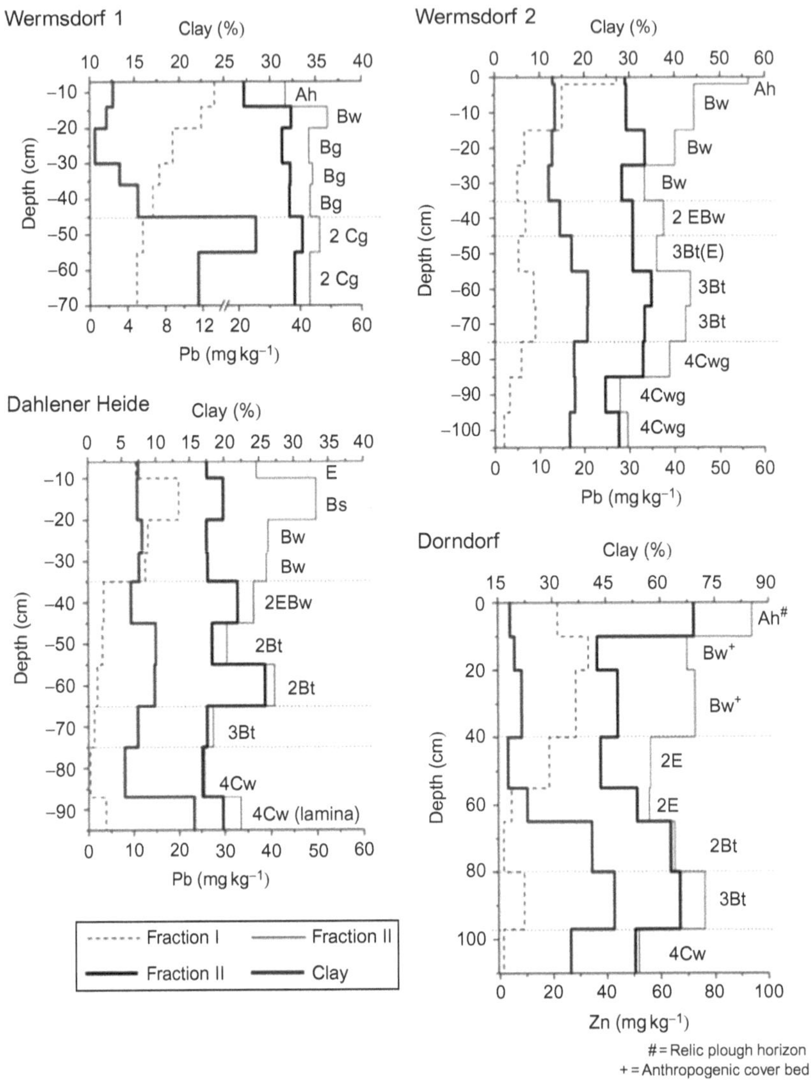

FIGURE 3.7 Fractions of lead and zinc in example profiles. For soil classification and source of data, see Fig. 3.5. The shortened sequential extraction used here includes the mobile and adsorbed fraction [I] (0.1 M NH_4 oxalate) and the total concentration [II] (microwave). From the difference between the total content and the sum of fractions I and II, a "quasi-lithogenic" concentration [III] was calculated (Lorz, 2008). The heavy metals used in the analyses depicted in this and following figures differ in origin: (this fig.) dominantly anthropogenic (Pb, Zn), with a distinct accumulation of mobile fractions in the organic and mineral top soil; (Fig. 3.8) dominantly lithogenic (Cu, Ni), with a clay-dependent accumulation throughout the entire profile; and (Fig. 3.9) dominantly pedogenic (Mn), with an accumulation within the solum. (For color version of this figure, the reader is referred to the online version of this chapter.)

bound fraction I (blurred contrast), although the lower limit of upper-soil enrichment sometimes coincides with the disconformity between a silty cover bed and the underlying material (profiles Wermsdorf 2, Dahlener Heide). Surprisingly, there is no gradient and no effect of the disconformity regarding the lithogenic fraction (III) in these profiles. The dependence of Pb on the clay content is obvious for profiles with clay illuviation. Their argillic horizons are zones of higher concentrations of Pb (fractions II and III). Effects of stratification are not recognizable from the depth gradients, but the accumulation may be explained by pedogenic effects alone. Remarkable exceptions are profiles with a strong contrast in clay content but a constant depth function of Pb concentrations (Wermsdorf 1). This phenomenon may be understood, if a geogenic difference in the geochemical character of the upper sandy loess and the lower till is assumed. In contrast, the gradients of the profile Wermsdorf 2 may be explained by a similarity of the involved layers and presumably similar sources. The consequent pedogenic alteration (clay illuviation) caused zones of higher concentrations. Sections of lower concentrations (4Cwg, 85–95 cm) are caused by hydromorphic patterns. For the profile Dahlener Heide, fly ash is likely influential (Fürst et al., 2007). The accumulation of Pb (fraction I) is restricted to the upper layer (intensified contrast), where there is an even distribution of Pb concentrations.

An additional effect may be observed for the disconformity between the anthropogenic layer and loess loam in the profile Dorndorf, where the disconformity affects the depth functions of Zn fractions I and II, but not fraction III (Fig. 3.7). We take this as an indicator for a topsoil source of the anthropogenic layer. An effect of stratification may be present for all zinc fractions between the loess loam cover, the loess loam with admixed terra fusca material, and the limestone debris with lower zinc concentrations (simple or blurred contrast). An atmospheric input of zinc is indicated by a decrease in fraction I concentrations with depth (down to 80 cm), crossing one disconformity (blurred contrast).

Copper and nickel distributions are only slightly affected by atmospheric deposition. For Cu fractions II and III, a simple contrast may be observed in profiles Wermsdorf 1 and Dahlener Heide for the disconformity between the loess-rich cover bed(s) and underlying materials (Fig. 3.8). In both profiles, the eolian input created a cover bed with lower concentrations (II + III) than the underlying material. For the profile Wermsdorf 1, the underlying material is till, which is known to be relatively rich in copper (Blume, 1981). The upper section of the profile Wermsdorf 2 does not show effects of stratification because all fractions are affected by clay illuviation. However, there is an effect at the disconformity between till and an unclassified transitional periglacial sediment at ~75 cm depth, as the Cu concentration is even higher in the sandy till, although its clay amount is substantially lower.

The profile Dorndorf shows a similar strong relationship between clay and nickel (Fig. 3.8). An increase in concentrations in all fractions with depth was found, probably caused by the increasing clay content. An effect of lithological discontinuity may occur at the boundary between the anthropogenic layer and

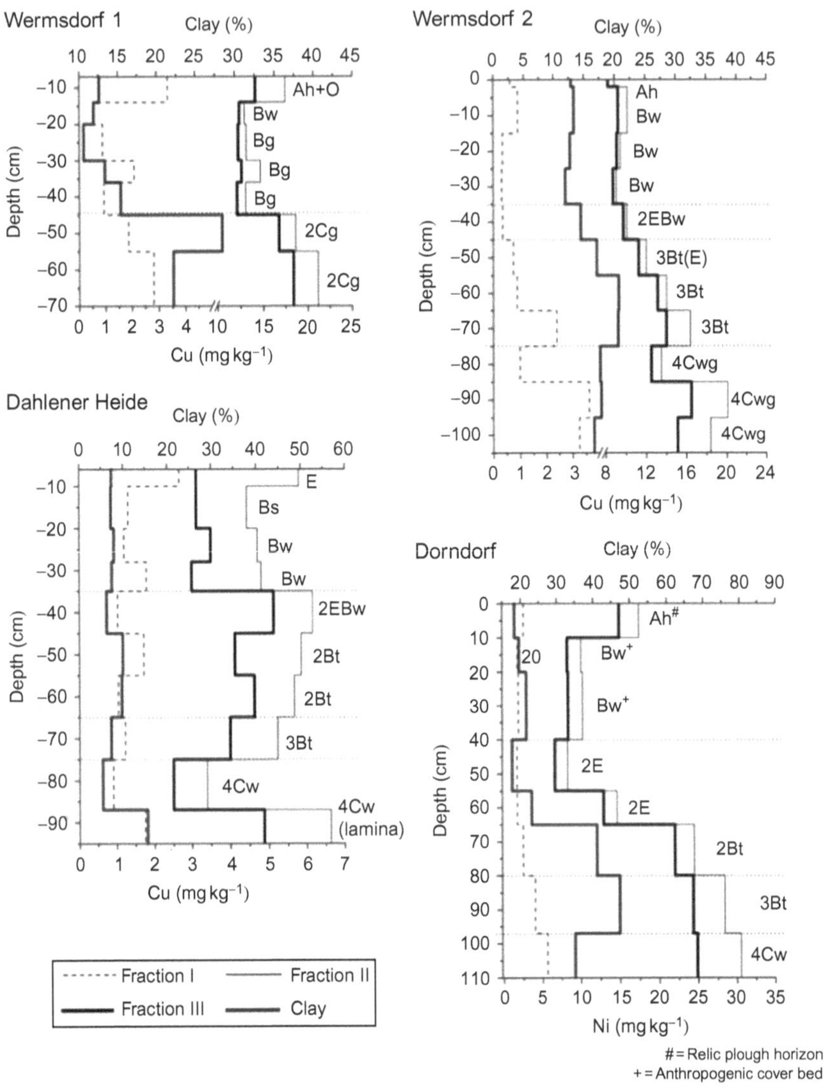

FIGURE 3.8 Fractions of copper and nickel in example profiles. For soil classification and source of data, see Fig. 3.5; for methods, see Fig. 3.7. (For color version of this figure, the reader is referred to the online version of this chapter.)

the buried profile, where the Ni concentrations for all fractions increase with clay content and depth. At the disconformity at a depth of 97 cm, the highest concentrations occur regarding all fractions, indicating a high geogenic content of the underlying, weakly weathered limestone.

The depth functions for manganese are characterized by strong pedogenic dynamics, as is to be expected (Fig. 3.9). Pedogenic manganese is enriched

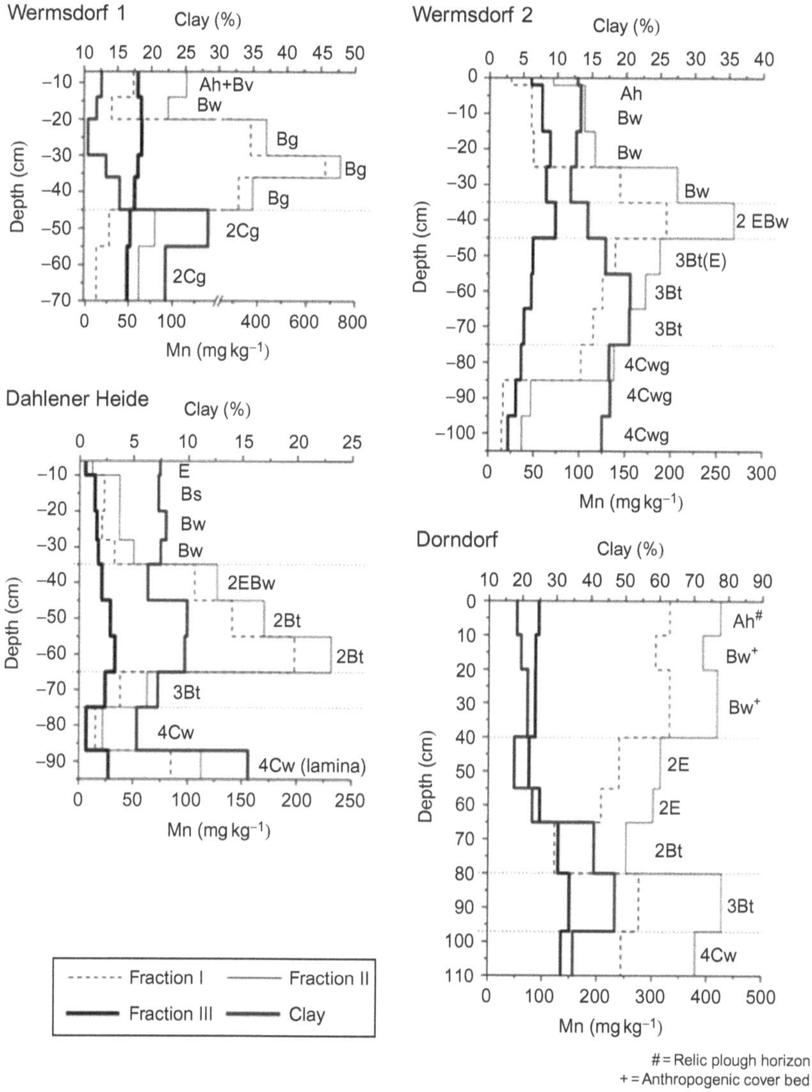

FIGURE 3.9 Fractions of manganese in example profiles. For soil classification and source of data, see Fig. 3.5; for methods, see Fig. 3.7. (For color version of this figure, the reader is referred to the online version of this chapter.)

down to the lower boundary of the silty cover bed with a sudden decrease in concentrations for fractions I and II below the disconformity as a secondary effect of stratification (profiles Wermsdorf 1 and Dahlener Heide). A similar pattern occurs in the profile Wermsdorf 2, where the contrast is blurred by a deep reaching (45–85 cm) influence of (paleo[?])pedogenesis. In the profile Dorndorf, the effect of stratification is obvious for the anthropogenic layer, as eroded topsoils from up-slope, rich in active manganese, are a substantial part of them. Below, the pedogenic pattern of the buried soil is retained.

3.4.3.3 Oxalate- and Dithionite-Extractable Iron

We use depth functions of oxalate-extractable iron and the Fe_o/Fe_d ratio as an indicator for soil formation (cf. Blume and Schwertmann, 1969), focusing on the effects of disconformities on the concentration/depth gradients.

All example profiles (Fig. 3.10) have depth-dependent gradients of oxalate-extractable iron and of the Fe_o/Fe_d ratio, reflecting the intensity of soil formation. Abrupt changes in the depth functions of the Fe_o/Fe_d ratio and the Fe_o concentration coincide with disconformities. The Fe_o/Fe_d ratio narrows at the disconformity

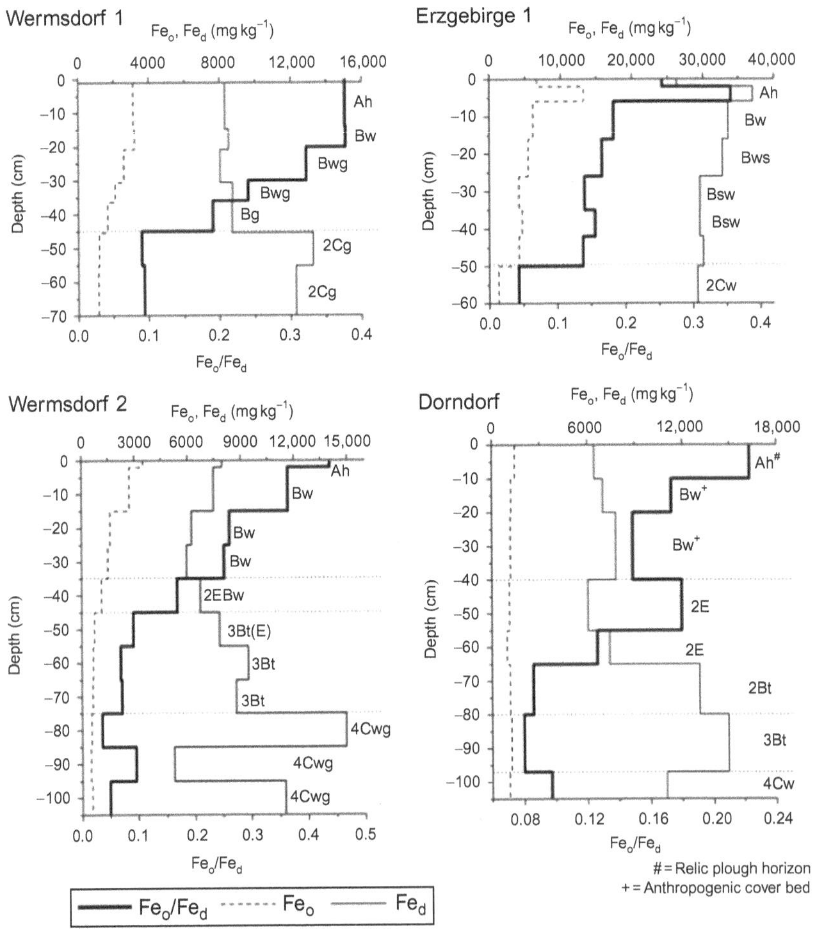

FIGURE 3.10 Oxalate- and dithionite-extractable iron (Fe_o, Fe_d) and Fe_o/Fe_d ratio of example profiles. For soil classification and source of data, see Fig. 3.5.

between silt-rich cover beds and underlying layers in all profiles. A distinct gradient for Fe_o may occur in the upper layer. Both features mark this zone as the solum of the surface soil. The higher ages of the lower layers in profiles Wermsdorf 1 and 2 are indicated by higher Fe_d concentrations. In the profile Dorndorf, a double pattern exists. The upper soil is an anthropogenic layer with a gradient that has developed after its sedimentation (at least 200 years ago; Lorz, 2008). Below 40 cm, the buried Alfisol with its (fossilized) depth gradient begins.

All soil profiles show a similar pattern of current soil formation, indicated by a decrease in the Fe_o/Fe_d ratio with depth. The pattern ends at a depth of 40–50 cm, coinciding with the disconformity of an upper cover bed (intensified contrast), except for the profile Wermsdorf 2, where a second shallow cover bed is crossed (blurred contrast). Although the depth gradient supports the idea of an *in situ*, top-down younger soil formation overprinting geogenic gradients, its dating remains rather vague. Pedogenesis must have started after sedimentation, and soil maturity presumably increased with wetter and warmer climate conditions during the Holocene.

3.4.3.4 pH Value and Acidification

The A–B–C model lets one expect a depth function of soil properties such as pH or base saturation due to top-down progression of soil acidification. In general, one may expect increases in pH with depth (Breemen van and Buurman, 2002; Scheffer and Schachtschabel, 2010), indicating the anisotropic effect (Johnson et al., 1987) of soil acidification as a top-down process.

The profiles Dahlener Heide and Erzgebirge 1 (Fig. 3.11) have perfect depth gradients of pH values, with a steady increase with depth, crossing disconformities (blurred contrast). The top-down process of acidification has created a pattern that is typical for humid regions with acidic topsoil and less acidic subsoil. The impact of stratification on the depth functions of the pH values (H_2O, KCl, $CaCl_2$) is obviously limited in acidic soil profiles with little geochemical contrast between the layers (pH [H_2O] 3.0–4.5). However, there are several deviations from this pattern. A less obvious irregular pattern occurs, if the pH gradient (KCl, $CaCl_2$) is affected by liming carried out with afforestation of the site in the early 1960s and/or hydromorphic processes—as is the case in the profile Wermsdorf 1 (blurred contrast). More obvious irregularities occur in profiles with young anthropogenic layers containing calcareous material superimposed on an older soil profile, as in the profile Dorndorf. Time and intensity of the pedogenesis were not sufficient to form a pedogenic gradient in the profile (simple contrast). A similar pattern occurs in the profile Erzgebirge 2, which has developed in young mining deposits from the thirteenth to the fourteenth century. A depth gradient may be present only in the upper topsoil (depth <10 cm), whereas no gradient has formed in the unstratified lower part of the profile.

Comparing the results of the profiles Erzgebirge 2 and adjacent Erzgebirge 1, we conclude that for the profile Erzgebirge 2 time was insufficient (about 500–700

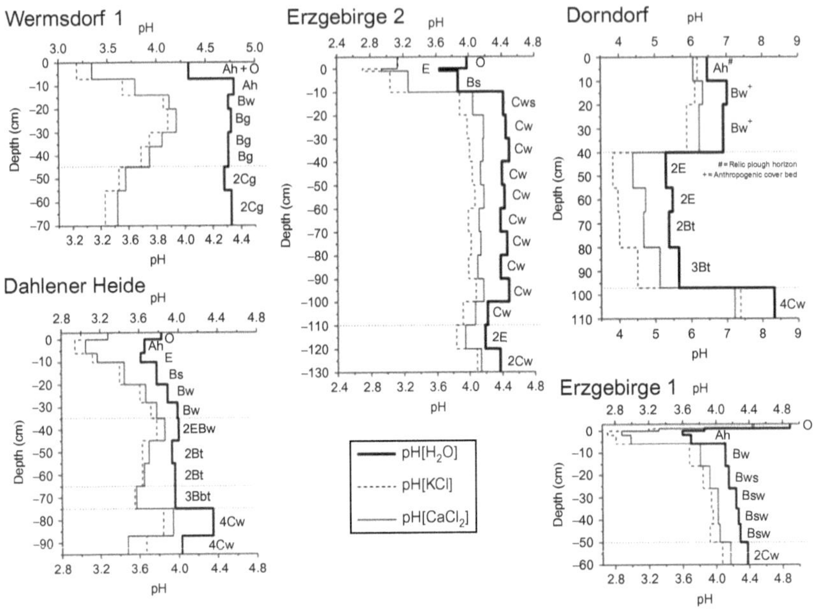

FIGURE 3.11 pH of example profiles. For soil classification and source of data, see Fig. 3.5.

years: Lorz, 2008) to form a typical pH gradient that is different from that of the profile Erzgebirge 1. The mining deposits are a mixture of regolith material and form a "neutral" parent material, that is, the material is not fresh, but the vertical organization by *in situ* pedogenesis started only after its formation.

Disconformities will also affect the depth functions of other parameters of soil acidification (e.g., base saturation). The crossing of the depth functions of base saturation and aluminum saturation—the so-called lower limit of acidification—often coincides with disconformities (simple contrast), as it is the case in the profiles Wermsdorf 2 and Erzgebirge 2 (Fig. 3.12). This coincidence may be explained by preferred rooting, prevailing uptake, and export of base cations from the upper layer with its lower bulk density (intensified contrast). This phenomenon occurs even if the geochemical differences between the layers are weak (profile Erzgebirge 1). However, the subsoil of the profile Wermsdorf 4 with its low buffer capacities (sandy till) does not show such a coincidence of disconformity and boundary of acidification. Rather, the gradient of base and aluminum saturation crosses the disconformity. The approximating gradients indicate decreasing acidification with depth. Due to an assumed lower acid neutralization capacity of the sandy till (sand ~80%), its resistance against acidification is weak (blurred contrast). The importance of disconformities and stratification for the formation of pedogenic depth functions may be shown by comparing a young soil profile in mining deposits (Erzgebirge 2) with the profile Erzgebirge 1. Pedogenesis has not formed a gradient in the former, but a depth function without a trend has formed due to mixing of

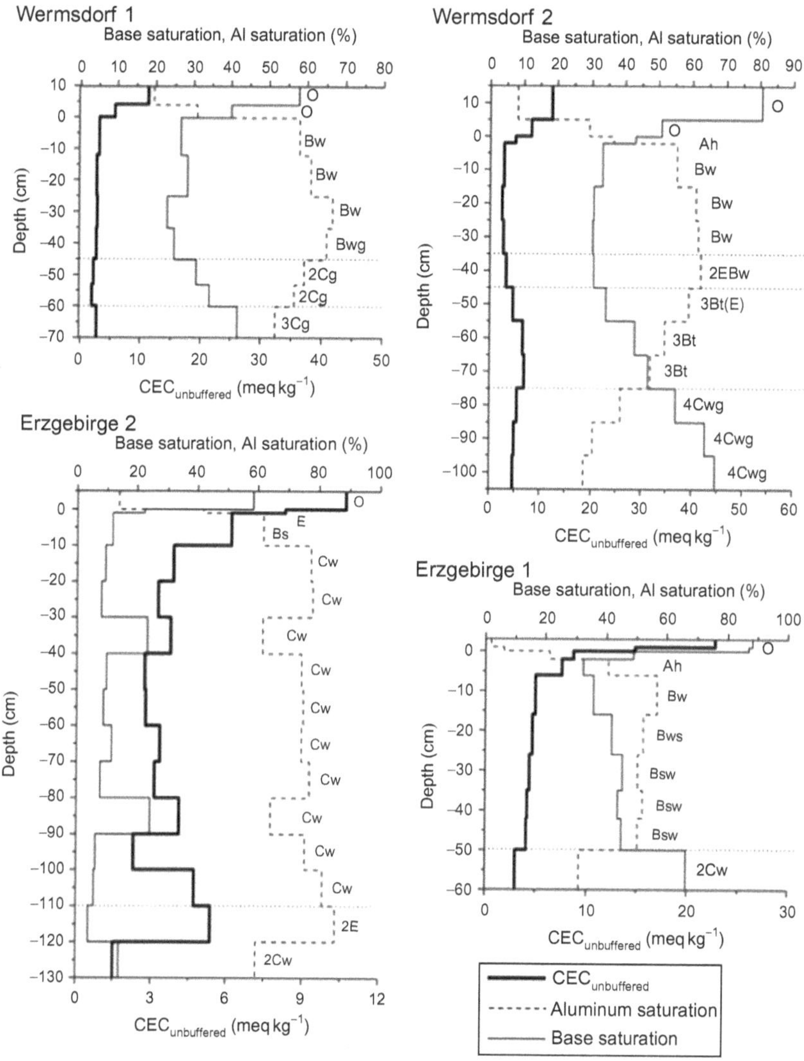

FIGURE 3.12 Base saturation, aluminum saturation, and unbuffered cation exchange capacity (CEC) of example profiles. For soil classification and source of data, see Fig. 3.5.

material. Aluminum saturation is predominant throughout the profile (below a depth of 10 cm), presumably because preweathered topsoil material is a substantial part of mining deposits.

As previously discussed, acidification that makes its way into the soil from top down is buffered in loess-rich substrates. In most soils, the buffering capacity of the substrates has been exceeded during Holocene pedogenesis so that there is a trend of decreasing pH values with depth. However, there are profiles

TABLE 3.2 Soil Profiles with Inverted pH Gradients (*Modified from Beck, 1998, Profile 1, and Mahr, 1998, Profile 2*)

Soil horizon	Cover bed	Depth (cm)	pH Value ($CaCl_2$)	Clay (%)	Silt (%)	Sand (%)
Profile 1						
A	LH	11	5.8	12	35	53
E		31	5.5	14	32	54
2Btg	LB1	60	4.0	38	19	43
3Cg	LB2	90	3.5	42	34	25
4Cg	PM	110+	3.5	28	22	50
Profile 2						
Bw	LH	25	4.92	23	38	39
2Bw	LM	50	5.12	23	35	42
3C1	LB	100	4.08	5	20	75
3C2		115	3.98	7	23	70
4C	PM	125	3.82	8	24	68
5C	PM	200+	3.84	5	13	82

Profile 1 is an Aqualf (Stagnic Luvisol)[1] from the Schönbuch Mountains, Southwest Germany, developed upon disintegrated Mesozoic sandstone, at 420 m asl, on a slope very slightly inclined in easterly direction. Profile 2 is a Dystrudept (Haplic Cambisol)[2] from the Bavarian Forest (Fig. 2.1 for location) developed upon weathered gneiss, at 750 m asl, on a steep slope inclined to the north. LH, upper layer; LM, intermediate layer; LB, basal layer, LB1 and LB2 for differentiation of basal layers; PM, underlying parent material.
[1]Soil with combined properties from clay translocation and hydromorphosis.
[2]Inceptisol with dystrophic topsoil.

that show pH values opposite to this tendency (see examples in Table 3.2), with pH values decreasing with depth, which is assumed to be due to remarkable admixture of buffering eolian silt.

3.4.3.5 Stratigraphic Approach to Alteration in Soils from Cover Beds

As outlined in the previous sections, quantifying postdepositional alteration of mineral soil substrates is complicated by the fact that most soils consist of layered parent materials, that is, cover beds with varying geogenic contents. Previous attempts at measuring the alteration of such substrates involved comparing the concentration of the component under study in the soil with that of the underlying parent material using a mass balance approach (for heavy metals, see Hindel and Fleige, 1990; Ruppert and Schmidt, 1987; Walk, 1982) along depth/concentration functions (Blume and Schwertmann, 1969; Filipinski and Grupe, 1990), referred to as autochthonistic approach in Section 3.2. This approach has an often overlooked prerequisite: the original parent material is required to be homogeneous or to have known consistent trends with regard to the component under study (Wang and Arnold, 1973). This requirement is rarely met, because most landscapes in the mid-latitudes are veiled by cover beds. Their composition is unpredictably different from underlying bedrock or saprolite even if they are of local, up-slope origin and even more so once eolian matter is admixed. This variability also holds true for the distribution of heavy-metal concentrations in soils (Eiberweiser and Völkel, 1993; Sabel, 1989; Spies, 1991).

Some other approaches to quantify postdepositional contamination of mineral soils, in particular by heavy metals, have been reported to have specific shortcomings. Continuous measurement of deposition (Schimming, 1992) does not allow for long-term conclusions because of the short duration of such measurements. Measurement of bonding strength (heavy metals enriched after deposition by pedogenic processes are bound weakly to the soil, whereas geogenic ones are bound strongly: Chlopecka et al., 1996; Filipinski and Grupe, 1990) reasonably indicates alteration but quantitative assignments and attribution to sources are difficult (Gerth, 1985; Kaupenjohann and Wilcke, 1995).

Kleber et al. (1998a) presented a procedure to overcome errors from vertical variability by comparing concentrations within the stratigraphically same deposit along a lateral gradient of exposition to surface-related processes. They used heavy metals, which are in the center of recent environmental concerns, to exemplify this approach.

They studied trenches in the Steigerwald Mountains, SW Germany (Fig. 3.13), dug into moderately cemented Mesozoic sandstone (5C horizon), and overlain by cover beds. This sandstone has silty and clayey interbeddings up-slope of the sites. The trenched sites cross boundaries of medieval agriculture: their lower portion was plowed and, accordingly, underwent anthropogenic soil erosion. Thus, just the same layer of sandstone at the base of the trench is overlain by more than 1 m of cover beds (Fig. 3.13, Profile A) or by a thin plowed layer (Profile B) within a distance of 1.5 m, with a >40° steep slope between; these two segments will be referred to as "protected" and "exposed" in the following, respectively. Sampling included Profiles A and B down to the uppermost layer in the exposed segment (which is weathered sandstone in the example trench, and basal layers in two other sampled trenches) and two to three lateral transects (each with six samples taken at approximately equal distances) within it to get an indication of gradients of heavy-metal concentrations. Farming at the sites was abandoned before 1554 A.D. (first mentioning of the devastated site in a historical document: Machann, 1972), allowing for forest growth. The entire site is now covered by approx. 60-year-old spruce (*Picea abies* (L.) Karst).

In the example trench, two of the basal layers overlying the protected range of the sandstone layer under consideration are rich in fine particles and, thus, impede vertical soil-water flow, which is also indicated by mottling in the

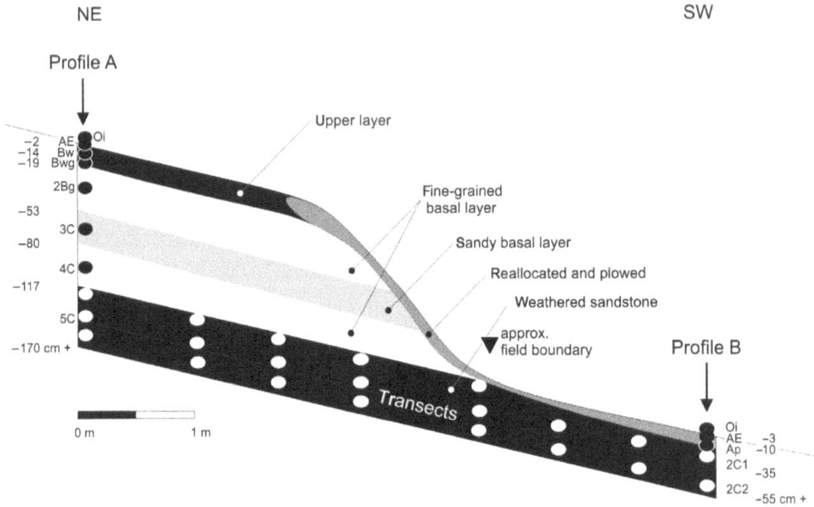

FIGURE 3.13 Example of a trenched section: deposits, soil horizons, and sampling design. Oval dots represent soil samples (*modified from Kleber and Mailänder, 2000*).

2Bg horizon. Because of their thickness, these layers are assumed to effectively shield what is beneath from most influences from the surface. Accordingly, gains of metals as well as losses through leaching are supposed to be unimportant in the protected range. The particle sizes within the particular transects, that is, within the same lithological layer, are essentially uniform.

The concentrations of soil organic matter, important for binding some heavy metals, are only high near the surface and do not vary much within the transects. The pH is very low near the surface but reaches a constant level at depth (Profile A). Along the transects, it decreases with decreasing protection.

By far the highest heavy-metal concentrations occur in the organic layers. This indicates that the site is polluted by atmospheric deposition, even though it is far from industrial centers and other heavy-metal point sources. However, it is close to highways, with traffic having been a major nonpoint heavy-metal source until the recent past (Kaupenjohann and Mai-Peter, 1998), and, accordingly, there is ubiquitous background contamination of heavy metals in Germany (Wilcke and Kaupenjohann, 1994).

The heavy-metal concentrations generally decrease with depth in Profile A (Fig. 3.14A), but this decrease is irregular because of the variability of the cover beds, indicated by particle sizes and by eolian admixture to the upper layer (derived from heavy-mineral analyses). If bonding strengths are considered, the differences at least beneath the topsoil are mainly concentrated to the two strongest bondings, which are not easily changed in the short term and, therefore, usually regarded as indicators of indigenous properties of geologic

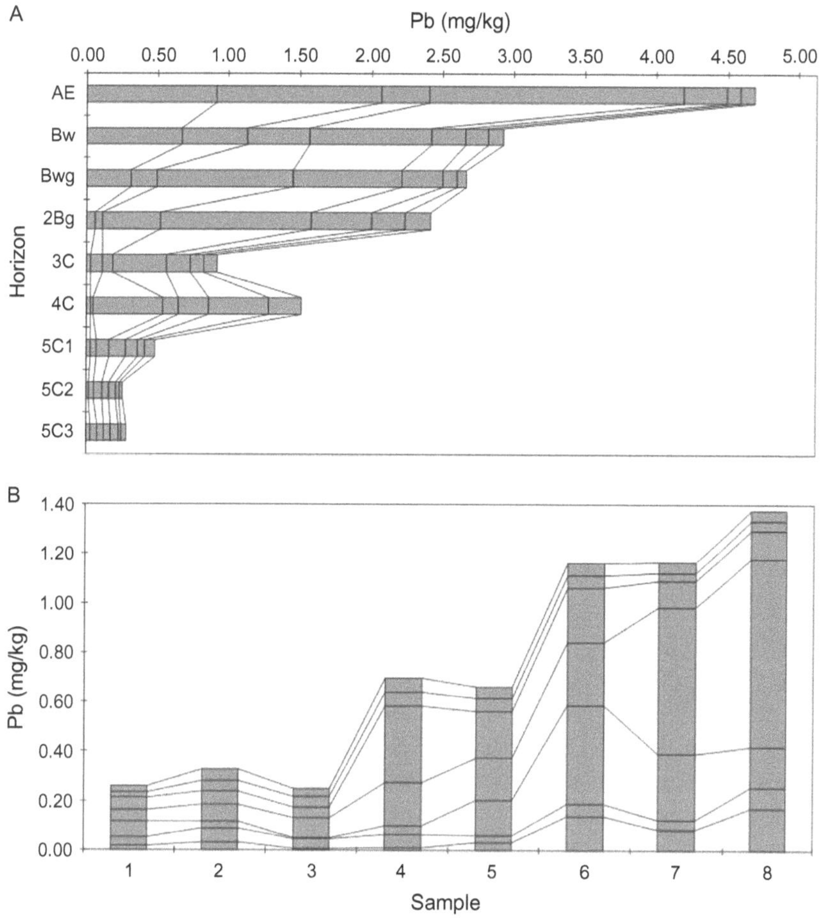

FIGURE 3.14 Pb concentrations in the example trench. (A) Vertical Profile A; (B) intermediate transect in the deepest trenched layer from most protected (1) to most exposed (8) sample. Segmented bars with connections indicate weak to strong bonding measured through sequential extraction, according to Zeien and Brümmer (1989) as modified by Wilcke and Amelung (1996); weak to strong = left to right in (A) and bottom to top in (B).

substrates (Wilcke and Amelung, 1996). The heavy-metal concentrations in Profile B also decrease with depth though with less irregularities (not shown).

The heavy-metal concentrations are generally higher in the exposed parts of the transects than in the protected ones, though there are differences between the studied metals. Cd probably has been hydrolyzed from strong bondings by acid percolating water and has migrated along the transect, where it has become concentrated in a weakly bound form (Kleber and Mailänder, 2000). In the exposed segments, total values, thus, are not significantly increased: there is no clear indication of an anthropogenic impact of Cd. However, an opposite example is Pb with constantly low concentrations in the protected segments (Fig. 3.14B).

With decreasing protection by overlying layers, Pb concentrations increase to double or more than triple the values of the protected segments. Increases in Pb with decreasing protection within the same, homogeneous deposit is a measure of the gross accumulation of these metals to the exposed segment since agriculture was abandoned, probably in the late Middle Ages but at least 60 years before the investigation started as indicated by the age of the oldest tree stands on the exposed segment.

Such balances may include effects of pedogenic and geomorphic processes besides the anthropogenic contamination. Heavy metals may be translocated down the steep erosion scarp to the area of Profile B either with dislocated organic material or with interflowing slope water. The first effect is assumed to be unimportant because the thicknesses of the organic layers and their heavy-metal concentrations are similar in Profiles A and B. The latter effect may be indicated by Fe and Mn concentrations, which are higher in the topsoil of down-slope Profile B than up-slope Profile A. However, Al, measured for comparison, is similarly different between the topsoils of A and B, though Al is unlikely dislocated in solution (Kleber et al., 1998a).

The increased Pb content with decreasing protection provides an overall measure of alteration. It demonstrates the feasibility of the approach to distinguish postdepositional modification from indigenous concentrations in mineral soils. This approach may also help understanding bonding preferences of particular metals under longer lasting, more natural conditions (compared with similar laboratory experiments), as demonstrated in Fig. 3.15. The well-known affinity of Pb to become bound to organic matter (fraction 4) or to remain dissolved (fraction 1) is well corroborated by this increase and by significant

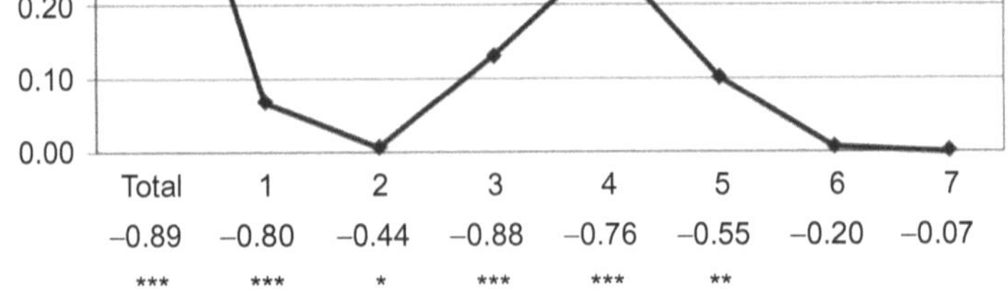

FIGURE 3.15 Bonding preferences of Pb. The maximum differences in Pb concentration between the best protected and the most exposed sample are shown for total Pb and for fractions (see Fig. 3.14) from the weakest bonding (1) to the strongest (7). Below, correlation coefficients r are given, calculated for all samples from three trenches, correlating Pb concentrations with sampling depths, and significance levels as asterisks (***: >99.9; **: >99; *: >95%); $n=37$.

correlation between the Pb contents and the degree of protection against influence from the surface, expressed as depth to sample. However, there is also significant binding of Pb to manganese (fraction 3) and to weakly crystalline iron (fraction 5) over time intervals of several decades.

This kind of approach has the potential to help quantifying various types of pedogenic alteration in stratified substrates, because the protected layer is nearly identical to the original parent material of the altered soil—to a degree higher than any other known technique is able to ensure—and casual lateral gradients of indigenous properties may well be controlled. Furthermore, the maximum time span involved may be confined through historical sources. Similar settings, though bound to certain constraints regarding historical and ensuing geomorphic evolution, are frequent throughout Europe in former arable land (Abel, 1976) so that subsequent studies could cover various environments.

3.5 CONCLUSIONS

Sharp boundaries within soil profiles, marking abrupt changes in soil physical and chemical properties, frequently originate from disconformities rather than exclusively from pedogenic processes. In soils and weathering profiles, simple stratigraphic principles do rarely hold, as discussed in detail by Paton et al. (1995) and Ollier and Pain (1996). Likewise, as discussed here, top-down pedogenesis operating in purely residual materials cannot be assumed either. Therefore, the interpretation of the vertical distribution of soil properties requires a concept of regolith *and* soil genesis, that is, an allochthonistic approach. As a crucial part of this approach, the original properties of the parent material and its pedogenetically formed alteration products have to be reconstructed (Blume and Röper, 1977).

Strong contrasts between layers will cause distinct stratification of geogenic properties and will decisively affect the vertical organization of the resultant soil properties. However, a vertical contrast in soils cannot be ascribed *a priori* to any particular dominant process or set of processes. Rather, any of a number of geological and pedological processes (or a combination thereof) may be responsible for intensifying or blurring vertical contrasts. Layering in soils is characterized by polygenesis and multiple causality. Thus, it is critical to evaluate property/depth functions in the light of an overall pedogeomorphic interpretation of substrates and processes.

Chapter 4

Influence of Cover Beds on Slope Hydrology

K.-M. Moldenhauer, K. Heller, P. Chifflard, R. Hübner and A. Kleber

4.1 INTRODUCTION

Slopes are critical elements of catchments (Graham et al., 2010), as they link the atmosphere with discharge systems. Therefore, watershed behavior is decisively controlled by slope characteristics (Wenninger et al., 2008). In the current research on Earth's critical zone, hydropedology, the science of the behavior and distribution of water in contact with subsurface solid materials, plays an important role (Lin, 2003). However, there still are many unresolved issues in physically based models of slope-hydrological systems (McDonnell et al., 2007), particularly for those having a large proportion of near-surface flow compared to other discharge components. This is caused by the high spatial variability of physical and hydraulic soil parameters (Reszler et al., 2006), incomplete understanding of hydrological processes (McDonnell et al., 2007; Sivapalan et al., 2003), and nonlinear interlinking between subsurface and base flow in space and time (Tilch et al., 2006; Uhlenbrook, 2005).

Flow beneath but close to the surface, approximately paralleling its inclination, independent of whether it is saturated or not, is in the following referred to as "interflow." Interflow research is crucial for at least two reasons: It is a major component of flooding and, thus, its knowledge is a necessary prerequisite for any flood forecast to rest on flood origin (e.g., Ward, 1978); second, interflow may quickly transport contaminants through ecosystems, avoiding filtering, which depends on the dwelling time (e.g., Beck, 2002; Lindemann, 1997).

As a result of differences in their genesis, upper layer, intermediate layer, and basal layer vary in their pedophysical and sedimentological properties such as bulk density, texture, and alignment of clasts. This brings about changes in hydraulic properties at certain depths, determining the permeability of subsurface materials. Therefore, and considering the omnipresence of periglacial cover beds in sloped terrain of the mid-latitudes, we regard cover beds as major triggers of slope-water paths. Next to slope inclination, they are the main factor influencing the further direction of water after penetration of topsoils, whether it

Developments in Sedimentology, Vol. 66. http://dx.doi.org/10.1016/B978-0-444-53118-6.00004-0

tends to seep to groundwater or to flow approximately parallel to the slope as an interflow. Despite this presumed hydrological importance, this aspect has not been studied systematically in the older literature (and in much of the current one either). Rather, slope-hydrological studies (e.g., Barsch and Flügel, 1988; Flügel, 1993; reviews in Anderson and Burt, 1990; Kirkby, 1985; Miyazaki, 1993) measured the water flow along slopes at constant, arbitrarily chosen depths, without taking into account the stratified character of the substrates, wherein the water flows, neglecting possible lateral connections. Also, slope depressions were rarely in the scope of such studies (with the exception of Anderson and Burt, 1978), although they collect surface and subsurface waters and, thus, likely are important guidelines of slope waters (see also Einsele et al., 1986; Frühauf, 1990; Semmel, 1994a).

As distribution and properties of cover beds are known or may be determined in the field, understanding fundamental relations between them and water flow would much improve forecasting the hydrological role of slopes. Improved knowledge about the interrelation between subsurface layering and runoff processes will certainly enhance current environmental issues such as flood forecasting or modeling pollutant translocation.

Soil classification in this section follows Soil Taxonomy (SSS, 2010) with nomenclature according to World Reference Base for Soil Resources (WRB, 2006) in parentheses.

4.2 BASIC HYPOTHESES

Figure 4.1 outlines in a very schematic manner cover beds that typically occur on a convex–concave slope segment and assumed water paths therein, derived from taking into account solely the differences in bulk densities of the particular

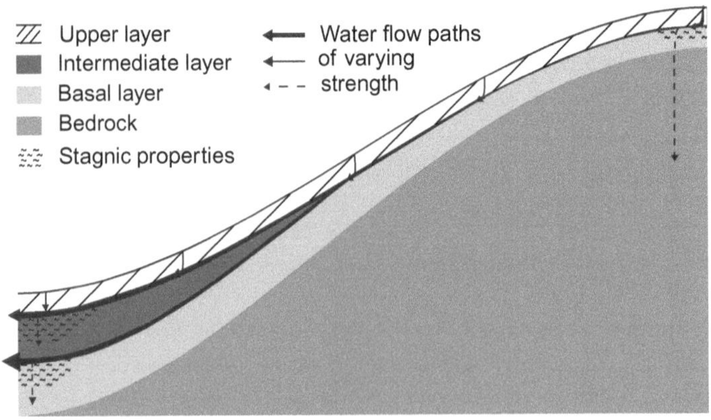

FIGURE 4.1 Basic hypothesis of water flow paths along a slope with cover beds (*reproduced from Kleber and Schellenberger, 1998*). (For color version of this figure, the reader is referred to the online version of this chapter.)

layers: Upper layers are of loose consistency, are intensely penetrated by roots, and are rich in macropores. These properties induce high, omnidirectional hydraulic conductivity. Consequently, precipitation water infiltrates fast, counteracting the origination of surface runoff (Casper, 2002; Schneider et al., 2001). On the flat summit, a mottled zone due to temporary soil-water stagnation is often observed within the basal layer, and corresponding properties of a perched water table occur in the overlying layers. One may expect that some of the water will penetrate into the underlying bedrock, despite the typical high bulk density of the basal layer, because the low slope angle impedes lateral flow. As slope inclination increases on the slope shoulder, much of the perched water may flow above the basal layer within the upper layer (Einsele et al., 1986; Semmel, 1994a) rather than tickle down to groundwater. The quantity of interflowing water will increase along this course, fed by further seeping from the surface.

Intermediate layers typically have a higher bulk density than upper layers, and they differ in clast content and thickness, according to location and region. Because of their density, they should in theory be less aquiferous than upper layers (Kleber and Schellenberger, 1998). On the back slope, the onsetting intermediate layer may act as an aquitard for water from the overlying layer, but as an aquifer for water perched upon the basal layer. Therefore, part of the water flowing above the basal layer may retain its direction and flow within the intermediate layer. The other part, as well as the water added from the surface along the remaining slope, will probably flow above the intermediate layer. As a consequence, two levels of interflowing water at different depths may exist with different quantities and velocities, depending mainly on the distribution of the intermediate layer, but above the basal layer anyway.

In general, basal layers are rich in clasts aligned parallel to the slope inclination, and they are characterized by high bulk density. Accordingly, a basic hypothesis on slope-water dynamics would interpret basal layers as aquicludes (Semmel, 1994a; Fig. 4.1). Several studies indeed noted water-blocking effects of basal layers (Damm & Terhorst, 2010; Körner, 1996; Schneider et al., 2001). They ascribed this finding to the high bulk density and low hydraulic conductivity of the underlying basal layer.

4.3 CASE STUDIES

4.3.1 Overview

The above basic hypotheses have been challenged in a number of current studies. Due to their heterogeneous characteristics, there are contradicting assertions regarding the hydraulic properties of intermediate layers. Abdank et al. (1999) and Scholten et al. (1999), using tracer tests, identified water-retardant and water-permeable zones within intermediate layers. Völkel et al. (2002a) described a high water permeability of intermediate layers from the Bavarian Forest.

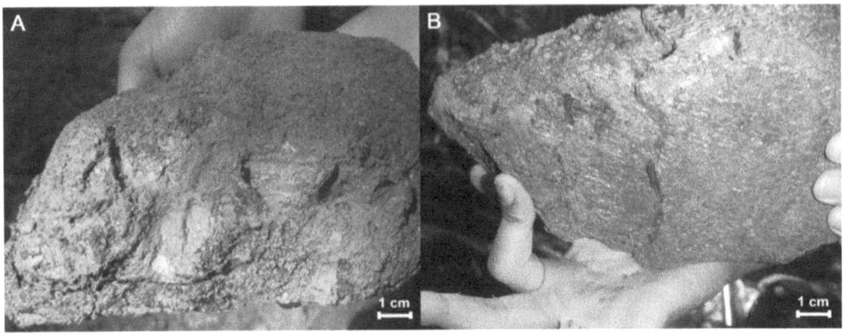

FIGURE 4.2 Upper side (A) and bottom side (B) of a clast with a silt cap (*photo M. Wegener*). (For color version of this figure, the reader is referred to the online version of this chapter.)

The sole hydrological role of basal layers as aquicludes in inclined terrain was put to doubt by Kleber and Schellenberger (1998, 1999), who conducted tensiometer measurements, and by Völkel et al. (2002b), who observed water leaking out of basal layers outcropping in road cuts and building pits. These authors concluded that basal layers may well function as aquifers. Another indicator for water flow in basal layers are silt caps (Fig. 4.2; see also Völkel et al., 2002a): loamy covers on the upper sides of clasts are opposed to bare bottom sides; percolating water dislocates fine particles into the basal layer, where they accumulate on the upper sides of clasts, whereas the bare bottom sides attest water flowage inside these layers.

As described in preceding sections, small-scale variations in composition and structure of cover beds are common. In theory, this will affect the conditions of the pore space and, herewith, the hydraulic conductivity, which in turn, governs seepage processes. Soil cracks act as preferential flow paths and disconformities between different layers favor interflow parallel to the surface (Kleber, 2004b). It is mainly for this reason that cover beds may be treated sufficiently neither as aquifers dominated by pores or fractures nor as aquicludes.

To overcome these difficulties and provide a deeper insight into hydrological processes in cover beds, several case studies were conducted in subdued mountains of Central Europe during the past decades, covering various types of bedrock. These research activities are characterized by a number of different approaches focusing on qualitative as well as on quantitative aspects of the topic. Data analyses obtained from experimental watersheds yielded hydrometeorological threshold values for the occurrence of interflow and allowed to derive interflow signatures by hydrograph separation. Field experiments using sprinkler irrigation and tracer techniques provide clues on infiltration processes and on preferential flow paths in cover beds (Table 4.1).

Although these case studies differ in many aspects, such as bedrock type, size of the study area, or applied techniques, some basic conclusions may be drawn from their results. Only three of the aforementioned studies report interflow in upper layers passing through its entire body (Casper, 2002; Körner,

TABLE 4.1 Case Studies on Slope Hydrology Focusing on the Influence of Cover Beds

Region	Bedrock geology	Area (km^2)	Applied technique	Interflow occurrence	Preconditions	Author
Schönbuch	Marl	1 × 1-m field plots	Sprinkler, tracer	Boundary between LH and LB, preferential flow	Field capacity plus 30 mm d^{-1}	Schwarz (1985)
Odenwald	Sandstone	0.005	Instrumental records	Preferential flow	Field capacity plus 30 mm d^{-1}	Barsch and Flügel (1988)
Taunus	Phyllite, slate		Field observation	Boundary between LH and LB		Semmel (1991a)
Westerwald	Slate, graywacke, diabase	0.1–0.3	Instrumental records, sprinkler	In LB	Field capacity exceeded	Ernstberger et al. (1992)
Odenwald	Granite	0.08	Instrumental records	Boundary between LH and LB	Field capacity plus 40 mm d^{-1}	Moldenhauer (1993)
Taunus	Phyllite, slate	0.14, 0.09	Instrumental records	In LM	Field capacity plus 40 mm d^{-1}	Moldenhauer (1993)
Schönbuch	Clay, sandstone	0.1–1.0	Instrumental records, tracer	In LH	Only winter half year	Körner (1996)
Taunus	Phyllite, slate	0.14, 0.09	Instrumental records	In LM	During snowmelt	Schneider et al. (1997)
Harz	Slates, graywacke	1.4	Tracer	In LM, LB, as preferential flow		Abdank et al. (1999) and Scholten et al. (1999)

Continued

TABLE 4.1 Case Studies on Slope Hydrology Focusing on the Influence of Cover Beds—Cont'd

Region	Bedrock geology	Area (km^2)	Applied technique	Interflow occurrence	Preconditions	Author
Odenwald	Sandstone	1 × 1-m field plots	Sprinkler, tracer	In LH, as preferential flow		Peters (2000)
Harz	Slate, graywacke	0.35, 1.4	Instrumental records	Yes	Field capacity exceeded	Wenk (2000)
Black Forest	Sandstone	7.0	Instrumental records	In LH, preferential flow		Casper (2002)
Hunsrück	Slate, limestone		Sprinkler	Boundary between LH and LB, as preferential flow		Schneider et al. (2001)
Westerwald	Slate, basalt			Boundary between LH and LB, in LB		Sauer et al. (2001)
Bavarian Forest	Granite		Instrumental records	In LB		Völkel et al. (2002a) and Huber and Völkel (2009)
Black Forest	Gneiss	1 × 1-m field plots	Instrumental records and tracer	Preferential flow		Wenninger et al. (2004)

LH, upper layer; LM, intermediate layer; LB, basal layer. Additional case studies will be reported in subsequent sections of this chapter.

1996; Peters, 2000). This always occurs as preferential flow, which presumably is due to atypical coarse textures of the upper layers in these particular studies.

Field experiments carried out in Sandstone Odenwald by Peters (2000), using sprinkler irrigation and Brilliant Blue as tracer, showed that dilapidated roots, small voids underneath larger rock fragments, and fissures in compacted soil act as preferential flow paths. Thus, infiltrated water passes by very rapidly without moistening the soil matrix (Fig. 4.3). These findings may explain why no correlation of soil moisture and interflow was found in some field studies (e.g., Schwarz, 1985). By contrast, in most cases interflow occurs beneath the upper layer and at its lower boundary, respectively, which matches the theoretical assumption that interflow is mainly conducted via sedimentary disconformities.

A closer look at the hydrometeorological preconditions that favor the occurrence of interflow reveals that macropores must be saturated or at least close to saturation due to preceding rainfall or snowmelt. Moreover, some of the studies exhibit that even as soon as field capacity is achieved, additional precipitation of 30–40 mm or thawing of a snow cover is required to produce appreciable quantities of interflow (Barsch and Flügel, 1988; Moldenhauer, 1993; Schneider et al., 1997; Schwarz, 1985). Such conditions are rarely met in most of the areas under study. Therefore, interflow in cover beds close to the surface is limited to rare events, connected to exceptional rainfall or to rapid thawing during winter.

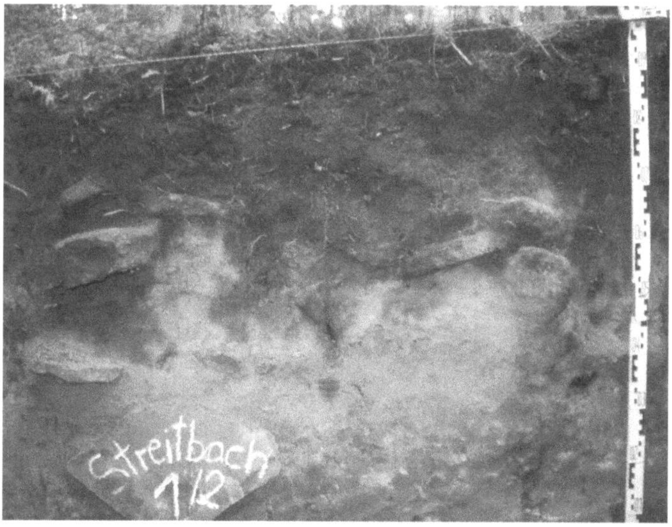

FIGURE 4.3 Tracer experiment using Brilliant Blue® conducted in the Sandstone Odenwald Mountains (*photo K.-M. Moldenhauer*). The upper layer down to 60 cm on the scale has almost ubiquitously been affected, whereas the blue color penetrates the intermediate layer only by approximately 10 cm. However, in the right part of the picture, there is clear indication of seepage down into the basal layer between 30 and 20 cm. (For interpretation of the references to color in this figure legend, the reader is referred to the online version of this chapter.)

However, from the viewpoint of flood prediction, it is just the exceptional case that matters.

The results of the selected studies are quite different and sometimes contradictory. This is supposed to be due to local peculiarities, different field methods, observation times, and catchments areas. In the subsequent sections, we present three studies that are mutually comparable with regard to the applied methods. They intend to lead to a more conclusive picture regarding the role of cover beds, especially basal layers, in slope-hydrological systems.

4.3.2 Frankenwald Mountains

4.3.2.1 Introduction

The study site reported here is a small, spruce-covered catchment of approximately 0.07 km^2 at an elevation of some 600 m asl (see Fig. 2.1 for location; Fig. 4.4). It is underlain by shale, which typically produces platy clasts upon physical weathering. This bedrock is essentially impermeable as derived from calculated water balances (Lindemann, 1997), cores drilled even directly beneath the spring bog (see below), and geoelectrical sounding (Kleber and Schellenberger, 1998, 1999); this is also known from hydrogeological

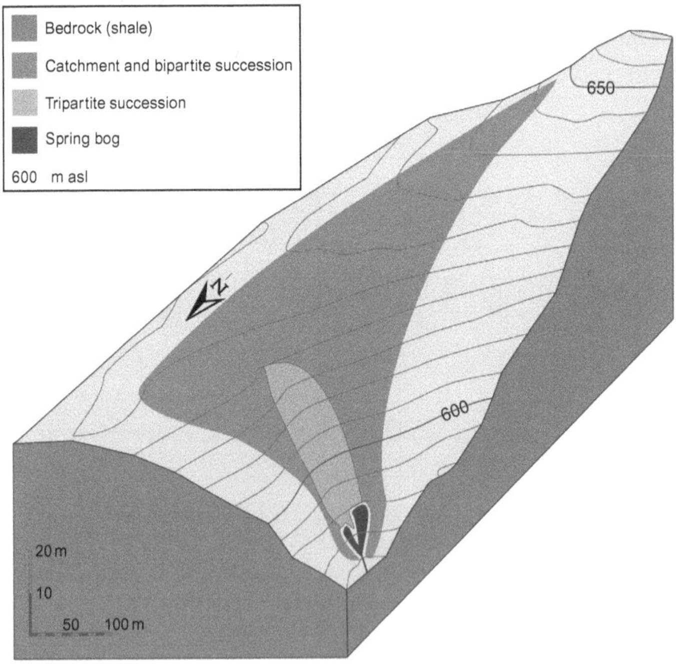

FIGURE 4.4 Block diagram of the studied catchment in the Frankenwald Mountains with contour lines every 10 m above sea level indicating slope curvature (*modified from Kleber et al., 1998b*). (For color version of this figure, the reader is referred to the online version of this chapter.)

surveying for water supply (Von Horstig and Stettner, 1976). The relief of the lower part of the catchment is formed as a slope depression draining into a bog. This bog was instrumented with a Thomson weir to register discharge. The distribution of the cover beds and some of their hydraulic properties was determined in three hand-dug pits (Kleber et al., 1998b), by coring hand-auger (1 or 2 m), by coring motor-auger (max. depth reached 4 m)—altogether 68 drill holes reached deeper than 1 m—and by seven Schlumberger-type geoelectrical profiles (Kleber and Schellenberger, 1998, 1999).

Water flow within the layers was assessed with 14 semipermanently (1-h interval) recording tensiometers arranged in four vertical profiles, and with manual pressure-transducer tensiometer measurements to control the recording stations and to extrapolate from them. The tensiometers, calibrated to register positive pressure, were placed specifically into particular layers. Here, we will discuss the results from four of the recording tensiometers arranged one above the other in a single borehole isolated from one another by bentonite filling:

- 1/1–0.50 (read: Profile 1/1, tensiometer located at 0.5 m depth) in the upper layer;
- 1/1–0.95 in the lower, denser part of the bipartite intermediate layer;
- 1/1–1.35 in the upper part of the basal layer, which part has no redox concentrations of iron; and
- 1/1–2.00 in the deeper basal layer, at which depth intensely oxidized interbeddings are frequent (later referred to as “deep interflow”).

The field-saturated hydraulic conductivity of the surficial materials was measured with a standard Guelph Permeameter.

Input variables were recorded every 10 min with a standard weather station somewhat outside the forest (precipitation, temperature, wind velocity, relative humidity, and radiation), and by 20 precipitation collectors (throughfall) within the forest. Ceramic suction cups were installed at different depths (0.30–2.20 m) along a transect from the terrestrial forest soils to the spring bog to collect soil pore water for chemical analyses once a week (Lindemann, 1997). We measured three times the $^{34}S/^{32}S$ ratio of sulfate in different compartments of the hydrological system (Kleber et al., 1998b).

4.3.2.2 Results

The underlying shale is ubiquitously covered by periglacial cover beds. The bedrock was determined by auger to be covered by less than 2.5 m but more than 0.7 m of cover beds on the straight slopes outside the central slope depression, as well as under the spring bog. Below the baseline of the slope depression, cover beds reach to a maximum of 6 m below ground level according to geoelectric sounding (see Fig. 2.5 in Chapter 2). Slope water becomes concentrated in the depression’s baseline and distributed down to the spring.

With respect to the hydraulic properties, the cover beds vary mainly in texture, especially regarding coarse clasts, and bulk density (Kleber et al., 1998b). An important property, which is unique to rock that weathers to platy clasts, is its imbricate structure. This is the result of the orientation of such clasts parallel to the slope by gelifluction. Such material may be more permeable in a direction parallel to the clasts than in any other direction, particularly in the basal layer where no eolian fines seal gaps between clasts; such deposits are assumed to own a hydraulic anisotropy. Despite this, water probably seeps into the basal layer in places, preferentially where periglacial disturbance (ice-wedge casts) or deep tree rooting occurs; in these places, the measured hydraulic conductivities in the basal layer are highest (up to 130 cm d^{-1}). Water seeped into such material is forced to follow the path predetermined by the imbricate structure rather than being able to easily move vertically. Considering the distribution of the cover beds in the catchment (see Fig. 2.5), it becomes evident that two levels of interflowing water may exist: one in the upper part of the soils upon the perching basal layer (not separated in Fig. 2.5) and the other within the basal layer (referred to as "interflow in basal layer" in Fig. 2.5), activated once water happens to seep into it. The velocity of the water flow as well as the size of the contributing catchment area may differ among the two.

Soil profiles in this catchment provide the first clues to the hydrological processes (Fig. 4.5). Below the upper layer, the basal layer often contains a gray horizon without hydromorphic properties or with introverted[1] mottling, indicating the absence of water surplus or seasonal saturation derived by perching of water percolating from the surface (stagnic properties), respectively. In deeper parts of the basal layer extroverted[1] iron mottles occur. Such properties are usually ascribed to changing groundwater levels (gleyic properties). However, based on the findings presented below, we assume that this is not a valid explanation in this case, because there is strong evidence that this seemingly gleyic horizon is occupied by interflowing water rather than groundwater (designated as "interflow in basal layer" in Fig. 2.5), which, however, may create similar effects as changing groundwater levels due to changes in the oxygen concentrations of the water at different times.

Figure 4.6 gives an example of tensiometer results, limiting the view to April 1995 and to the tensiometers of Profile 1/1. The 1/1–0.50 recorder in the upper layer quite sensitively follows precipitation most of the time (note that April started with a snowmelt event, melting 10 cm of snow). Except during some major events (see below), there is no saturated flow in it. The 1/1–0.95 recorder in the lower intermediate layer is less sensitive, indicating that there is a hydraulic contact between the two, though it is imperfect. The 1/1-1.35 recorder in the upper (mostly dry) part of the basal layer shows a fidgety curve that does not perceptibly react to meteorologically induced pressure changes in the overlying

1. Mottles of iron (hydr)oxides in the interior of soil aggregates; extroverted mottles on aggregate outside.

FIGURE 4.5 Soil profile with stagnic (light gray) above gleyic (yellowish brown) properties in the basal layer beneath the upper layer (dark gray) (*photo C. Beierkuhnlein*). (For interpretation of the references to color in this figure legend, the reader is referred to the online version of this chapter.)

layers; there seems to be no hydraulic contact with the surface, except for the event starting around April 20 (discussed below). The 1/1-2.00 recorder in the deeper basal layer within the intercalated intensely oxidized zones shows the largest pressure amplitude; it reaches excess pressure after each major precipitation and/or snowmelt event (as during the rain storm starting on April 17). On the other hand, this tensiometer drains down to a very low pressure. In all, this tensiometer always reacts strongly with the same signals recorded in the upper two layers, though with a remarkable time lag. Furthermore, the figure displays one strong rain storm. During this event, the deepest tensiometer recorded the highest (positive) pressure, indicating that there is no downward seepage possible against the pressure gradient. The inset of the figure shows the rise to this peak in more detail. The lowest tensiometer records the event first, followed by the next higher ones successively; the highest is the last one to react: the profile fills up from the bottom. During the 1995 hydrological year, six strong events that size occurred, whereas none of this strength may take place in other years.

Whether the observed phenomena in the basal layer may be defined as interflow events is important, because another plausible explanation would be that there was groundwater deeper in the profiles below the tensiometers; the instruments would have registered just the rise of the level of this persistent

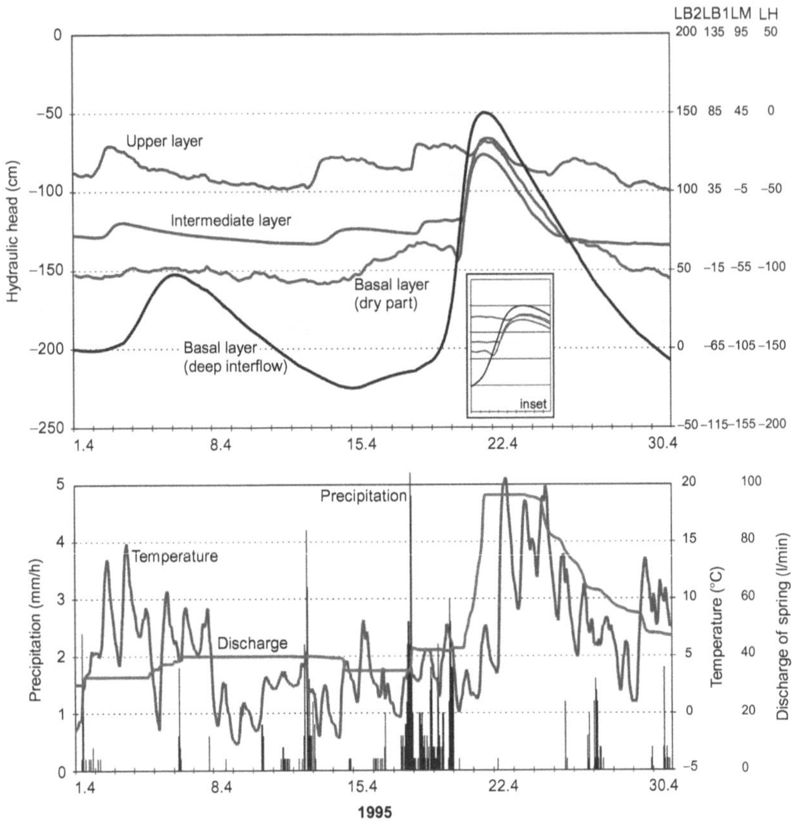

FIGURE 4.6 Tensiometer measurements at station 1/1, spring discharge, and weather data. Timescale of inset diagram with details of the major rise is 10 h per abscissa unit, starting on April 19, at 3 am (*modified from Kleber and Schellenberger, 1998*). (For color version of this figure, the reader is referred to the online version of this chapter.)

groundwater. The suction cups occasionally reached the permanent water level below the depth recorded by tensiometers: except for a transitional zone, explained by an occasional mixture of ground- and interflow waters, both water bodies are remarkably different in their chemical and isotope composition. The $\delta^{34}S$ values support that finding: Fig. 4.7 shows a schematic section of the watershed near the spring. The sulfate of the atmospheric input has $\delta^{34}S$ values of 4%. In the forest soil profile (rightmost in the figure), enclosing the upper layer, intermediate layer, and upper basal layer, the $\delta^{34}S$ values range from 2% to 3%. In the transition zone (middle in the figure), the upper horizons show essentially the same values. But at a deeper level, in the deep aquifer, the soil solution has negative $\delta^{34}S$ values. The $\delta^{34}S$ values of the bog's profile display a trend from a positive value of 1% in the uppermost horizon to less than −11% at 1.4 m, fed by groundwater. A difference in sulfate reduction is the only process

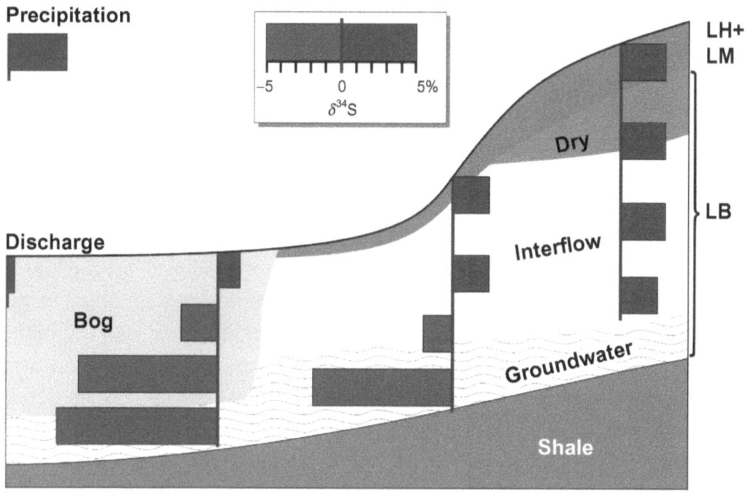

FIGURE 4.7 $\delta^{34}S$ of dissolved sulfate in pore waters from different depths of the spring bog and its surroundings, in precipitation water, and in spring discharge water (*modified from Kleber et al., 1998b*). Schematic sketch of topography, not to scale. Bars represent values in per thousand of dissolved sulfate relative to Standard Mean Ocean Water (zero value in legend). LH+LM, upper and intermediate layer and reworked materials derived from them; LB, basal layer. (For color version of this figure, the reader is referred to the online version of this chapter.)

in such ecosystems that may induce a significant shift in $\delta^{34}S$, but this is always positive (Thode, 1991). However, when comparing the sulfate input (precipitation) and output (discharge), there is a negative shift. This indicates an additional source of sulfur with a negative $\delta^{34}S$ value, which must be located in the deeper soil profile; the most likely additional source of sulfate is geogenic pyrite in the shale debris making up the deeper basal layer. Thus, $\delta^{34}S$ values imply the presence of water with very different sources of their dissolved sulfate: one in groundwater and others at higher levels.

The chemical composition of ground- and interflow waters confirms that both types of subsurface water are remarkably different, which conclusion is based on several thousands of individual measurements of ion equivalent concentrations and pH values. These data show that the average pH in surface including interflowing waters is much lower than in groundwater (4.7 compared to 6.1); the HCO_3 anion typically has a more than 10-fold concentration in groundwater than in interflow water; the mean concentration of Al cations in groundwater is lowered by a factor of more than 20, substituted by mainly Ca, Mg, and especially Fe ions; these Fe ions are almost absent in surface waters (Lindemann, 1997).

The likely explanation for all these findings is that interflowing waters have a significantly shorter retention time than groundwater so that chemical reactions with surrounding materials are less well possible. This indicates that the water identified as an interflow is an individual water body overlying the persistent groundwater at times of high seepage through the slopes of the catchment under study.

4.3.2.3 Conclusions

Water flow in the studied catchment in the Frankenwald Mountains takes place at different levels of depth, during most of the time confined to certain cover beds. The basal layer supports lateral flow, whereas vertical flow is impeded. This is attributed to the slope-parallel alignment of small, platy shale clasts and to their imbricate structure, which causes anisotropy. Thus, cover beds mainly derived from shale act as aquitards for seeping water, but as aquifers for water once entrapped in the cover bed. However, during events of high discharge, saturation becomes exceeded in this aquifer, leading to some upward relieve of water, and then to saturated flow in the overlying layers, while the aquifer in the basal layer retains excess hydrostatic pressure, indicating that it is still confined.

The three-level hydrogeology of the catchment allows the water in the deep basal layer either to reach strata rich in pyrite or to dwell and flow within such material for a longer distance and time, enhancing the total uptake of geogenic sulfate. Thus, the deep basal layer contributes to the spring bog interflowing water, containing sulfate that is isotopically different from the atmospheric one. Because sulfate reduction is only evident in a few, distinct zones in the spring bog and because sulfate-reduction rates are low (Kaiser, 1996; Müller, 1996), ^{34}S is regarded as a well-suited natural tracer for water of strongly differing retention times in this system.

Three types of discharge events may be distinguished:

- Minor events induce increased soil-water potentials only in the upper part of the soil with each consecutive tensiometer lagging behind the one above, indicating approximately vertical seepage. This response is limited to the upper part of the BL at most and does essentially not reach the deep aquifer.
- Events of intermediate strength, particularly those accompanied by snowmelt, lead to similar effects in the upper part of the soil. A large wave passes the basal layer (leaving its upper, dry zone unaffected) much later, which, however, has no feedback to the surficial tensiometers.
- Strong events lead to water from the deep aquifer breaking upward into the higher parts of the soil in the lower part of the slope depression near the spring bog.

During minor and intermediate events, the upper part of the soil is characterized by unsaturated flow. Flow rates are much lower under such conditions, and therefore, a blocking effect of the BL to seepage may be assumed. During type 3 events, the flow is saturated in the IL and BL, and blocking effects are moderate at most, leading to a faster release of water from the deeper interflow level up into overlying layers.

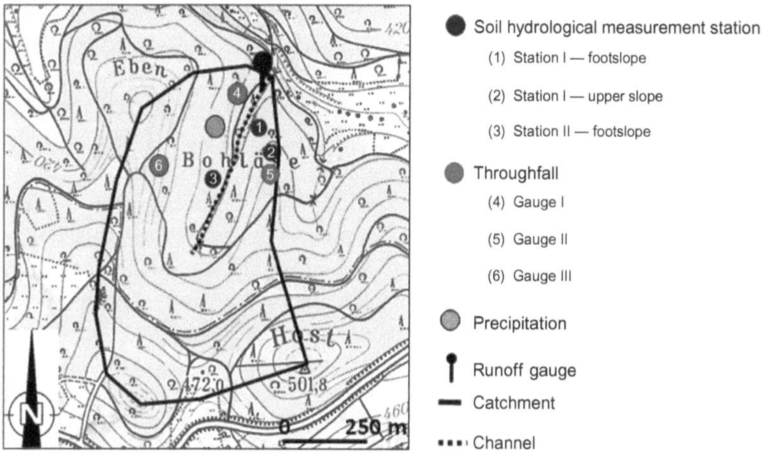

FIGURE 4.8 Bohläse study area (*modified from Chifflard et al., 2008*).

4.3.3 Sauerland

Similar results were obtained by Chifflard et al. (2008) regarding the small forested catchment "Bohläse" (0.7 km^2, North Rhine-Westphalia, Germany; see Fig. 2.1 for location and Fig. 4.8 for details), in which up to three layers of periglacial cover beds are widespread. The underlying basal layer is composed of weathered silty shale covered by an intermediate and an upper layer. Periglacial cover beds form the parent material of the soils in the catchment, which are mostly Inceptisols[2] (Cambisols). At the top of the slopes, Entisols[3] (Leptosols) are widespread, and at the footslopes, Aquepts[4] (Stagnic Cambisols) dominate.

To describe and understand the hydrological dynamics of the catchment, three measurement stations were equipped with tensiometers, which were installed in the various layers. Hydrochemical and tracer-hydrological methods were applied as well.

The results obtained at different spatial scales show that the periglacial cover beds have strong influence on both the soil-water dynamics at the plot scale and the runoff dynamics at the catchment scale. These results confirm the importance of cover beds for runoff generation in subdued mountains. It has to be emphasized that the basal layer shows a temporally variable influence on soil-water fluxes, dependent on preexisting soil moisture, the latter being defined as initial soil moisture at the onset of a runoff event. On one hand,

2. Soil type characterized by a cambic (Bw) subsurface horizon with evident oxidization of iron-bearing minerals.
3. Immature soil type characterized by slight to absent weathering and thin to discontinuous humus accumulation.
4. Soil having combined cambic properties and properties induced by seasonal water logging.

the basal layer functions as a barrier for vertical water flow when soil moisture is low; on the other hand, it is a preferential pathway for lateral subsurface run-off in the case of high soil moisture.

At the plot scale, the temporal distribution of the soil moisture at two measurement stations plotted for 1 year shows that the basal layer at station II (upper foot slope), which has a high bulk density, acts as a barrier for vertical water flow, the effects of which may be seen in Fig. 4.9. This leads to saturation of the lower intermediate layer over a long time, whereas in the subjacent layer the soil-water potential increases (the soil-water tension becomes negative). In this case, the vertical distribution of soil moisture may be described by hydraulic anisotropy according to Kleber et al. (1998b). However, at station I (lower foot slope), the vertical soil moisture is remarkably different because in comparison with station II the basal layer has a lower bulk density and there is no intermediate layer.

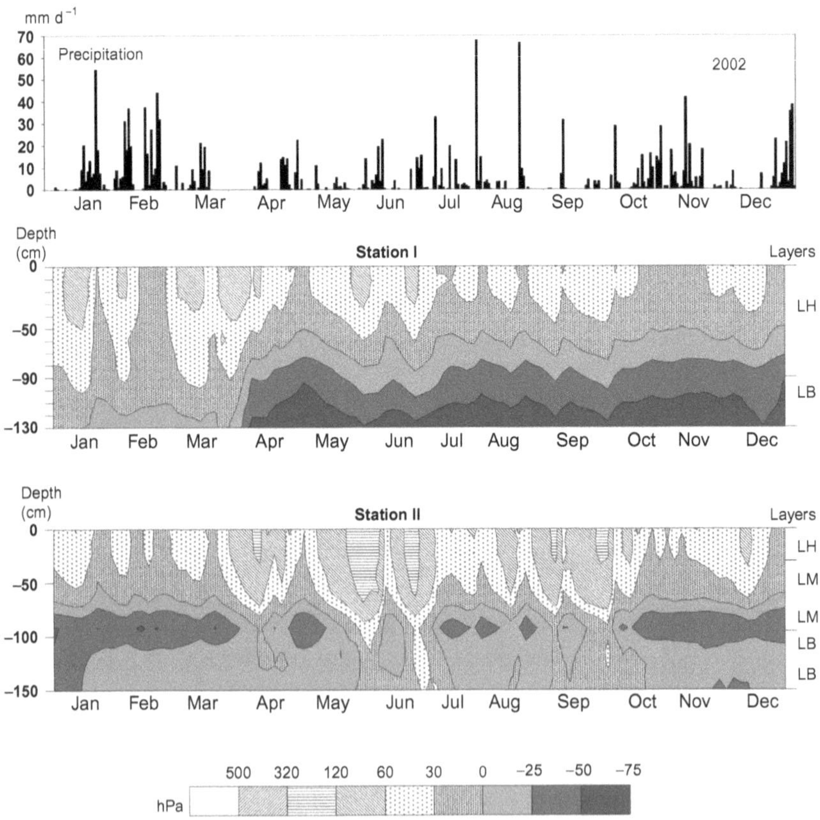

FIGURE 4.9 Isopleth diagram of soil-water potentials at Stations I (lower foot slope) and II (upper foot slope), precipitation, and periglacial layers (*modified from Chifflard et al., 2008*). LH, upper layer; LM, intermediate layer; LB, basal layer.

At the catchment scale, the combination of tracer-hydrological and hydrometric measurement during different rainfall/runoff events shows that the intermediate and basal layers influence soil-water dynamics and runoff generation in different ways, depending on preexisting soil moisture. The observed events have total rainfall amounts of 51.1 mm (E1), 18.1 mm (E2), and 35.6 mm (E3), which lead to peak runoffs of 45, 12, and 19, respectively (Figs. 4.10 and 4.11).

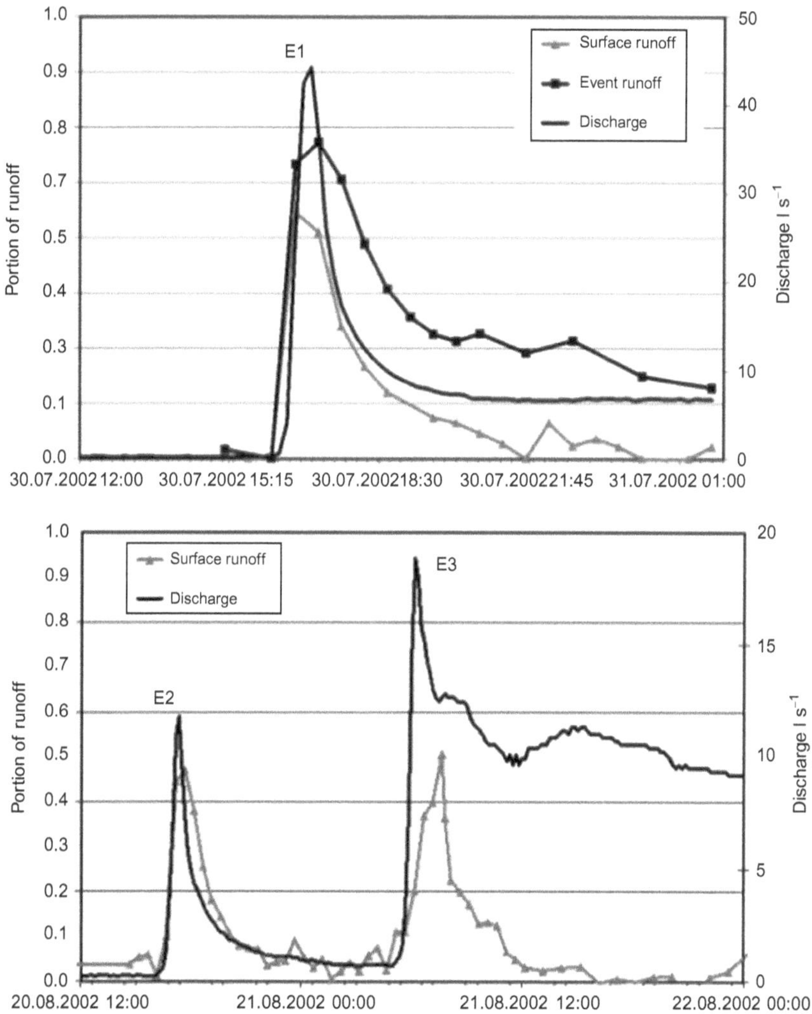

FIGURE 4.10 Event-runoff and surface-runoff coefficients based on hydrograph separation by using the stable isotope ^{18}O and dissolved silica during the various rainfall/runoff events (*modified from Chifflard et al., 2008*). The event runoff coefficient is the fraction of rainfall which becomes direct runoff during a particular runoff event.

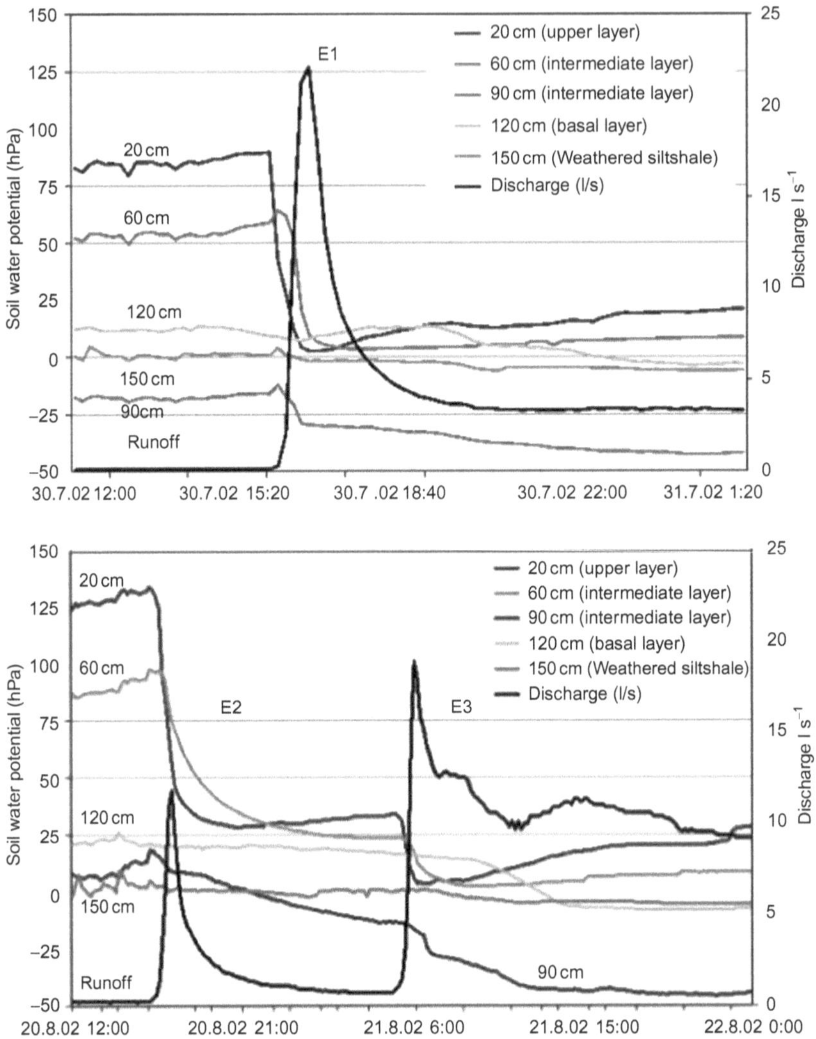

FIGURE 4.11 Soil-water potentials at Station II and channel runoff measured at the catchment outlet during the various rainfall/runoff events (*modified from Chifflard et al., 2008*).

With respect to the dynamics of the three rainfall/runoff events, we conclude that the basal layer acts as a barrier for vertical soil-water flow during dry soil moisture conditions, which leads to high soil-water potential in the intermediate layer (event E2 in Fig. 4.11). This results in lateral subsurface runoff, which has attenuated the contribution to channel runoff. The upper layer and the saturated areas near the channel may be identified as sources for shallow subsurface runoff (Fig. 4.12).

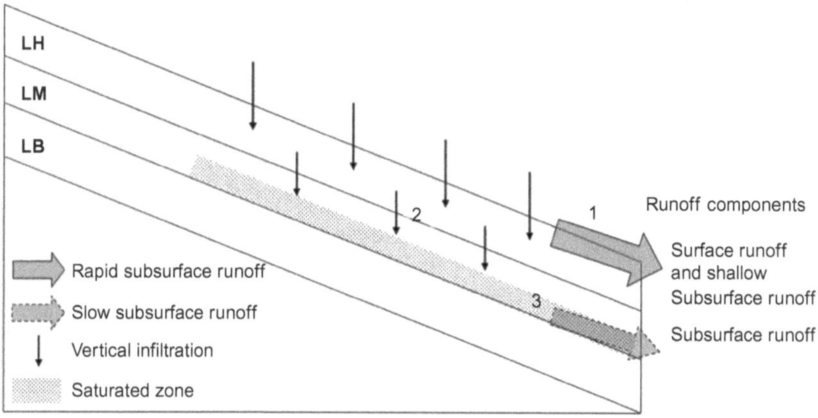

FIGURE 4.12 Runoff generation with low preexisting soil moisture, schematic (*modified from Chifflard et al., 2008*). LH, upper layer; LM, intermediate layer; LB, basal layer. Numbers illustrate the temporal sequence of the activation of the various source areas and flow paths.

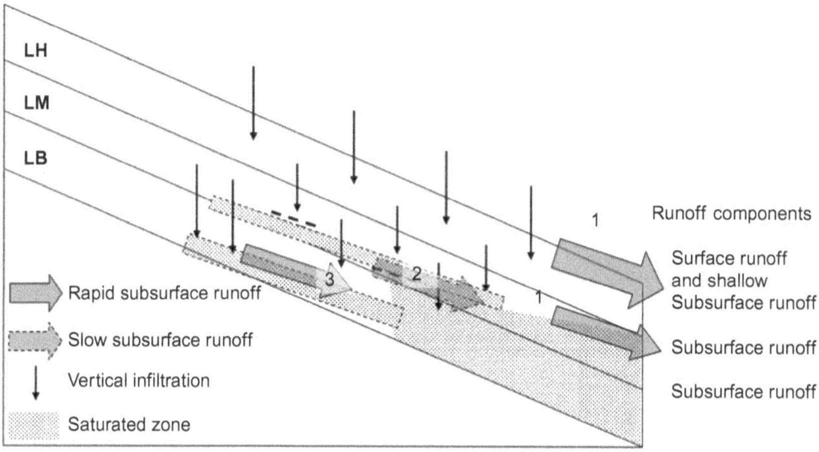

FIGURE 4.13 Runoff generation with intermediate preexisting soil moisture, schematic (*modified from Chifflard et al., 2008*). See Fig. 4.12 for explanation.

With increasing soil moisture, soil-water potentials in the intermediate layer increase considerably so that soil water infiltrates gradually into the basal layer, which so far had been unsaturated (E1 in Fig. 4.11). This leads to steady increase in soil-water potential, indicating saturation (Fig. 4.13).

If the basal layer is saturated prior to the runoff event (E3 in Fig. 4.11), there appears to be a threshold the surpassing of which leads to a higher lateral hydraulic conductivity and to rapid subsurface runoff in the basal layer (Fig. 4.14). Obviously, in addition to subsurface runoff in the intermediate layer, interflow in the basal layer contributes to channel runoff. This leads to

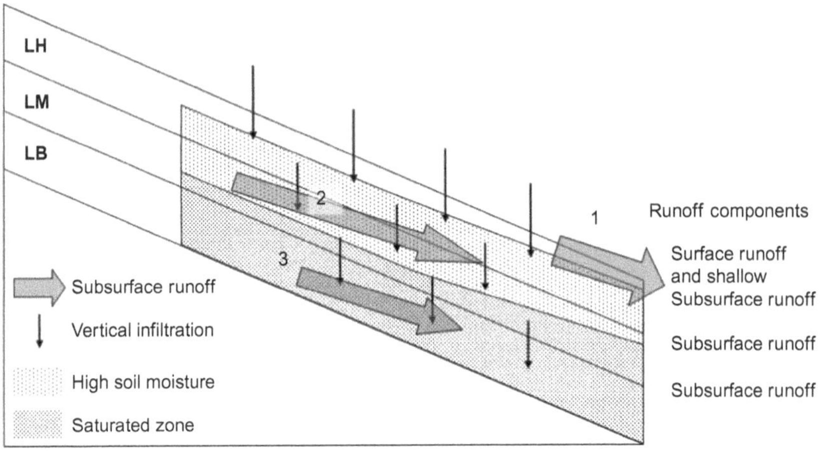

FIGURE 4.14 Runoff generation with high preexisting soil moisture, schematic (*modified from Chifflard et al., 2008*). See Fig. 4.12 for explanation.

a delayed peak in runoff on August 21, 2002 at around 15:00 (3 pm) at the catchment outlet (Figs. 4.10 and 4.11). This delayed peak is not caused by additional precipitation, as the surface runoff coefficient remains constant during the entire event (E3, Fig. 4.10).

Chifflard et al. (2008) calculated a runoff volume of 30 m^3 per 500 min for this delayed event. Based on Darcy's equation (Kutílek and Nielsen, 1994), they estimate lateral hydraulic conductivities of 84 cm d^{-1} in the intermediate and basal layers being necessary to generate this second runoff peak. Due to the orientation of the rock fragments parallel to the slope inclination, high lateral hydraulic conductivities of the basal layer may be expected (Kleber and Schellenberger, 1998). Therefore, Chifflard et al. (2008) take this 84 cm d^{-1} as a more reasonable value for conductivity rather than 5–35 cm d^{-1} as estimated using the approach of AGB (2005).

4.3.4 Erzgebirge

Approach and most methods used in an ongoing study in the eastern Erzgebirge, Saxony, Germany are similar to those applied in the Frankenwald and Sauerland Mountains. The goal of the project is expanding this type of research on runoff processes to an area with another type of bedrock because both studies depicted in the previous sections were on shale with its ability to create pronounced platy clasts upon physical weathering, possibly inducing some bedrock-related bias into the results (Heller, 2012).

The study site "Mulda" (see Fig. 2.1 for the location referred to as Ore Mountains; see Fig. 4.15 for details) is a small, forested spring catchment (0.06 km^2) with an elevation of 521–575 m asl. The watershed is formed as a

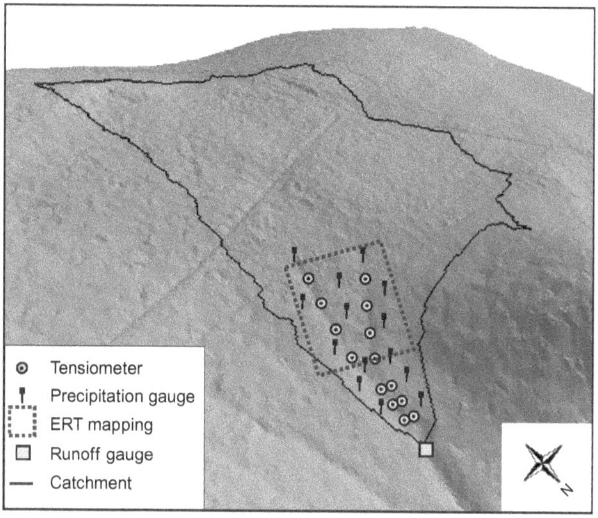

FIGURE 4.15 Study area Mulda and sampling design (not to exact scale).

TABLE 4.2 Measured Properties of Cover Beds of an Example Profile

Layer	Soil horizon	Color (moist)	Texture	Clasts (%)	Bulk density (g cm^{-3})	Conductivity (cm d^{-1})[a]
Upper layer	A/Bw	10YR/ 5/8	Loamy-sandy silt	36	1.2	27
Intermediate layer	2Bg	10YR/ 5/4	Loamy-sandy silt	43	1.5	9
Basal layer	3CBg	10YR/ 5/3	Very loamy sand	56	1.7	52

[a]Field-saturated hydraulic conductivity measured using the Compact Constant Head Permeameter (CCHP) method (Amoozegar, 1988).

slope depression and has a mean slope angle of approximately 7°. The bedrock is gneiss, which is overlain by periglacial cover beds comprising an upper layer, occasionally an intermediate layer, and a basal layer. Some soil-physical and sedimentological properties of the layers are specified in Table 4.2.

The bulk density and the clast content increase from the upper to the basal layer. The texture of both the upper layer and the intermediate layer is similar, but the substrate change between the intermediate layer and the basal layer is substantial. The field-saturated hydraulic conductivities of the layers are quite different. The moderate hydraulic conductivity of the upper and the low conductivity of the

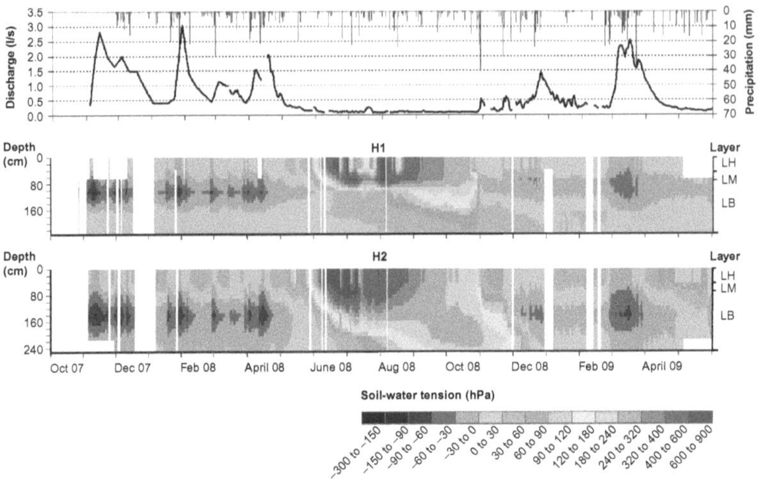

FIGURE 4.16 Throughfall precipitation, spring discharge, and mean daily soil-water tension at Stations H1 and H2 from November 2007 to May 2009. (For color version of this figure, the reader is referred to the online version of this chapter.)

intermediate layer are as expected. The high values in the basal layer are hardly explained by soil-physical and sedimentological properties but are assumed to be an effect of anisotropy (see below). In fact, measurements conducted in the Frankenwald Mountains with a standard Double-Ring Infiltrometer, which is able to prevent much of the anisotropy effects, yielded much smaller conductivities.

Since November 2007, soil-water tension within the layers has been measured using 76 recording tensiometers (10-min intervals) arranged in 14 profiles along the slope at five to seven different depths. A V-notch weir and a pressure meter have been installed to quantify the spring discharge. The rainfall has been measured with 14 precipitation gauges. Furthermore, electrical resistivity mapping was carried out. To investigate the petrophysical relationship between resistivity and saturation with pore water, soil probes were analyzed in the lab.

During the major study period from October 2007 to May 2009, the spring discharge fluctuated between 0.08 and 3.04 l s^{-1} (Fig. 4.16) with the highest discharges occurring in the winter. In summer 2008, the spring reacted only slightly to precipitation events. The diagrams of two representative tensiometer stations with the daily average of soil-water tension related to depth and time (Fig. 4.16) indicate the impact of the soil moisture on the spring discharge.

During summer, increased evapotranspiration causes the soil to dry out. Therefore, rainfall could not at all drain to the spring. In contrast, in the winter season at all depths low tensions (<90 hPa), that is, high water contents, were measured, and after rainfall or snowmelt a rapid runoff response was observed. Furthermore, the influence of the layered subsurface on soil moisture may be concluded. In the upper basal layer, a saturated region develops in contrast to the lower basal layer with higher tensions.

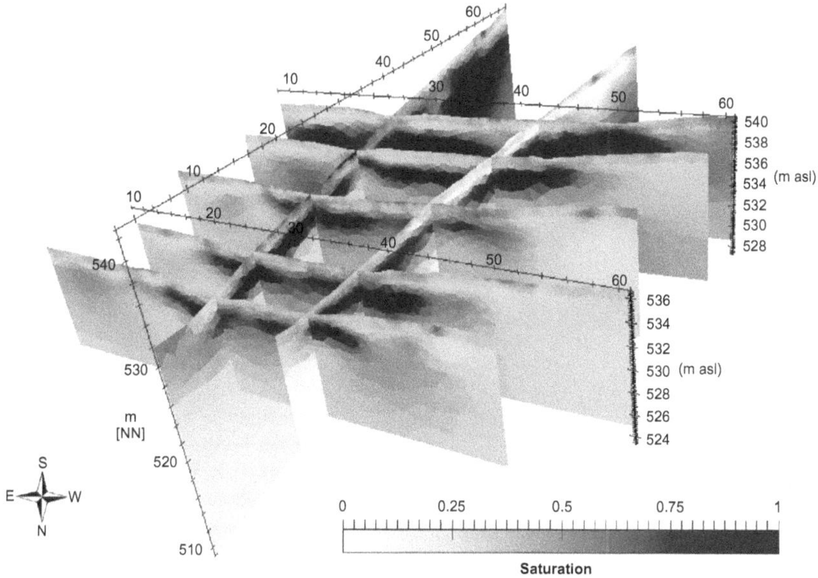

FIGURE 4.17 Saturation calculated from electrical resistivity mapping. (For color version of this figure, the reader is referred to the online version of this chapter.)

We interpret these findings as due to the fact that the lower basal layer acts as an aquitard, whereas the upper basal layer acts as an aquifer under moist conditions. Referring to the high hydraulic conductivity measured in the basal layer, the moisture distribution is inconsistent. Therefore, the hydraulic conductivity of the basal layer is assumed to be anisotropic; that is, water drains faster parallel to the slope than vertical (Kleber and Schellenberger, 1998).

Electrical resistivity mapping corroborates the results of the hydrometric measurements (Fig. 4.17). Saturation values calculated from resistivity, using a small ERT-measuring system on representative, successively water-saturated soil samples in laboratory, are shown in Fig. 4.17. At about 1.5–2 m depth, a saturated zone is discernible. This zone lies in the baseline of the slope depression and confirms the runoff concentration in the basal layer measured by the tensiometers.

Analyses of the hydraulic head and the spring discharge also support this hypothesis (Fig. 4.18). The data selected are characteristic of autumn, with medium soil-moisture conditions. Besides the spring discharge, three tensiometer stations were selected along the slope (H1, H2, and D2) to represent the spatial connections in a slope depression.

At the beginning of the charted period, the soil moisture was moderate. Within 5 days, there were four precipitation events with breaks of only some hours and a total amount of 82 mm. The spring discharge increased from 0.08 to a maximum of 0.65 $l\ s^{-1}$. This distinctive rise of the discharge on October 30 corresponds to the increased hydraulic head in the upper and the intermediate layer. This probably indicates interflow in these layers. The second

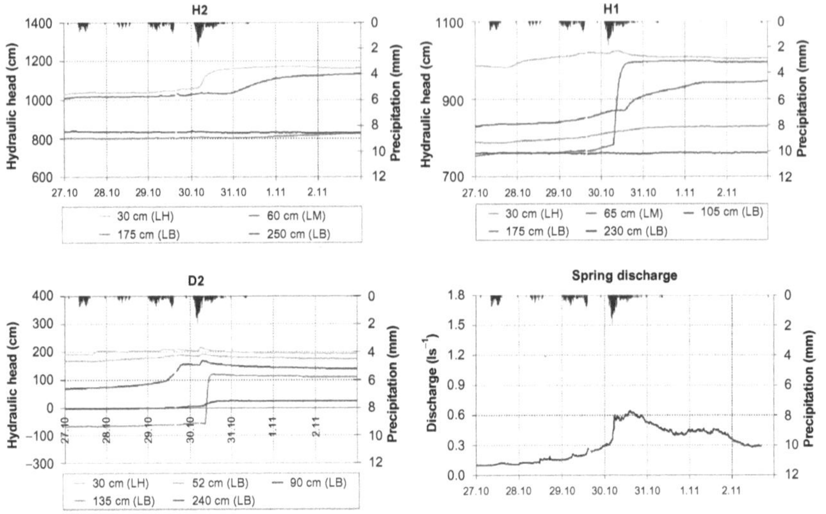

FIGURE 4.18 Hydraulic head at tensiometer stations H1, H2, D2, and spring discharge, October 27 to November 2, 2008. (For color version of this figure, the reader is referred to the online version of this chapter.)

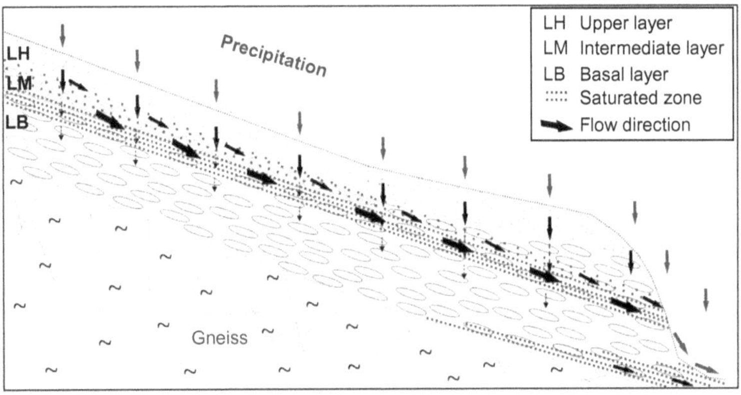

FIGURE 4.19 Water flow paths with medium preexisting soil moisture along a slope in the eastern Erzgbirge, schematic.

maximum discharge on November 1 correlates with the increase in soil-water potential in the upper basal layer (105 and 120 cm). Simultaneously, the hydraulic head remained almost constant in the lower basal layer (150–250 cm). This zone considerably impedes seepage through its matrix. Therefore, we assume that interflow in the upper basal layer caused the second peak in discharge long after precipitation had ceased (cf. similar findings by Chifflard et al., 2008 and Section 4.3.3). Figure 4.19 summarizes these results. Precipitation water infiltrates into the upper and intermediate layers. Interflow

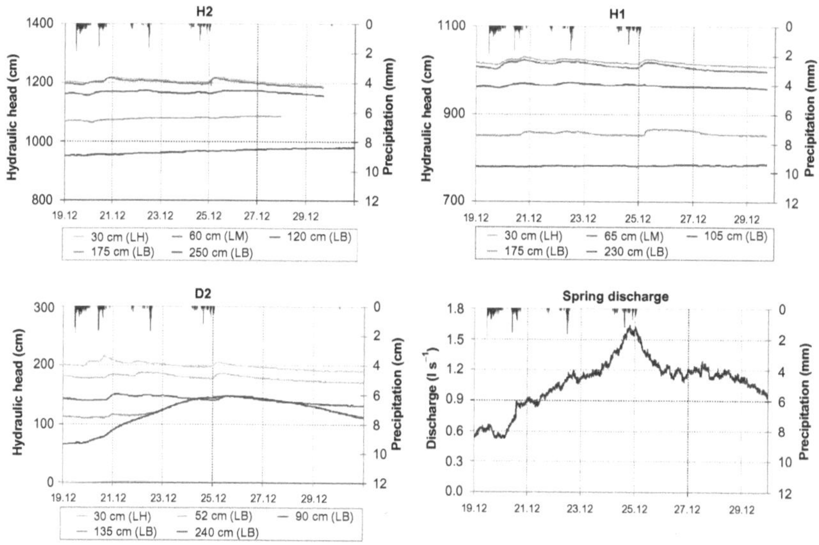

FIGURE 4.20 Hydraulic head at tensiometer stations H1, H2, D2, and spring discharge, December 19 to 28, 2008. (For color version of this figure, the reader is referred to the online version of this chapter.)

occurs in both layers and, with a short delay, in the upper basal layer. A rainfall/runoff event with high preexisting soil moisture, characteristic of the winter and spring seasons, occurred from December 19 till December 28, 2008 (Fig. 4.20). With onsetting precipitation (68 mm), the discharge continuously increased from $0.6\,\mathrm{l\,s^{-1}}$ and reached a maximum of $1.7\,\mathrm{l\,s^{-1}}$ on December 25. The hydraulic heads rose in the upper layer, intermediate layer, and upper basal layer, consecutively. The deeper basal layer did not react, indicating that compensation of hydraulic potentials took place in a lateral direction. This became apparent by a continuous rise of the hydraulic potentials in the deepest basal layer close to the spring. Furthermore, the spring discharge increased.

On the slope water infiltrated through the upper and intermediate layers into the upper basal layer (Fig. 4.21). Close to the spring, it reached the deeper basal layer as an interflow and led to water rise into the overlying layers (return flow, cf. Kleber and Schellenberger, 1998).

The analyses of rainfall–runoff events show that, depending on pre-event soil moisture, different water-flow pathways become activated in the subsurface. Furthermore, stratification exerts a decisive influence on these pathways: the basal layer with its high bulk density hinders vertical seepage through its matrix but prefers lateral water flow because of the clasts that are adjusted parallel to the slope inclination. This finding indicates that the results of previous studies from shale bedrock (see preceding sections), viz., that the clast orientation in basal layers triggers its hydrological behavior, may be transferred to other bedrock types rather than shale, which are able to deliver oblate clast as well.

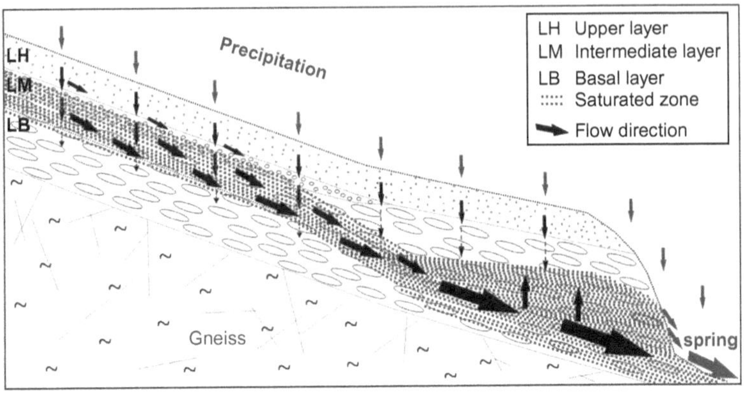

FIGURE 4.21 Water flow paths with high soil moisture conditions along a slope in the eastern Erzgebirge, schematic.

4.4 CONCLUSIONS

The reaction of the studied catchments to precipitation and snowmelt largely depends on the different properties of the cover beds. Water paths may be expected to be completely different if these deposits are absent or homogeneous.

The observations reported here from several regions are quite similar. The conclusion of all presented studies is a significant interrelation between subsurface layering and runoff-generation processes. Exact knowledge of these interrelationships will be helpful to improve rainfall–runoff modeling. Future work has to test these hypotheses on the basis of observations in other experimental watersheds to separate local anomalies from regularities (McDonnell, 2003). New measurement techniques such as geophysical sounding and monitoring will facilitate systematic approaches at lower cost.

A major conclusion from the studies presented here is that discharge from a particular catchment may react in nonlinear ways; that is, the system output following minor to moderate events may correlate with rainfall/snowmelt intensities, whereas strong events may produce unexpected output, maybe much larger, faster or delayed in its timing than forecasted from this simple correlation. Results of such strong events, possibly connected with high preexisting soil-water contents, may be unpredictable if using only observed system reactions on weaker events. This may significantly influence the quality of hydrological modeling when it comes to forecasting extreme floods. However, the predictable properties of the particular layers promise that this drawback for hydrological modeling might be overcome in the future, helping to transfer models between scales, which is one of the major challenges modern hydrology is facing (Finke et al., 1998).

Chapter 5

Geotechnical Properties of Cover Beds

B. Damm, B. Terhorst and F. Ottner

5.1 INTRODUCTION

The occurrence and spatial distribution of cover beds are decisive for modern hillslope morphodynamics. This aspect is strongly associated with the geotechnical properties of cover beds and especially with their anisotropy as discussed in Chapter 4. The impact of periglacial cover beds on the occurrence of mass movements, in particular landslides in subdued mountains, will be discussed in this section.

In general, examples from Central Europe show a predominance of shallow landslides with a thickness of 2–3 m, whereas deep-seated landslides with slide planes up to 10 m below the surface play a minor role (Damm et al., 2008; Semmel, 1987; Terhorst et al., 2009; Thein, 1998). Due to the topography, composition, and thickness of Quaternary slope deposits, by far, most landslide masses in Central European subdued mountains comprise volumes of 100–1000 m^3. The failure areas are mostly 10- to 30-m wide and 20- to 50-m long (Fig. 5.1). Even though landslides like these are comparably small, they frequently affect transport infrastructure as well as residential areas (Damm, 2005; Krauter et al., 2004).

The case studies (Section 5.3) show that soil-physical and soil-mechanical properties significantly influence the forces in soils and periglacial cover beds and, thus, directly control the slope stability. Aside from the physical characteristics of the coarse fraction ($\geq$2000 μm), those of the fine fraction (<2000 μm) are of substantial relevance in this context (Terhorst and Damm, 2009). In detail, grain size, grain structure, pore volume, water content, and permeability control the shear strength and affect parameters such as friction angle, cohesion, and deformability (Terzaghi, 1950; Zaruba and Mencl, 1982).

In addition, parameters like the degree of consolidation, particle-size structure (clay content, coherence, etc.), and the rate of permeability in cover beds and unconsolidated rocks are factors which control the stability of hillslopes (Damm et al., 2009; Terhorst, 2001). In contrast to long-lasting, stable

Developments in Sedimentology, Vol. 66. http://dx.doi.org/10.1016/B978-0-444-53118-6.00005-2

FIGURE 5.1 Representative landslide in the subdued mountains of Central Europe. The landslide occurred in cover beds overlying Early Triassic sandstone at Neustadt/Main in northern Bavaria (February 14, 2002) (*Reproduced from Damm and Terhorst, 2010*). (For color version of this figure, the reader is referred to the online version of this chapter.)

geomorphological factors, such as slope gradient and slope curvature, the aforementioned factors may induce abrupt instability. Abrupt instabilities in periglacial cover beds particularly come into being when the soil-water content increases and, consequently, both the pore water and the hydrostatic pressures rise. These changes in hydrological conditions reduce the cohesion and friction angle (e.g., Bromhead and Ibsen, 2004; Terhorst and Kreja, 2009). In particular, the latter work shows that landslides to a large extent occur in slope deposits, which are weakly consolidated and sensitive to water penetration.

Furthermore, there is also a strong impact of bedrock stratification on the soil-mechanical instabilities in periglacial cover beds via subsurface water flow (see Cardinali et al., 1994; Damm et al., 2010b; Kovacik and Ondrasic, 1991; Montrasio et al., 2009). Studies in landslide areas of subdued mountains show that the discharge from subterraneous bedrock sources into basal layers may saturate the substrates independent of the surface infiltration, which is directly derived from precipitation (see also Damm, 2005; Govi et al., 1985). In such cases, instable sediment becomes mobilized after long-lasting rainfall periods. Thus, a relationship between the dip of the bedrock layering and the occurrence of landslides in cover beds is well documented (see Cardinali et al., 1994; Damm et al., 2010b).

Soil terminology in this chapter follows Soil Taxonomy (SSS, 2010) with the classification according to the World Reference Base for Soil Resources (WRB, 2006) in parentheses. The study areas discussed in this chapter are depicted in Figs. 2.1 and 6.1.

5.2 INTERNAL STABILITY OF COVER BEDS DERIVED FROM THE "INFINITE MECHANICAL SLOPE MODEL"

The internal stability of cover beds on natural slopes may be evaluated using generally accepted criteria and models of slope stability analysis (Knoblich, 1967; Terzaghi, 1950; Zaruba and Mencl, 1982). The stability of cover beds against sliding may be expressed as a factor of safety η, based on the equation

$$\eta = \frac{H}{T}$$

with H representing the sum of resisting forces and T being the sum of the driving forces. In the case that H and T are in balance, slope stability is defined as an "instable equilibrium." Among the resisting forces, shear strength is the dominant parameter. The shear stress at a slide plane, frequently an interface between layers with different shear strengths or permeabilities, is determined in accordance with the Mohr–Coulomb failure criterion.

By far, the most common types of slope failure in cover beds are shallow landslides of one or more layers of cover beds. In general, these types of landslides are connected with the vertical structure of cover beds and the occurrence of discontinuities, for example, at the interface "bedrock—basal layer" or "basal layer—intermediate/upper layer." In this case, slope failure produces a shallow planar slide of great width and considerable length compared to its depth (Damm, 2005; Gudehus et al., 1985), which may be described by the infinite mechanical slide model (Fig. 5.2). This slide model may be applied in particular to soil and debris complexes with layering relevant for soil-mechanical

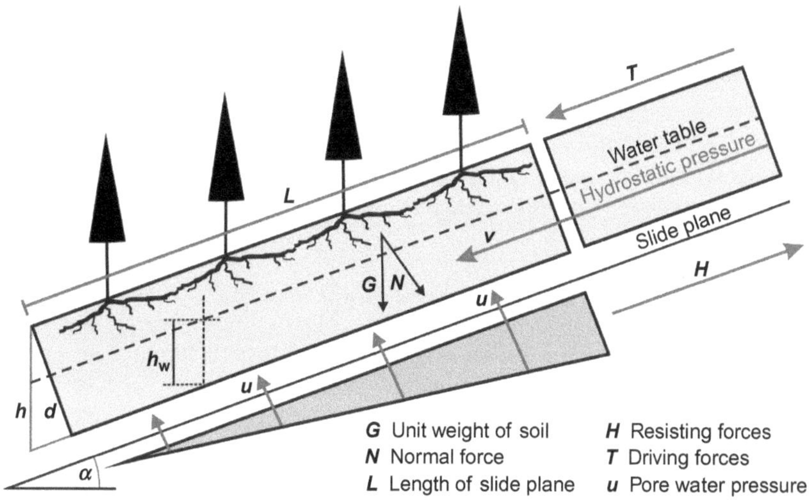

FIGURE 5.2 Parameters and physical interrelationships of the "infinite mechanical slide model" for the calculation of slope stability in planar hillslope sediments (*Modified from Terzaghi, 1950, Knoblich, 1967,* and *Damm, 2005*).

properties and, thus, with the occurrence of disconformities. However, the model ignores possible lateral effects.

The resisting forces are the effective shear strength of the soil, whereas the driving ones are the gravitational forces tending to move the soil downward, supported by the tendency of water to reduce the soil strength and to cause a buoyancy effect within the soil. Driving and resisting forces are described by the unit weight of soil (F_G), the length of the sliding mass, respectively, slide plane (L) and through base variables (angle of friction φ, cohesion c, pore water pressure derived from the unit weight of water u, hydrostatic pressure v, angle of slope, respectively, slide plane inclination α). The mechanical model includes driving and resisting forces as follows:

$$T = F_G \sin\alpha + v \left(\mathrm{kN/m^2}\right)$$

and

$$H = \tan\varphi(F_G \cos\alpha - u) + cL \left(\mathrm{kN/m^2}\right)$$

The unit weight of soil, F_G, of the sliding mass is calculated according to

$$F_G = \gamma_B L d \left(\mathrm{kN/m^2}\right)$$

where γ_B is the specific weight of the (moist) soil and d is the vertical depth of the sliding mass. In the case that measured values are not available, the pore water pressure u and the hydrostatic pressure v are calculated. This procedure may be applied, if there is adequate knowledge of the formation of the slide mass and its basement.

The effective pore water pressure u' results from

$$u' = uL = \gamma_w h_w L \left(\mathrm{kN/m^2}\right)$$

where γ_w is the specific weight of water and h_w is the vertical height of the water table above the slide plane as a part of the soil thickness above the plane. The hydrostatic pressure v is calculated in assumption of a slope parallel flow through

$$v = Ld\gamma_w I = Ld\gamma_w \sin\alpha \left(\mathrm{kN/m^3}\right)$$

where the hydraulic inclination is $I = \sin\alpha$ (methodological approach according to Terzaghi, 1950 and Zaruba and Mencl, 1982).

5.3 CASE STUDIES

During the past few years, analyses related to the geotechnical stability of Quaternary cover beds were carried out in the German subdued mountains and the eastern Austrian Prealps. In both—geologically different—areas, landslides in Quaternary slope deposits are widespread.

It was found that slope stability in general is closely connected to long-term landscape formation through factors like weathering, erosion, rock properties, and tectonics. Furthermore, the studies aimed to assess geotechnical parameters and geomorphological processes to develop a base for susceptibility modeling.

5.3.1 Stability of Cover Beds in the Flysch Zone of the Vienna Forest (Austria)

5.3.1.1 Study Area

The northern Vienna Forest (Fig. 5.3) is part of the eastern Prealps and represents an undulating landscape with an altitude ranging from 300 to 500 m a.s.l. The area belongs to the Rhenodanubian Flysch Zone and adjoins the Vienna Basin at the periphery of Vienna City.

The flysch formations in the study area, with focus on the Hagenbach Valley, predominantly consist of sandstones, marls, and clay shists (Wessely, 2006). Marly sandstone, in particular, tends to be profoundly decomposed by percolating water. In general, the bedrock is covered by periglacial sediments, like cover beds and loess. Both the petrography of the bedrock and the soil-mechanical properties of the slope deposits control the present-day slope dynamics, as changes in permeability are numerous and abrupt (Poisel and Eppensteiner, 1986; Terhorst and Damm, 2009).

Based on historical sources and engineering reports (Brix, 1969; GBA, 2006; Götzinger, 1943; Plöchinger and Prey, 1993; Wessely, 2006) as well as on field surveys, it is obvious that mass movements are widespread in the northern Vienna Forest. A current database comprises about 400 datasets on mass movements of which 82% are related to shallow sliding processes mostly bound to cover beds (see Damm and Terhorst, 2010).

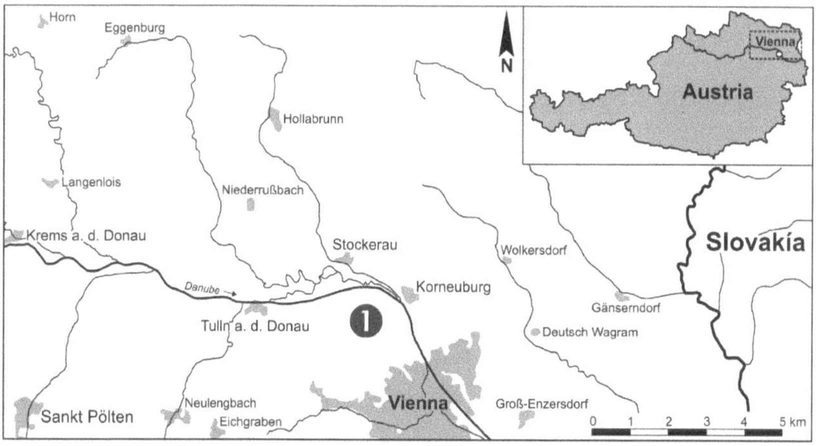

FIGURE 5.3 Position of the Hagenbach Valley and surroundings (①) in the Vienna Forest northwest of Vienna (*Modified from Frank et al., 2011*).

FIGURE 5.4 Vertical structure of Quaternary hillslope sediments at the "Pfarrwald III" landslide site in the Hagenbach Valley. The entire section is about 3.5 m high and comprises an upper layer (Ah–Al/HL), loess (Bt–C), and basal layer (C2). The complete sedimentary succession was exposed by a landslide in 1996 (*photo B. Damm, 2010*). (For color version of this figure, the reader is referred to the online version of this chapter.)

5.3.1.2 Distribution and Composition of Slope Deposits in the Hagenbach Valley

The slopes of the Hagenbach Valley and its surroundings are characterized by several types of sediments (for details, see Terhorst et al., 2009). The bedrock consists of calcareous and marly sandstones, marly shales, calcareous marls, and clayey schists. On the upper slopes, periglacial cover beds (Fig. 5.4) and loess are important factors smoothening the terrain as they cover bedrock, old landslide scars, and sliding blocks (Fig. 5.5). On the lower slopes, sandstones, marls, and basal layers are exposed at the surface due to erosion of overlying cover beds. Numerous landslide scarps in unconsolidated Quaternary sediments expose sediment successions and slide planes, indicating a disturbed relief situation.

Most of the upper slopes are covered by undisturbed soils and cover beds. Figure 5.4 depicts a complete succession of slope deposits, with an Alfisol[1]

1. Soil-type characterized by an argillic (Bt) subsurface horizon, enriched by translocated clay, usually overlain by a depleted E horizon.

FIGURE 5.5 Pleistocene sediments smoothen the terrain of an old, semicircular landslide scarp and corresponding sliding blocks (*photo B. Damm, 2009*). (For color version of this figure, the reader is referred to the online version of this chapter.)

(Haplic Luvisol, Ah–E–2EBt–2Bt–3C horizons) developed in the upper and intermediate layer. Below, unweathered loess is present, of which the basal part is characterized by redoximorphic features such as thin rusty and grayish iron bands and mottles due to water perched above the underlying basal layer. The latter exclusively consists of flysch components of clay, that is, marl and sandy debris. It is densely bedded and exhibits wavy layer boundaries.

Special emphasis should be placed on the fact that the basal layer frequently has strong perching effects due to its high bulk density (Damm et al., 2008). In contrast to the upper slopes, soils and cover beds in the middle and lower slope sections are eroded to a large extent and basal layers as well as bedrock have been uncovered completely (Fig. 5.6).

The upper layer consists mainly of loess and of sandstone debris as a secondary component. It shows silt contents between 56% and 63% (Table 5.1), similar to nearby pure loess deposits. The clay content of the upper layer may rarely reach 25%. Intermediate layers contain less silt (25–50%) and more clay (around 30%), largely due to pedogenic clay enrichment in the Bt. The silt fraction of loess samples is comparably high with values between 57% and 62%, whereas their clay content is significantly lower (7–10%). Basal layers are characterized by intense variations in all grain-size fractions. In most samples, the clay content is relatively increased and may reach values up to 36%. In material derived from decomposed flysch sandstone, the sand fraction is dominant (50–80%), whereas the clay and silt contents are low. The relevance of the clay minerals and the grain size for the stability properties of cover beds was studied in a characteristic and complete profile exposed in a landslide scarp (partially shown in Fig. 5.7).

FIGURE 5.6 In middle and lower slope sections, soils and periglacial sediments have been eroded to a large extent and basal layers as well as bedrock have been unveiled completely by slide processes (*photo B. Damm, 2007*). (For color version of this figure, the reader is referred to the online version of this chapter.)

Table 5.2 shows some results of mineralogical analyses. The higher contents of dolomite than in all other layers demonstrate that the loess consists of allochthonous material, which is not present in the flysch bedrock. Most samples contain smectite and 18 Å vermiculite. However, in the basal layer, the content of smectite is exceptional and corresponds to a high proportion of swellable clays compared to the total clay content. All other samples contain much less swellable clay.

The analyses indicate that in particular the swellable clay minerals smectite and 18 Å vermiculite are relevant for the stability of the studied slope deposits. Due to swelling of the clay minerals, the basal layer tends to be impermeable and, therefore, forms a natural slide plane. Cover beds and loess sediments overlying the basal layer may move due to increased hydrostatic pressure. All together, the difference in permeability and water stagnation on top of the basal layer are important controlling factors for landslides in the study area.

5.3.1.3 Soil-Mechanical Stability of Cover Beds and Bedrock

Laboratory analyses followed the European standards of engineering geology (ISO/DIN), which are well comparable to the analogous recommendations of the Unified Soil Classification System (USCS; see ASTM, 1985). Stability-relevant attributes of the substrates were derived from laboratory data combined with relevant equations using standard parameters and are specified according to USCS (see Bell, 2007; Damm, 2005; Damm and Terhorst, 2010; Mitchell, 1988). The shear strength was measured using a shear strength tester. The results are based on about 80 laboratory analyses and 140 *in situ* measurements.

TABLE 5.1 Laboratory Analysis of Characteristic Sediments and Soil Horizons of Landslide Sites in the Hagenbach Valley, Northern Vienna Forest *(Modified from Terhorst et al., 2009)*

Sample	Layer/Horizon	$CaCO_3$ (%)	Silt, total (%)	Sand, total (%)	Clay, total (%)
PWII-H1	Upper layer, E	0	62.3	18.7	19.0
PWII-H2	Upper layer, EBt	0	56.5	18.1	25.4
PWII-H3	Intermediate layer, Bt	0	49.7	20.8	29.5
PWI-A2	Intermediate layer, Bt	0	25.7	44.6	29.7
PWII-E5	Loess, C	36.5	57.5	32.9	9.6
PWII-H4c	Loess, C	36.2	59.5	33.1	7.4
PWII-H4b	Loess, C	36.2	62.0	30.2	7.8
PWI-A3	Basal layer, Bw	0.2	26.0	55.5	18.5
PWII-H5	Basal layer, C	31.9	53.8	14.4	31.8
PWII-E9	Basal layer, C	18.7	53.4	10.6	36.0
12	Decomposed sandstone, C	1.5	40.3	49.3	10.4
10	Decomposed sandstone, C	2.8	17.4	79.7	2.9
9	Decomposed sandstone, C	4.5	29.3	60.2	10.5
8	Decomposed sandstone, C	6.0	31.7	61.2	7.1
11	Decomposed sandstone, C	41.7	19.9	76.1	4.0

$CaCO_3$ was determined with the Scheibler apparatus, and grain sizes were determined by the combined sieve and pipette method according to Köhn.

The grain-size distributions of loess and loess-rich cover beds classify them as low-plasticity silts in reference to soil-mechanical properties. The bulk density is moderate (ρt 3). The *in situ* shear strength, measured under low-field moisture, ranges from 50 to 85 kN/m^2 (Table 5.3). As a result, loess and loess-rich cover beds are stable at slope gradients up to 40° and more when

FIGURE 5.7 Profile "Pfarrwald I" showing the upper layer (HL), the intermediate layer (AlBt–Bt), loess (Cvc–C(g)), and the basal layer below (BL/Cv). The basal layer is composed of clay (left-hand side) and sandstone fragments (right-hand side). The layering of its matrix is very dense (*photo B. Terhorst, 2010*).

dry. However, upon moistening, they lose cohesion, skip from solid to supple conditions, and reduce stability until their friction angle of 27.5° is reached (see Gudehus et al., 1985).

Based on grain size, clays and marls of the basal layer are classified as moderate-plasticity clays. Their bulk density is predominantly high, corresponding to level ρt 5. The span of the shear strength under low-field moisture is between 85 and 190 kN/m^2 (Table 5.3). The soil-mechanical stability is comparable to that of the loess sediments mentioned earlier. Clayey basal layers are stable under dry conditions up to a slope gradient of 45° and more. With increasing moistening and at the transition from solid to supple consistency, their strengths rapidly decrease. Thus, the soil-mechanical stability is reduced to that of a friction angle under supple conditions and total loss of cohesion appears.

The sand/silt mixture of the decomposed sandstones, in turn, is not cohesive and is characterized by a moderate to high carbonate content. The soil-mechanical property was determined as a sand/silt aggregate. Depending on the compactness of the bedding, different shear strengths were measured under low-field moisture. For moderately dense bedding, the values range from 11 to 15 kN/m^2, and the range for dense bedding is between 20 and 45 kN/m^2 (Table 5.3). Due to

TABLE 5.2 Some Mineralogical Results from a Typical Sediment Succession in the Northern Vienna Forest (*Modified from Damm et al., 2008*)

Layer	Texture	$CaCo_3$ (%)	Calcite (% of total sample)	Dolomite (% of total sample)	Smectite (% of clay fraction)	Vermiculite 18 Å (% of clay fraction)	Clay fraction (<2 μm) (% of fine earth)	Swellable clay min. (% of total sample)
Upper layer	Silty loam	0.4	n.d.	n.d.	n.d.	tr.	22.8	tr.
Upper layer	Silty loam	0.6	n.d.	n.d.	n.d.	n.d.	30.2	n.d.
Intermediate layer	Silty loam	0.6	tr.	1.3	tr.	23	30.7	7.1
Loess	Sandy silt	34.2	8.2	13.7	27	tr.	10.7	2.9
Loess	Sandy silt	33.8	7.8	15.1	32	tr.	9.1	2.9
Basal layer	Silty loam	36.0	30.9	tr.	66	tr.	37.3	24.6
Flysch bedrock	Very silty clay	8.1	6.8	n.d.	32	40	8.1	5.8

tr., traces; n.d., not detectable. The mineralogy was processed with a PW 1710 X-ray diffractometer. The clay content was determined by combined sieve- and sedimentation-rate analyses using a sedigraph.

TABLE 5.3 Soil-Mechanical Classes and Characteristics Following USCS (see ASTM, 1985; Bell, 2007) and DIN 1055-2, and Shear-Strength Values of Cover-Bed-Related Sediments and Loam Beds of Landslide Sites in the Vienna Forest Flysch Zone

Sediment unit	Class symbol (USCS)	Soil-mechanical specification (USCS)	Consistency, compactness	Bulk density (ρt)	Specific weight (kN/m^3)	Cohesion (kN/m^2)	Friction angle (°)	Shear strength (kN/m^2)
Loess, moderate compact	ML	Low plastic silt	Supple	3	20.0	0	27.5	50–85
			Solid		21.0	5		
Clays and marls of cover beds, compact	CL	Moderately plastic clay	Supple	5	19.0	0	22.5	85–190
			Solid		20.5	10		
Sand and silt mixture of decomposed (calcareous) sandstones, earth moist	SM	Sand–silt aggregate	Moderately compact	3	20.0	–	32.5	11–15
			Compact		22.0	–	35.0	20–45
Marls of interbedded flysch strata, compact	CL	Moderately plastic clay	Supple	5	19.0	0	22.5	70–110
			Solid		20.5	10		

The results are based on about 80 laboratory analyses and 140 *in situ* measurements, predominantly from sediments of landslide sites in the Hagenbach valley.

leaching of calcareous cement out of the sandstones, and to decomposition of the rock structure, the density increases with depth. Sand/silt aggregates are almost free of clay, and thus cohesion is absent. Therefore, the stability is controlled by the friction angle which, depending on the bedding density, ranges from 32.5° to 35°. A short-term increase of the stability may result from increasing moistening, defined as "apparent cohesion" (Zaruba and Mencl, 1982).

Marls of interbedded flysch bedrock are classified as moderate-plasticity clays. Its bulk density is predominantly high (ρt 5), and its shear strength is within 70 and 110 kN/m^2 (Table 5.3), which is similar to that of the clays and marls of basal layers. The moderately plastic marls of flysch bedrock interbedding are stable under dry conditions up to slope gradients of 45° and more. With increasing moistening and at the transition from solid to supple consistency, the shear strength rapidly decreases. Thus, the soil-mechanical stability is reduced to that of the friction angle for supple conditions and a total loss of cohesion appears.

5.3.2 Stability of Cover Beds on Early Triassic Sandstones of Southern Lower Saxony (Germany)

5.3.2.1 Study Area

The study area is part of the Solling anticline, a saddle structure between the Rhenish Massif and the Harz basement complex (Fig. 5.8). The Solling anticline is predominantly characterized by bedrock complexes of the Early Triassic Buntsandstein. These bedrock formations consist of sandstones and interbedded siltstones, which occasionally alternate with thin layers of claystones that act as aquitards. In the southern Solling area, the main valleys are 200–300 m incised into a pre-Oligocene sandstone plateau. For the most part, the solid bedrock is covered by thin layers of Quaternary slope deposits.

About 160 landslides, most of them developed in cover beds (Fig. 5.9), have occurred in the study area during the past 140 years. Due to topography, composition, and properties of the slope deposits, the landslides are mainly shallow with thicknesses of 2–3 m (see Damm et al., 2009).

5.3.2.2 Distribution and Composition of Slope Deposits

Quaternary slope deposits are widespread on landslide sites in the study area. To a large extent, loess-rich deposits are absent on steeper slopes. However, in slope depressions and foot-slope areas, there are loess-rich cover beds and anthropogenic cover beds with thicknesses between 3 m and occasionally more than 6 m. The grain-size maxima are generally in the coarse-silt fraction. Clay and silt make up 80–90% of the fines. Loess has been found to be less susceptible to landslides, which fact is closely related to its occurrence on slopes with low gradients (Damm, 2005). Furthermore, cover beds in the area under study

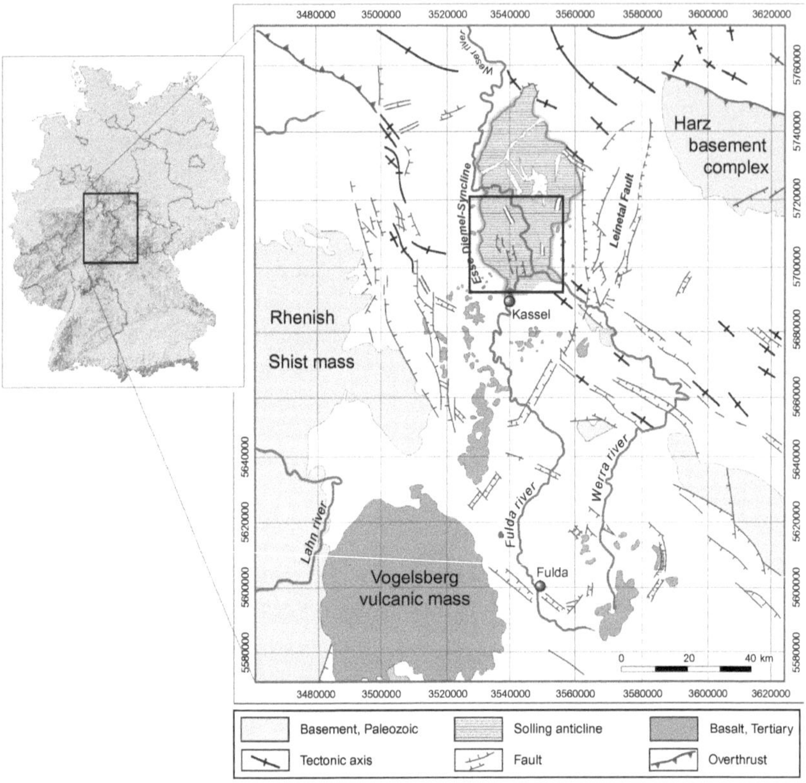

FIGURE 5.8 Location of the Solling anticline in the central part of Germany between the Rhenish Massif and the Harz basement (box). The Solling anticline is predominantly characterized by bedrock complexes of the Early Triassic Buntsandstein (*Modified from Damm et al., 2010b*).

contain much loam and usually are difficult to distinguish. Their total thickness, averaging 1–3 m, is determined by local topography and the degree of weathering of the bedrock (see Section 2.3).

The ubiquitous upper layer is 20- to 60-cm thick and may be kept apart from the basal layer by its higher silt content. Furthermore, it can be distinguished both macroscopically and microscopically from the basal layer. In general, it consists of loess and deeply weathered bedrock, including rock fragments.

The basal layer is largely composed of angular debris ranging from the stone to the coarse block fraction (diameter >630 mm). The coarse fraction amounts to 30–80%. The grain-size distribution of the fines strongly depends on the composition of the bedrock. On average, the basal layer contains 69% sand, 20% silt, and 11% clay.

The saprolite zone that underlies the basal layer is derived from the Early Triassic Buntsandstein bedrock. The latter was affected by deep and for the most part chemical decomposition during the Tertiary. Decomposition may reach a maximum depth of 60 m. On average, the saprolite zone is 5- to 10-m

FIGURE 5.9 Characteristic landslide in cover beds overlying Buntsandstein rock at "Roter Stein" in the southern Solling area. As a consequence of long-lasting and modern creep processes, the cover beds move toward the edge of the rock wall and break successively. The failure zones are 2–3 m high (*Reproduced from Damm, 2010*). (For color version of this figure, the reader is referred to the online version of this chapter.)

thick and characterized by bleaching, rock disintegration, decalcification, as well as neoformation of kaolinite within the Buntsandstein rocks. Water infiltration may reach very deep in this porous media and, thus, may advance the weathering of the underlying bedrock.

5.3.2.3 Soil Mechanic Characteristics and Stability of Slope Deposits

Cover beds and decomposed bedrock are particularly important for the occurrence of mass movements in the study area. Characteristic geotechnical values are compiled in Table 5.4. The sediments are largely unstable, due to their specific grain-size distribution and their sensitivity to water supply. The stability of cover beds is directly related to moisture penetration. The cohesive fine-earth fraction of these sediments quickly responds to moisture penetration with changes in consistency and reduction of strength properties (Damm et al., 2009) and change from semi-compact or compact to plastic or viscous (Table 5.4). Thus, sliding in Pleistocene cover beds and soil sediments frequently alternates between slide and flow processes.

TABLE 5.4 Characteristic Soil-Physical Parameters and Atterberg Limits of Hillslope Sediments Covering Buntsandstein Rocks in the Southern Solling Anticline

Consistency limit value	Symbol	Minimum value	Maximum value	Property
Specific weight of soil	γ (kN/m^3)	20.0	21.4	Earth moist
Natural water content	W (%)	9.2	26.3	Normal to excessive
Water permeability	k_f (m/s)	1.0×10^{-6}	1.5×10^{-6}	Semipermeable
Plastic limit	W_P (%)	12.6	22.0	Weakly cohesive to cohesive
Liquid limit	W_L (%)	22.5	32.6	Weakly cohesive to cohesive
Plasticity index	I_P	7.8	17.5	Weakly cohesive to cohesive
Consistency limit	I_C	0.27	0.84	Pappy to stiff plastic
Dry density	ρ_d (t/m^3)	1.63	1.86	High
Frost sensitivity	F	3	–	High-grade frost sensitive

Analyses of 34 samples from landslide materials.

5.3.2.4 Calculation of Soil Mechanic Stability of Slope Deposits

The soil-mechanic stability is significantly controlled by stagnant and percolating slope water. The stability is reduced if the friction angle, φ, and the cohesion, c, decrease, and if both the pore water pressure and the hydrostatic pressure increase. In the case that the fine fraction changes from solid to supple or pappy conditions, a total loss of cohesion must be assumed. Based on the experiences with landslides in cover beds of southern Lower Saxony, the impact of the pore water pressure as a landslide trigger is of secondary importance subsequent to that of the hydrostatic pressure (Damm, 2005).

The geotechnical stabilities of slope deposits may be calculated by backward projection of former mass movements, if failure criteria are known (see Gudehus et al., 1985; Knoblich, 1967). However, the friction angle and the cohesion are unknown for the larger part of the studied cases. Based on the experience that landslides occur commonly if the slide masses are highly water saturated and lose their coherence during movement, it must be assumed that a loss of cohesion is the most important parameter for failure.

TABLE 5.5 Characteristic Soil-Mechanical Values of Hillslope Sediments Covering Buntsandstein Rocks in the Southern Solling arch

Soil-mechanical parameter	Decomposition zone of bedrock	Basal layer (earth moist)	Mixed layer (earth moist)	Mixed layer (water saturated)
Specific weight (kN/m^3)	21	21	20	10
Friction angle (°)	35	27.5	22	22
Cohesion (kN/m^2)	50	5–15	5–10	0

Averages from laboratory analyses of 65 samples and mean values according to DIN 1055-2.

TABLE 5.6 Safety Against Sliding in Mixed Layers of Backward Projected Altmündener Wand Landslides due to Cohesion and Slope Gradient

Cohesion (KN/m^2)	Slope gradient (°)					
–	30°	32°	34°	36°	38°	40°
10	**1.47**	**1.37**	**1.28**	**1.20**	**1.12**	**1.06**
8	**1.37**	**1.27**	**1.19**	**1.11**	**1.04**	0.98
6	**1.27**	**1.18**	**1.10**	**1.02**	0.96	0.90
4	**1.16**	**1.08**	**1.00**	0.94	0.88	0.82
2	1.06	0.98	0.91	0.85	0.79	0.74
0	0.96	0.89	0.82	0.76	0.71	0.66

Numbers in boldface indicate conditions defined to be stable.

The geotechnical stability criteria of slope deposits were calculated using the infinite mechanical slope model (see Section 5.2), using the example of six landslide sites at the Altmündener Wand in southern Lower Saxony (Damm, 2005). During the past decades, landslides repeatedly occurred at this place. As a consequence of long-lasting and present-day creep processes ranging from 0.1 to 1.6 cm/a, with an average of 0.2 cm/a (see Damm, 2005), the various cover-bed layers typically cannot be differentiated anymore. Hence, the slope deposits are composed of mixed layers, and landslides in material like this frequently occur at the disconformity between mixed slope deposits and decomposed bedrock.

Landslide masses at the Altmündener Wand, used for backward stability projection, vary between 1.5 and 2.5 m in depth and between 6 and 20 m in length. These landslides are related to slopes inclined from 35° to 38°. The soil-mechanical parameters were applied according to Table 5.5, where cohesion is 10 kN/m^2 for earth moist and 0 kN/m^2 for water-saturated mixed layers. The level of the water table, significant for the calculation of the hydrostatic pressure (see Fig. 5.2), was assumed at a low level of 0.1 m above the slide plane due to experiences from field surveys (see Damm et al., 2010b).

The backward projection of the slope stability results in factors of safety, η, of 0.7–0.8 for water-saturated mixed layers under humid weather conditions and low cohesion (see Table 5.6). Under mean conditions, the safety increases up to 1.2–1.3 for earth moist mixed layers. However, increased safety against sliding can be attained if the consistency of the slope deposits tends to stiff-solid conditions with c between 5 and 10 and a slope between 35° and 40°.

5.4 PERSPECTIVES

Studies concerning the geotechnical properties of cover beds and substrates near to the surface were already initiated and performed by Semmel (1986, 1987, 1991a, 2000). However, few investigations have focused on this topic since then. The latest studies show that the properties of the cover beds may have significant relevance for mass movements, in particular, for landslides (Damm, 2005; Damm and Terhorst, 2010; Damm et al., 2008; Terhorst, 2007; Terhorst and Damm, 2009). In connection with topography, geological framework, and tectonics (see Bíl and Müller, 2008; Damm et al., 2009, 2010b; Margielewski, 2006; Neuhäuser and Terhorst, 2006; Schmanke, 1999), the cover-bed properties decisively affect the slope stability and safety of the infrastructure in subdued mountains of Central Europe.

The mentioned studies also highlight the relevance of process-related analysis for recognition and evaluation of factors that control the stability of, and dynamics triggered by, cover beds and other weekly consolidated slope deposits such as loess (Semmel, 1989). As, for the most part, the present geomorphologic, geologic, and structural framework of hillslope systems and their sediments developed over the course of the Quaternary, the regional susceptibility to landslides may be traced back to comparably long-lasting processes and dynamics. As these processes continue to date, analysis of Quaternary landscape formation is an important tool to understand the present-day dynamics of slope deposits (Terhorst et al., 2009).

Chapter 6

Transferring the Concept of Cover Beds

A. Kleber, M. Leopold, C. Vonlanthen and J. Völkel

6.1 INTRODUCTION

Previous chapters of this book demonstrate the research density on slope deposits in Central European subdued mountains since the 1960s. If at all, similar research in other parts of the world is in its beginnings (Jacobs et al., 2012; Kleber, 1997). Slope deposits, as reviewed by, among others, De Wolf (1988), Daniels and Hammer (1992), and Selby (1993), have often been recognized as soil parent materials (e.g., Birkeland and Burke, 1988; Molloy, 1988; Sokolov, 1989), but few publications focus on their genesis, distribution (Paton et al., 1995), or whether they display recurring sequences, that is, on their internal structure (Veit, 1987, 1993; Veit et al., 2011). Especially in the mid-latitudes outside of Central Europe, evidence for cover beds is scarce.

This chapter examines successions of soil parent materials on slopes in several mid-latitude areas outside Central European subdued mountains, located between 37° and 56°N (Fig. 6.1), a zone mainly characterized by Pleistocene periglacial conditions. Its intention cannot be a comprehensive overview, laying claim to covering a great variety of, or even all, possible constellations. Rather, this chapter presents case studies from various, but deliberately chosen, relief and climatic settings.

Section 6.2 has its focus on low-lying relief (basins and plains) and juxtaposes examples from a humid area with case studies from arid continental climates.

Section 6.3 attempts to pursue cover beds from basins and subdued mountains up into alpine heights, starting with the European Alps. It commences with case studies reaching intermediate elevation and is followed by examples from the higher Alps. A similar approach was chosen in the following section, which first summarizes studies from the Rocky Mountains, USA, and adjacent mountain ranges in an attempt to trace the cover beds from basin up to the timberline and then depicts cover beds at alpine altitudes of the Colorado Front Range.

Soil terminology in this section follows Soil Taxonomy (SSS, 2010).

Developments in Sedimentology, Vol. 66. http://dx.doi.org/10.1016/B978-0-444-53118-6.00006-4

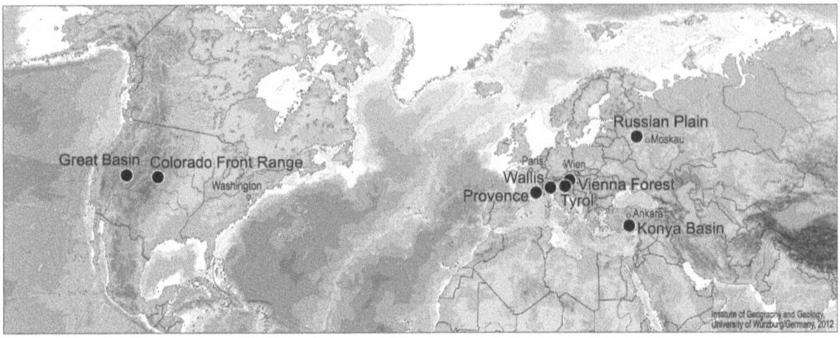

FIGURE 6.1 Location of study areas in the mid-latitudes of the northern hemisphere. Some areas will be depicted in more detail in the appropriate sections. (For color version of this figure, the reader is referred to the online version of this chapter.)

6.2 BASINS AND LOWLANDS OF THE MID-LATITUDES

6.2.1 Northern Russian Plain

6.2.1.1 Successions of Cover Beds

In the area of Ploskosh' (Fig. 6.2), till and glaciofluvial deposits of Moskva age (penultimate glaciation) are ubiquitously overlain by cover beds (Kleber and Gusev, 1998): their oldest one, referred to as "basal layer" in the following, solely consists of local components, but two properties distinguish it from the underlying glacigenic materials: the clasts are well-oriented down-slope (not so in the till or outwash) and it has a high bulk density. Except on upper slopes, this layer is overlain by a loess-rich cover bed (coarse silt typically increases from <20% to >35% and the heavy-mineral contents also point to

FIGURE 6.2 Location of study areas on the Russian Plain. Dashed line = approx. maximum extent of the last ("Valdai") glaciation (according to Svendsen et al., 1999; Svendsen et al., 2004; Velichko et al., 2006); dot-dashed line = approx. maximum extent of the penultimate ("Moskva") glaciation (according to Astakhov, 2004; Velichko et al., 2006).

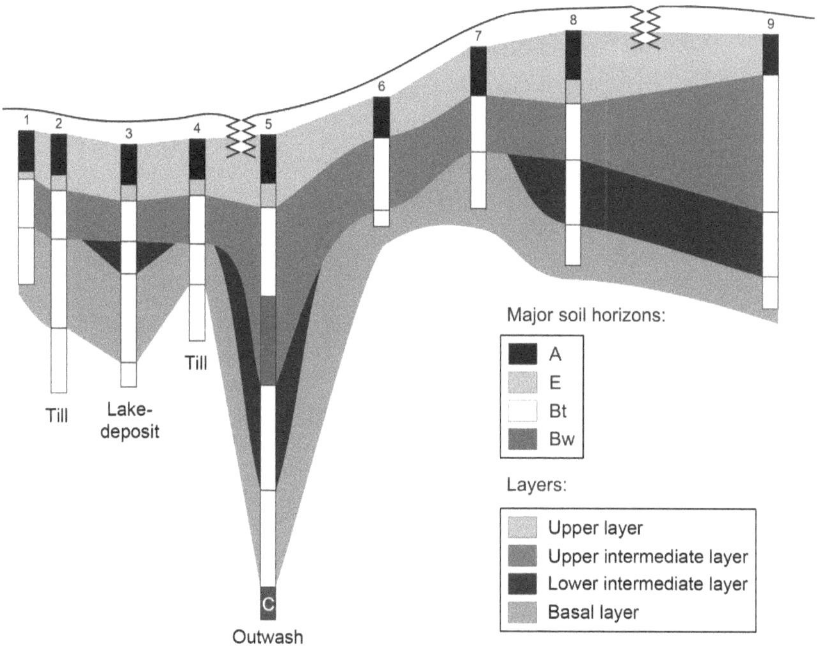

FIGURE 6.3 Composite catena of cover beds and soils in the area of Ploskosh', Russian Plain. Zigzag indicates composition of the catena (*reproduced from Kleber and Gusev, 1998*). Horizontally not to scale.

allochthonous components) with very mature argillic[1] and often stagnic[2] properties superimposed on the argillic ones. This layer may reach a thickness of 3 m on foot slopes (Fig. 6.3); then pedogenetically unaffected sediment eventually causes bifurcation of the argillic horizons. This indicates that these horizons belong to two different phases of soil formation, each developed in a separate parent material. These parent materials are referred to as "lower intermediate layer" and "upper intermediate layer," respectively. This finding renders the lower argillic horizon a paleosol, probably of the last interglacial (Mikulino) as it is younger than the underlying till of Moskva age. If this is the case, its parent material, overlying the Moskva deposits, dates from the late Moskva glaciation. Having observed this bifurcation in a few profiles, it becomes possible to keep track of the disconformity between the two argillic horizons in stacked profiles without bifurcation, because of notable, recurring changes in soil color (especially of clay films: cutans of the deeper argillic horizon are darker probably because of a larger content of clay-complexed organic matter), structure, and often texture and mineral contents (Kleber and Gusev, 1998; Fig. 6.4).

1. Subsurface soil properties formed by clay illuviation.
2. Soil properties due to temporary soil-water stagnation.

FIGURE. 6.4 Boundary between grayish paleosol (where the ruler is) and upper soil. The disconformity is extremely sloping here (*photo by A. Kleber*). Inset: sketch depicting the three-dimensional position of the photo, which was taken within a trench (hatched line) with a horizontal plane (light gray) and two vertical ones. The position of the photo is indicated by a dotted line, the paleosol body by dark gray color. (For color version of this figure, the reader is referred to the online version of this chapter.)

All landforms, except where anthropogenic soil erosion has taken place, are covered by another, loose layer of constant (40–50 cm) thickness, referred to as the "upper layer." Compared to the underlying deposits, this is further enriched in loess (approx. 50% coarse silt), whereas coarse rock fragments decrease (on average from >10% to <5%). Its heavy-mineral suite is often somewhat closer to that of the till than to the cover beds directly beneath. This layer hosts the Ap[3] and E[4] horizons of the surface soils. However, on flat relief, modern pedogenesis is additionally influenced by the effects of water perched on the denser layers below. The soils qualify as "sod-podzolic soils" in the Russian classification (Glossaqualfs and Glossudalfs in Soil Taxonomy: SSS, 2010)[5]; they are relatively uniform, because all three layers are widespread with only the intermediate ones varying considerably in thickness but rarely missing entirely.

As a result, there are recurring variations in soil parent materials, which comprise several cover beds: the upper layer has a loose consistency, usually the largest content of coarse silt, and the smallest content of sand and gravel of all layers. It is assumed to be Late Glacial in age (Kleber and Gusev, 1998). The upper intermediate layer is yellowish brown, has an intermediate bulk density, and is usually rich in coarse silt, though to a lesser degree than the upper layer. It is separated by a disconformity, occasionally by pedogenetically less or not affected substrates, from the lower intermediate layer, which

3. Surficial soil horizon with properties dominated by tillage.
4. Soil horizon characterized by depletion of clay or of metals and humus substances.
5. Soil types characterized by combined argillic and stagnic properties with tonguing of the depleted domains into the enriched horizons.

hosts a mature paleosol. A basal layer is of high bulk density, poor in silt, and mostly has reddish color derived from underlying or up-slope materials. This color is typical for Moskva till (Sudakova, 1990).

The number of profiles studied was insufficient for sound statistical analyses. However, some preliminary trends may be seen: the amounts of certain grain-size classes tend to rise in particular layers parallel to those in underlying layers. This trend is weak ($r \approx 0.4$ for coarse sand and clay) at the ends of the grain-size spectrum but significant (>99%) in the center ($r \approx 0.7$ for fine sand and medium silt), indicating inheritance from underlying deposits (see Section 2.6.4). The only class to show a very poor correlation ($r \approx -0.1$) is coarse silt. Based upon a much larger set of data from the western USA, such correlations have been interpreted as typical for loess-mixed cover beds (see Section 6.2.2.3). Eolian admixture mainly introduces coarse silt, thereby weakening the correlation for this size class. Heavy-mineral analyses by Kleber and Gusev (1998) also indicate remarkable eolian influence.

The maximum extent of the Valdai glaciation on the Russian Plain was intensively disputed until the 1980s (Chebotareva and Makarycheva, 1982, cited after Rusakov and Korkka, 2004). However, since then consensus has apparently been reached on the maximum possible extent, approximately as drawn in Fig. 6.2 even farther north (Kalm, 2010; Svendsen et al., 1999, 2004; Velichko et al., 2004). From this, there is no reasonable doubt that the glacigenic sediments underlying the cover beds under study belong to an earlier, probably the penultimate glaciation (Moskva glaciation: Astakhov, 2004; Velichko et al., 1984, 2004). If the older interpretation held true, the Ploskosh' site would be located on Valdai, the sites discussed in the next section on Moskva sediments.

Based on these findings, it should be possible to distinguish different soil parent materials through the identification of cover beds, especially through the disconformity characterizing a major interruption in sedimentation through very strong pedogenesis, and attempting a sequence-stratigraphic approach (Catuneanu, 2006; Catuneanu et al., 2010) to connect layer/soil successions.

6.2.1.2 Sequence-Stratigraphic Approach

The profiles (Fig. 6.2) studied for this approach are located near Krasnozawodsk (56°27′N, 38°13′E, 195 m asl) upon a Moskva till; near Dmitrow (56°19′N, 37°32′E, 150 m asl; Fig. 6.5) upon a Moskva outwash plain; near Rostow (57°17′N, 39°17′E, 260 m asl) on the highest shoreline of Lake Nero, which has been TL dated at about 150 ka (Gunova, 1972, cited after Wohlfarth et al., 2006), and which was an ice-dammed lake during Moskva times; near Sara (57°05′N, 39°17′E, 140 m asl) on a lower shoreline of Lake Nero, inferred to be of Valdai age (Kerzum et al., 1991); Dolgopolka 3 (57°55′N, 39°29′E, 155 m asl) on the Moskva river terrace of River Volga; and Dolgopolka 2 (57°54′N, 39°29′E, 130 m asl) on the Valdai terrace (Novsky, 1975). Results of the analyses are presented in Tables 6.1 and 6.2.

FIGURE 6.5 Dmitrov profile depicting the reworked till (5C horizon) identified by its reddish cast. The overlying loess-rich layers, especially the uppermost layer, are well identified by their whitish cast due to deplenishment (*photo A. Kleber*). (For color version of this figure, the reader is referred to the online version of this chapter.)

Though they cover a wide area, the sequences of soils and layers are very similar to those reported from Ploskosh' (Kleber and Gusev, 1998, and previous section). The disconformity between the 1st and 2nd layers and, especially, the paleosol buried by the 2nd layer are widespread on older, but absent (or restricted to redeposited material) on the younger river terrace. This corroborates the tentative correlation of Kleber and Gusev (1998) of this paleosol with the Mikulino (last interglacial) paleosol previously reported mainly from the Russian loess province south of the study area reported here. This correlation was mostly based on the notable maturity of this soil, which is even outraging the modern soil. The upper parts of the profiles also contain periglacial deposits, as indicated by the clast fabric and a typical richness in loess. The latter is barely in accordance with a Holocene age because of the continuous forest cover (Velichko et al., 1997) and the general lack of eolian activity (Ilyuchev and Targulian, 1981) during the Holocene in this area. On the other hand, the buried humic substrates have been described and dated as early Holocene (Aleksandrovsky and Chichagova, 1995; Aleksandrovsky et al., 1991); the datings, however, were from humic acid, which is not very reliable in a

TABLE 6.1 Grain-Size Data, Color, and pH of Profiles on the Russian Plain

Profile	Deposit	Horizon	Depth (cm)	ccl	fcl	cS	mS	fS	cSi	mSi	fSi	Clay	Color		pH
								(%)					*Dry*	*Moist*	
Krasnozawodsk															
		Ap	−20												
	u-l	Eg	−40	0	1.5	0.4	0.9	4.1	54.3	25.0	9.3	6.1	10YR7/2	10YR6/2	4.22
	ui-l	2Bt	−105	0	11.7	0.3	0.7	2.2	37.7	22.6	10.9	25.6	7.5YR7/6	10YR5/6	3.25
	li-l	3Btg	−170	0	16.0	2.3	2.4	2.5	19.2	25.2	12.9	35.5	10YR7/4	10YR5/4	3.00
	b-l	4Crt	−255	0	9.5	5.7	25.6	28.2	17.0	5.9	4.7	12.9	7.5YR5/6	7.5YR4/6	4.25
	Till	5Cr	−675+	0	3.2	5.6	22.9	30.8	20.9	6.1	3.3	10.3	7.5YR6/6	7.5YR4/6	7.50
Dmitrow															
		A	−3												
	u-l	Eg	−30	0	1.8	0.6	1.7	4.6	48.2	27.5	11.0	6.3	10YR7/3	10YR5/3	4.05
	ui-l	2Etg	−40	0	5.7	1.0	1.3	3.0	34.0	23.1	8.8	28.8	10YR7/4	10YR5/4	3.67
		2Bt	−90	0	7.3	0.4	0.6	4.2	44.6	20.7	7.2	22.4	10YR6/4	10YR5/4	3.62
	li-l	3Bt	−180	0	14.5	1.9	6.5	9.2	17.7	17.0	13.8	33.8	10YR6/4	10YR5/4	3.56
	Slope wash	4Cr	−205	1	14.6	2.7	9.4	15.3	16.6	14.8	11.6	29.6	10YR6/4	10YR5/4	4.20
	b-l	5Cr	−250	1	10.1	4.6	15.1	23.8	19.0	6.3	4.9	26.3	7.5YR6/6	7.5YR4/6	4.13
	Outwash	6Cr	−520+	5	14.6	30.5	60.6	6.5	0.6	0.6	0.6	0.5	10YR6/4	10YR6/4	7.85

Continued

TABLE 6.1 Grain-Size Data, Color, and pH of Profiles on the Russian Plain—Cont'd

Profile	Deposit	Horizon	Depth (cm)	ccl	fcl	cS	mS	fS	cSi	mSi	fSi	Clay	Color		pH
								(%)					*Dry*	*Moist*	
Rostow															
	u-l	Ap	−30	0	6.4	1.9	2.3	5.9	48.3	21.8	10.1	9.8	7.5YR6/2	7.5YR4/2	5.95
	ui-l	2Bt	−95	0	11.9	1.0	1.1	5.9	45.4	19.1	7.4	20.0	7.5YR6/6	7.5YR4/6	5.00
		3Bt	−130	0	14.0	1.2	1.3	2.8	31.6	21.0	9.9	32.2	7.5YR6/6	7.5YR4/6	4.37
		4Crt	−190	0	14.3	1.3	1.3	2.0	31.2	20.9	11.0	32.4	7.5YR6/6	7.5YR4/6	4.52
		5Cr	−240	0	15.8	1.1	2.1	3.2	22.2	22.3	14.6	34.6	7.5YR6/6	7.5YR4/6	4.82
	li-l	6Btg	−325	0	14.1	2.6	8.6	10.0	14.4	17.9	12.0	34.4	7.5YR6/6	7.5YR4/6	4.82
		6Btg	−375	0	19.4	2.5	7.1	8.5	10.6	16.4	15.0	39.9	7.5YR7/4	7.5YR5/4	5.50
	b-l	7Bt	−400	2	9.2	5.3	17.9	22.2	19.7	7.9	6.3	20.7	2.5YR6/6	2.5YR4/6	5.70
	Lacustrine	8Cr	−417	0	13.5	9.0	44.5	28.6	9.4	0.5	1.3	6.7	10YR5/4	10YR4/4	5.76
	Till	9Cr	−700+	0	12.9	6.7	16.5	21.9	22.4	9.8	6.4	16.4	2.5YR6/4	2.5YR4/4	7.45

Continued

TABLE 6.1 Grain-Size Data, Color, and pH of Profiles on the Russian Plain—Cont'd

Profile	Deposit	Horizon	Depth (cm)	ccl	fcl	cS	mS	fS	cSi	mSi	fSi	Clay	Color		pH
								(%)					*Dry*	*Moist*	
Sara															
		Ap	−25												
	u-l	Eg	−40	0	<u>3.4</u>	0.5	2.5	6.8	<u>48.9</u>	<u>22.6</u>	8.6	10.2	7.5YR7/4	7.5YR5/4	5.50
	ui-l	2Bt	−80	0	15.3	0.6	1.4	7.6	58.6	2.1	2.7	27.1	7.5YR6/4	7.5YR4/4	5.65
		3Bt	−185	0	16.4	0.1	0.2	9.0	44.9	16.4	5.9	23.6	7.5YR6/4	7.5YR4/4	4.25
		4Btg	−330	0	9.1	1.6	1.3	2.9	<u>40.4</u>	18.8	12.4	22.6	7.5YR6/4	7.5YR4/4	7.07
	li-l	5Bg	−520	1	8.9	1.3	3.7	7.3	<u>25.4</u>	23.1	14.0	25.2	7.5YR6/4	7.5YR4/4	7.17
		6Btg	−570	<u>0</u>	12.0	<u>4.9</u>	<u>10.7</u>	7.4	12.4	<u>16.5</u>	<u>14.4</u>	<u>33.7</u>	7.5YR6/4	7.5YR4/4	6.87
	b-l	7Cr	−610	2	12.3	25.6	<u>52.7</u>	7.0	7.2	1.9	1.9	3.5	10YR6/6	10YR5/6	7.45
		8Bg	−680	<u>4</u>	6.1	24.1	<u>42.8</u>	7.9	6.8	4.8	3.8	9.8	10YR6/3	10YR5/3	7.35
	Alluvium	9Bg	−700+	1	12.6	18.9	69.6	4.7	5.3	0.2	0.1	1.2	7.5YR6/6	7.5YR5/6	7.65

Continued

TABLE 6.1 Grain-Size Data, Color, and pH of Profiles on the Russian Plain—Cont'd

Profile	Deposit	Horizon	Depth (cm)	ccl	fcl	cS	mS	fS	cSi	mSi	fSi	Clay	Color		pH
								(%)					*Dry*	*Moist*	
Dolgopolka 3															
		Cr	+30												
		Ap	−55												
	u-l	Eg	−65	0	0.3	0.1	1.2	28.0	<u>52.2</u>	13.9	3.6	1.0	10YR7/3	10YR5/3	5.50
	ui-l	2Bt	−190	0	9.1	0.2	8.0	<u>19.6</u>	37.7	10.8	4.7	19.0	7.5YR6/6	7.5YR4/6	4.05
		2Cr	−270	0	<u>5.9</u>	0.3	7.4	<u>13.0</u>	<u>37.2</u>	15.8	8.2	18.1	7.5YR6/4	7.5YR4/4	7.22
	li-l	3Bt	−310	0	15.5	0.4	8.1	<u>30.2</u>	17.5	12.6	10.7	20.5	7.5YR6/4	7.5YR4/4	7.22
		4Bt	−360	0	10.2	0.8	7.4	18.9	<u>21.4</u>	14.1	13.2	24.1	7.5YR6/4	7.5YR4/4	7.10
		6Btg	−380	0	<u>15.9</u>	3.9	<u>9.1</u>	<u>15.0</u>	5.4	<u>21.1</u>	<u>17.8</u>	<u>27.7</u>	10YR6/3	10YR4/3	7.00
		7Bt	−425	1	4.5	4.3	<u>44.2</u>	30.0	10.7	3.6	2.5	4.7	10YR6/4	7.5YR4/4	6.80
	b-l	8Crt	−495	<u>5</u>	<u>6.3</u>	3.4	15.1	23.4	19.1	9.6	6.2	23.2	7.5YR5/6	7.5YR4/6	6.72
	Alluvium	9Bt	−540+	0	16.0	4.9	13.8	22.2	23.7	11.9	6.5	17.0	7.5YR6/6	7.5YR4/6	7.45

Continued

TABLE 6.1 Grain-Size Data, Color, and pH of Profiles on the Russian Plain—Cont'd

Profile	Deposit	Horizon	Depth (cm)	ccl	fcl	cS	mS	fS	cSi	mSi	fSi	Clay	Color		pH
								(%)					Dry	Moist	
Dolgopolka 2															
		A	−8												
	u-l	Eg	−40	1	1.9	0.1	6.0	49.4	38.0	3.9	1.9	0.9	7.5YR6/4	7.5YR4/4	5.04
	ui-l	2SwBt	−55	1	10.1	11.0	25.0	26.6	13.1	4.5	2.7	17.3	7.5YR5/4	7.5YR4/4	n.d.
		3BgBt	−75	1	3.9	9.7	38.5	24.6	10.9	3.5	1.6	11.1	7.5YR5/4	7.5YR4/4	6.07
	b-l	4Bt	−165	5	20.7	21.1	21.5	12.4	16.3	10.7	5.0	12.9	10YR6/4	10YR5/4	7.25
	Alluvium	5Cr	−300	3	40.7	40.3	45.6	6.3	3.2	1.8	1.6	1.3	10YR6/4	10YR5/4	8.05
	Landslide	6Bg	−360	0	12.8	34.7	36.0	18.3	7.2	1.4	1.4	1.0	10YR6/4	10YR5/4	7.80
	Landslide+			0	5.1	7.7	11.6	20.7	23.5	11.2	9.7	15.6	7.5YR6/6	7.5YR4/6	7.05
	Alluvium	7Bgr	−490	0	7.5	1.3	5.7	11.3	22.5	18.8	14.9	25.4	7.5YR6/6	7.5YR4/6	6.80
		8Cr	−520	0	5.1	15.9	47.1	21.0	6.9	2.0	2.5	4.6	10YR6/3	10YR5/3	6.77
		9Bg	−550	5	27.3	23.9	50.6	14.3	6.5	1.3	1.5	1.9	10YR6/6	7.5YR4/6	6.85
	Till	10Cr	−580+	2	5.2	7.0	24.0	24.2	18.4	8.5	5.1	12.8	7.5YR6/6	5YR4/6	6.07

Horizon: designation according to SSS (2010); depth measured from the surface to the top of the horizon; grain-size classes: clay (<2 μm), fine, medium, coarse silt (fSi 2 to <6.3, mSi 6.3 to <20, cSi 20 to <63 μm), fine, medium, coarse sand (fS 0.063 to <0.2, mS 0.2 to <0.63, cS 0.63 to <2 mm), fine clasts (fcl 2 to <20 mm), coarse clasts (ccl ≥ 20 mm); ccl = vol.% of whole soil (field estimate); fcl = wt.% of whole soil after removal of ccl; other particle classes = wt.% of fines <2 mm. pH measured in $CaCl_2$. Underlined = break assumed to indicate a disconformity. Landslide +, landslide deposit with redeposited patches derived from a cover bed, presumably the 3rd layer, analyzed independently from surrounding landslide materials. u-l = upper layer; ui-l = upper intermediate layer; li-l = lower intermediate layer; b-l = basal layer. Here, we do not use the acronyms by AGB (1994, 2005) because the area under study is outside where they have been defined for.

TABLE 6.2 Heavy-Mineral Content of Profiles on the Russian Plain

Horizon	Hornblende	Garnet	Epidot	Staurolite	Titanite	Disthene	Mica	Zircon	Augite	Tourmaline	Rutile	Others
Krasnozawodsk												
Eg	44.8	17.4	22.7	5.3	0.5	2.9	1.0	1.0	0.5	1.0	1.0	2.0
2Bt	<u>47.9</u>	12.0	20.4	5.6		3.5	1.4	0.7	0.7	3.5		4.2
3Btg	59.7	14.9	14.9	0.9	0.4			0.4	2.6	1.8		4.4
4Crt	58.0	10.8	21.2	3.7	0.7	0.4		0.7	1.1	0.7	0.4	2.3
5Cr	62.0	10.8	14.2	2.2	1.7	1.3	0.9	0.9	0.4	2.2	0.9	2.5
Dmitrow												
Eg	53.9	15.0	20.4	3.1	0.4	1.6	1.2		1.2	1.6	0.4	1.2
2Egt	57.7	10.3	18.6	4.9		3.0	0.5	1.5	0.5	1.5	1.0	0.5
2Bt	50.6	14.7	14.1	2.9		5.9	2.9	2.4	2.9	0.0	0.6	3.0
3Bt	<u>54.0</u>	13.4	20.6	2.1	0.4	2.9	0.4	0.8	0.4	1.8	0.4	2.8
4Cr	63.9	9.5	17.0	0.8	0.4	2.4	0.8	0.8	0.8	0.4	0.4	2.8
5Cr	60.1	8.4	19.3	1.4	1.4	2.3	0.9	0.5	0.5	2.3	0.9	2.0
6Cr	53.6	11.3	20.6	3.8	1.9	2.8		0.9	0.9	2.3	0.5	1.4

Continued

TABLE 6.2 Heavy-Mineral Content of Profiles on the Russian Plain—Cont'd

Horizon	Hornblende	Garnet	Epidot	Staurolite	Titanite	Disthene	Mica	Zircon	Augite	Tourmaline	Rutile	Others
Rostow												
Ap	41.8	12.8	32.2	1.9	2.7	1.5	0.4	0.8	0.4	0.8	2.3	2.4
2Bt	33.2	16.2	24.6	4.3	0.7	2.8	4.9	1.4		1.4	3.5	7.0
3Bt	44.7	12.7	31.1	2.8	0.5	2.4	1.4			1.9		2.5
4Crt	48.6	14.3	8.6	5.7		5.7	14.3					2.8
5Cr	47.6	13.8	24.5	0.4	0.4	2.6	1.8		0.4	1.1	1.1	6.3
6Btg	52.4	11.8	25.0	1.5	0.4	1.1	0.4		1.5	0.8	0.4	4.7
6Btg	56.4	9.8	18.6	2.5	0.4	2.1	0.4	0.8	0.8	1.2	0.8	6.2
7Bt	52.1	10.1	22.5	2.9	0.4	2.5	0.4	1.4	1.8	1.1		4.8
8Cr	39.7	19.4	23.9	3.4	1.1	1.9	0.7	0.7	1.1	1.1	0.7	6.3
9Cr	47.2	10.0	29.4	1.4	0.7	2.1	1.1	0.3	1.7	1.7	0.3	4.1
Sara												
Eg	43.4	10.1	27.2	4.3	0.9	3.9		0.9	0.9	1.3	0.4	6.7
2Bt	57.6	5.9	24.1	3.2	0.4	0.9		0.4		1.4	0.4	5.7
3Bt	56.9	3.1	23.8	1.1	0.4	5.1	1.5	0.4		1.5	0.8	5.4
4Btg	55.5	6.9	22.6	3.6		1.1	2.2	0.4	0.4	1.1	0.4	5.9
5Bg	47.8	10.0	29.2	0.8	0.8	2.6	0.8	0.8	0.8	0.8		5.6
6Btg	61.2	5.2	17.3	1.7	0.8	1.3	0.8	0.8	1.8	1.7	0.8	6.6

Continued

TABLE 6.2 Heavy-Mineral Content of Profiles on the Russian Plain—Cont'd

Horizon	Hornblende	Garnet	Epidot	Staurolite	Titanite	Disthene	Mica	Zircon	Augite	Tourmaline	Rutile	Others
7Cr	50.9	9.7	24.2	1.5	0.5	2.4	1.0	1.0	2.8	1.0	0.5	4.0
8Bg	<u>56.4</u>	10.3	18.9	2.5	0.4	1.2	0.4	1.2	2.5	0.8	0.4	5.0
9Bg	45.0	16.3	19.2	6.3	1.2	3.0	0.3	0.9	1.8	0.9	1.8	3.3
Dolgopolka 3												
Eg	53.4	11.5	20.6	2.4	1.0	0.3	3.4			2.0	0.7	4.7
2Bt	44.8	11.8	29.2	2.6	1.0	0.3	0.6	1.4	1.0	2.6	0.6	4.1
2Cr	48.4	10.9	24.7	4.0	1.1	1.1		1.5	0.7	0.7	1.1	5.8
3Bht	45.2	11.5	29.6	1.9	0.6	0.6		1.6	1.0	1.3	1.0	5.7
4Bt	47.0	8.5	33.4	2.4	2.1	0.3	0.3	0.3	0.3	1.1	0.3	4.0
6Btg	<u>55.3</u>	5.3	27.5	2.1	1.2	0.4	0.4	0.4		0.8	0.4	6.2
7Bt	44.8	11.4	34.0	0.6	1.3	1.3	0.3		0.6	0.6	0.6	4.5
8Crt	51.4	8.7	32.5	0.4	1.1	0.7	0.4	0.4	1.1	0.7	0.4	2.2
9Bt	54.2	7.2	29.4	2.4	0.8	0.4	0.8	0.4	0.8	0.8		2.8

Continued

TABLE 6.2 Heavy-Mineral Content of Profiles on the Russian Plain—Cont'd

Horizon	Hornblende	Garnet	Epidot	Staurolite	Titanite	Disthene	Mica	Zircon	Augite	Tourmaline	Rutile	Others
Dolgopolka 2												
Eg	55.5	11.3	18.7	2.3	1.2	0.8	0.4	1.2	0.8	1.9	0.4	4.7
2Btg	47.6	10.4	28.4	2.9	0.9	0.6	0.3	0.6	1.4	0.9	0.6	5.4
3Btg	<u>48.6</u>	7.3	29.8	2.5	0.3	1.7	0.3	0.7	1.1	1.7	0.3	5.7
4Bt	62.2	2.2	22.7	3.2	0.4	0.8	0.4	0.4		0.4		7.3
5Cr	52.2	10.2	20.5	1.6	0.8	1.2		0.8	4.2	0.4	0.8	7.3
6Bg	52.6	5.3	27.8	3.0	1.1		1.1		1.1	0.8	0.4	6.8
Landslide+	54.4	<u>7.8</u>	26.1	1.9	2.3	0.4			0.4	1.2		5.5
7Bg	<u>38.2</u>	21.8	27.2	1.3	2.3	0.7	0.3	1.7		1.3	1.0	4.2
8Cr	50.9	12.7	25.6	0.4	1.7	0.4		1.3	0.4	0.8	0.8	5.0
9Bg	47.0	13.3	25.8	2.3	1.6	0.4	0.4	1.6	1.2	0.8	0.4	5.2
10Cr	48.6	10.6	29.7	1.0	1.4	0.7	0.4	0.4	1.4	0.4	0.4	5.0

Minerals were determined by microscope using 200 nonopaque fine-sand grains at least, etched with 15% HCl. Bromoform was used for separation and synthetic resin (refraction index 1.67) for enveloping. Underlined = break assumed to indicate disconformity. Analyses by R. Schill, Univ. of Bayreuth.

carbonate-free soil environment (Anderson and Paul, 1984; Eckmeier et al., 2007), and there are dates of around 20 ka from similar material (Rusakov et al., 2007) and from artifact layers in the same stratigraphic position (Kurenkova et al., 1995). The reliability of soil dating in this area and age range may also be doubted considering the Holocene age derived from material beneath more than a meter of loess (Aleksandrovsky and Chichagova, 1995). Therefore, a Pleistocene periglacial origin of the upper layer is the favored interpretation here.

The findings at Sara may challenge the last-interglacial age suggestion of the paleosol (Kleber and Gusev, 1998), which at Sara is even buried by some less pedogenetically altered substrate, whereas the age of the underlying landform is believed to be Valdai. However, Kerzum et al. (1991; see also Davydova and Servant-Vildary, 1996) imply an age succession of lake shorelines with the highest being the oldest and so forth. They do not take into account the possibility of lake shorelines formed during lake advances (e.g., Godsey et al., 2005). However, pedogenesis during a Valdai interstadial would be an alternative age assignment for the paleosol. Soils of this age (for instance the so-called Bryansk paleosol: Little et al., 2002) are known from the Russian loess province, but only few occurrences have been described from upon Moskva material; thereafter, an Endoaquoll,[6] radiocarbon dated as approx. 33 ka, formed at that time (Rusakov and Korkka, 2004; Rusakov et al., 2007) with thin, discontinuous cutans at most (Gerasimenko, 2006; Rusakov et al., 2007) rather than mature argillic properties. The Mikulino paleosol of the Russian loess province, however, closely resembles the paleosol described in this section (e.g., Panin, 2007). The same holds true for the area of Moscow close to the sites reported here, located upon Moskva till (Velichko et al., 2006). Thus, the most likely age estimate for the paleosol is Mikulino (last interglacial), which would place parent and underlying cover beds (lower intermediate and basal layer) into the late Moskva, calling for redrawing of the loess-distribution map of Velichko et al. (2006).

6.2.1.3 Comparison to Central European Successions

In a comparison of cover-bed successions and associated soils of various areas, Kleber (1997) distinguished two major types (see Section 6.4). One is characterized by cover beds which differ in a predictable manner in their sediment properties: they have a loess-free layer at their base and loess-bearing layers above, whereas paleosols are not ubiquitous. At first glance, the described profiles belong to this kind. However, there is a remarkable difference with the Central European successions, as they are depicted in Chapter 2 (see Table 6.3): the recurrent record of a buried soil of probably last-interglacial age. This may be useful for paleoenvironmental reconstructions in the northern Russian Plain and for keeping apart Valdai from Moskva deposits.

---

6. Soil type with a surface horizon rich in humus and some properties due to temporary or permanent reduction by water.

TABLE 6.3 Comparison of Layers and their Soils on the Russian Plain with Central European Successions

Russian Plain	Central European subdued mountains
Upper layer	**Upper layer**
Maximum loess content	Loess content similar or smaller compared to intermediate layer
Strong pedogenic overprint due to soil-water stagnation	Yellowish cast of primary loess often still recognizable
Ubiquitous distribution, constant thickness of 40 cm (average on the Russian Plain) or 50 cm (average in Central Europe), eluvial soil horizon	
No age constraint yet, rather than Valdai	Age constrained to late glacial
Upper intermediate layer	**Intermediate layer**
Ubiquitous distribution	Patchy distribution
Somewhat less loess content than 1st layer	At lower elevation typically peak loess content
Rich in loess, pedogenic overprint by clay translocation	
No age constraint yet rather than Valdai	No certain age constraint known
Lower intermediate layer	–
Broad distribution, host for last-interglacial paleosol, age probably constrained to late Moskva (penultimate) glaciation	No widespread counterpart
Basal layer	**Basal layer**
Strong pedogenetic overprint (argillic)	Usually unaffected by major pedogenesis
Poor in rock fragments	Typically rich in rock fragments
Broad distribution, exclusively local components, rock-fragment long axes oriented down-slope, high bulk density	
Age constrained to late Moskva	Variable ages

Parallelism is based on facies, but does not imply synchronicity of the compared layers.

Complete successions at lower elevations of Central European subdued mountains often have their silt maximum in the intermediate layer and usually increased gravel and sand contents in the upper layer, whereas silt in the profiles of the present section peaks in the uppermost layer. Loess deposition rates were probably low in Central Europe during the Late Glacial, when the upper layer formed. Presumably, loess deposition still lasted or even peaked in the northern Russian Plain when the 1st layer evolved; with this stated, we do not imply synchronicity of the upper layers in both areas. Another difference is that a 2nd (and often 3rd) loess-rich layer was found in all profiles of appropriate age of the Russian Plain, whereas similar deposits, that is, intermediate layers, are spatially restricted in Central Europe.

Altogether, stratigraphic handles especially on older cover beds appear more frequent on the Russian Plain than in Central Europe, calling for the use of cover beds in terms of sequence stratigraphy.

6.2.2 Northern Great Basin, USA

6.2.2.1 *Buried Cover Beds*

The internally drained northern Great Basin is characterized by numerous mountain ranges and intervening basins (Fig. 6.6). The ranges mainly consist of sedimentary and acidic igneous rocks, whereas the basins are dominated by Quaternary alluvial and lacustrine deposits. The climate of the basins is arid to semi-arid, inducing shrubland vegetation, whereas in the ranges precipitation increases with elevation, giving room for conifer forests.

The basin floors of the north-eastern Great Basin were covered by a paleolake, Lake Bonneville, several times during the Pleistocene. Its lake deposits are

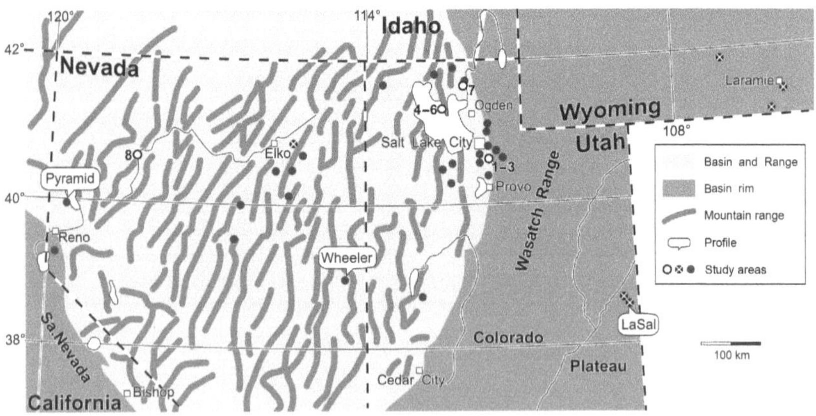

FIGURE 6.6 Location of profiles in and near the northern Great Basin, USA (*modified from Kleber, 2001*). Open circles are study areas of this section, numbers refer to Fig. 6.7; circles with criss-cross stroke point at study areas of Section 7.3; filled circles indicate areas and sites of Section 6.3.2.1, where named profiles are discussed in detail.

separated by paleosols, many of which were distinguished by Morrison (1965a–c), with later modifications by McCoy (1981) and Scott et al. (1983). The most prominent paleosol is the Promontory soil between the Little Valley lake deposits (penultimate deep-lake cycle of Illinoian age; Kaufmann and Broecker, 1965; Scott et al., 1983) and the sediments of the Bonneville lake cycle (last deep-lake cycle of Late Wisconsinan age; Currey and Oviatt, 1985; Currey et al., 1984; Kaufmann and Broecker, 1965; McCalpin, 1986). Scott and Shroba (1980) and Scott et al. (1983) presented evidence that Morrison's (1965a) Dimple Dell soil displays virtually the same degree of soil development and shares the stratigraphic position of the Promontory soil.

Oviatt et al. (1987) described an immature soil, the Fielding soil, on top of the early Wisconsinan (Kaufmann et al., 2001) Cutler Dam lake deposits. The Hansel Valley deposit of McCalpin (1986) and Robison and McCalpin (1987) is probably correlative (C.G. Oviatt, Kansas Univ., pers. commun., 1990) to the Cutler Dam. The Fielding soil is buried by Lake Bonneville beds.

The Eetza lacustrine deposits of paleolake Lahontan (north-western Great Basin) are broadly contemporary with the Little Valley deposits (Kaufman and Broecker, 1965; Lao and Benson, 1988; McCoy, 1981). They are overlain by subaerial deposits of Morrison's (1964) Wyemaha Formation, in which the Churchill soil is developed, which is in turn overlain by the Sehoo deposits of the Late Wisconsinan last deep-lake cycle (Benson, 1993). Thus, the Churchill soil is also correlative to the Promontory soil.

These paleosols were basically assumed monogenetic by the cited workers, whereas Kleber (1993, 2000) inferred some as multistoried, developed within chronologically different parent materials. He identified these parent materials as cover beds.

The paleosols of the profiles in Fig. 6.7 have most of the following horizons in common (from top to bottom):

1. an argillic horizon, developed in a cover bed (referred to as 2nd cover bed in Fig. 6.7);
2. a calcic[7] horizon, developed in a cover bed or in loess, having calcic properties typically decreasing with depth; this horizon is absent in some profiles and reaches a thickness of 30 cm in others;
3. a compound calcic and argillic horizon, developed in a cover bed (1st cover bed);
4. a lower calcic horizon, mostly developed in lake deposits.

At least between (1) and (2) and between (3) and (4), there are disconformities, indicated by textural differences and by the mineralogical composition (Kleber, 2000).

7. A subsurface soil horizon dominated by carbonate enrichment.

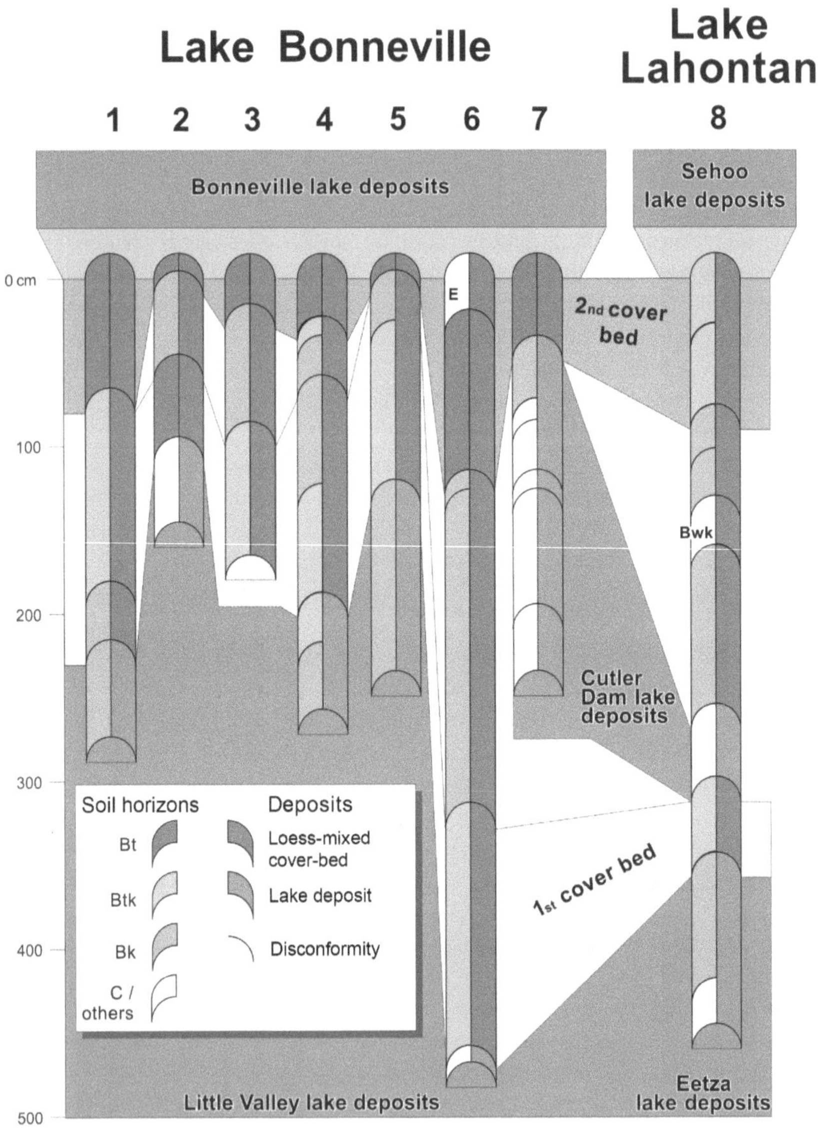

FIGURE 6.7 Correlation of paleosol parent materials in the northern Great Basin. Columns indicate soil horizons (left half) and deposits (right half) and start beneath the overlying Late-Wisconsinan lake deposits. Profile 8 was composed of several laterally connected profiles and is, thus, not to scale regarding the sections between the B horizons (*modified from Kleber, 2000*). The 1st and 2nd cover bed will be referred as lower and upper Wyemaha layers in the following, respectively. (For color version of this figure, the reader is referred to the online version of this chapter.)

Carbonate in soils is usually regarded as pedogenic if there is evidence of pedogenic processes, such as water percolation through the soils. In the soils under study, this evidence is provided by features such as nodules, filaments, and coatings on rock fragments (cf. Bachmann and Machette, 1977; Gile et al., 1966). Many possible sources of pedogenic carbonate are discussed in the literature (Birkeland, 1999; Neymark et al., 2005). After reviewing these possibilities, Kleber (2000) concluded that the majority of the carbonate in these soils was enriched through vertical percolation during pedogenesis.

The carbonate in the compound horizons (3) was deposited after the clay cutans of the argillic horizons had been developed, because carbonate enrichment and clay illuviation that is not restricted to vertical cracks or channels are mutually exclusive and, thus, cannot form simultaneously in the same horizon (Birkeland, 1999; Curtin et al., 1994). Clay translocation is only possible during or after carbonate depletion because clay tends to flocculate as long as Ca^{2+} ions are abundant; for the same reason, argillic horizons cannot form within or below calcareous material, except for clay translocation under the influence of exchangeable sodium (Natrargids). Except where mentioned otherwise, soils discussed in this section do not show such natric properties.

The carbonate in the upper calcic horizon (2) could stem from the parent material of the uppermost argillic horizon (1), if it was leached from upper into lower horizons before the formation of the argillic properties above. If no significant carbonate has later been added to the soil, the argillic horizon (1) is preserved unaltered. However, in the subjacent horizon with compound argillic and calcic properties, the carbonate must be a later overprint of a preexisting argillic horizon. It is unlikely that sufficient carbonate had remained above the compound horizon to account for this. The most likely conclusion from these theoretical considerations is that the parent material of the upper part of the soil (1) had not yet been deposited when the argillic properties in the deeper part (3) formed; thereafter, parent material (1) became deposited introducing new carbonate, which in the next step was leached during pedogenesis to overprint underlying materials (3) (Kleber, 2000). The parent material of horizon (2) is sometimes the lower part of the new, or the upper part of the overprinted layer; in other cases, it is an intercalated third deposit (as in the case exemplified in Table 6.4), or this horizon is absent at all.

One such profile (profile 6 in Fig. 6.7) is exemplified in Tables 6.4 and 6.5. Lake deposits in this area are often dominated by silt, so that it is difficult to conclude from grain-size analyses alone whether there is loess in the cover beds (Table 6.4). In terms of texture, it is mainly the clast content that allows keeping different layers apart. The allochthonous component is better identified by heavy-mineral analyses, which indicates varying sources contributing to the deposits (Table 6.5). The disconformities recognized in these analyses corroborate the previous conclusions from pedogenetic considerations. There is further evidence of a major disconformity between horizons (1) and (3): there occasionally is fresh material intercalated between them, which neither is

TABLE 6.4 Texture and Carbonate Content of a Paleosol in Little Valley, Utah (*Modified from Kleber, 2000*)

Depth (cm)	Soil horizon	Clay	fSi	mSi	cSi	fS	mS	cS	fcl	ccl[a]	Carb. (%)
		(%)									
	C[b]	9.0	13.1	23.8	40.6	12.7	0.3	0.5	0.0	25	28.5
−35	2E	23.0	9.2	18.2	31.6	13.5	2.2	2.3	26.7	2	0.3
−110	3Bt	31.8	6.5	11.5	28.0	13.4	2.0	6.8	9.9	4	0.4
−125	4Bk1 (III)	25.2	11.4	15.9	21.0	11.5	8.4	6.6	48.0	40	41.6
−315	4Bk2 (I+)										11.0
−460	5Btk (I+)	27.0	16.5	20.0	15.1	6.2	7.0	8.2	68.5	60	6.6
−480+	6Bk (II)[b]	No fine material								100	30[a]

Depth: measured from the top of the paleosol. Horizon: designation according to SSS (2010), in parentheses carbonate morphology *modified from Bachmann and Machette (1977)* where appropriate. Grain-size classes: clay (<2 μm), fine, medium, coarse silt (fSi 2 to <6.3, mSi 6.3 to <20, cSi, 20 to <63 μm), fine, medium, coarse sand (fS 0.063 to <0.2, mS 0.2 to <0.63, cS 0.63 to <2 mm), fine clasts (fcl 2 to <20 mm), coarse clasts (ccl ≥20 mm). ccl = vol.% of whole soil (field estimate); fcl = wt.% of whole soil after removal of ccl; other particle classes = wt.% of fines <2 mm; carb = total carbonate. Underlined = breaks assumed to indicate disconformities.

[a]Lake deposit; all other soil horizons are derived from cover beds.

[b]Estimated in the field.

TABLE 6.5 Heavy-Mineral Content of a Paleosol in Little Valley, Utah (*Modified from Kleber, 2000*)

Horizon	Augite	Brown hornblende	Cummingtonite	Garnet	Green hornblende	Hypersthene	Zircon	Accessories[a]	Total counts
C	<u>5.0</u>	12.8	0.5	0.5	<u>59.8</u>	5.6	1.7	14.1	179
2E	<u>31.6</u>	<u>11.5</u>	0.7	1.4	<u>43.4</u>	<u>8.7</u>		2.7	288
3Bt	<u>3.8</u>	23.8	–[b]	0.7	<u>68.0</u>			3.7	290
4Bk	17.3	15.2	–[b]	3.5	42.0	3.0	6.9	12.3	231
5Btk	11.6	19.9	–[b]	2.5	51.1	0.4	6.3	8.2	241

Numbers are the percentage of the nonopaque grains. Underlined notify breaks in sediment properties assumed to indicate disconformities.
[a]Minerals not assumed to be indicative for stratification.
[b]Mineral not found in a count of the entire sample (>1000 heavy-mineral grains).

affected by strong carbonate accumulation (sufficiently strong to destroy evidence of clay illuviation) nor contains illuvial clay (profiles 4, 6: 4Bk2 horizon, and 8 in Fig. 6.7). It is unlikely that argillic properties could form in a deeper and an upper part of the soil but left an intervening part untouched.

Another line of argumentation on the separation of the parent materials of this paleosol comes from profile 7 (Fig. 6.7), which represents the most mature paleosol found upon probable early Wisconsinan Cutler Dam lake deposits (Kleber, 2000). This paleosol consists of just one layer of parent material with a soil rather similar to horizon (1). This indicates that this soil is younger than the early Wisconsinan and, accordingly, horizon (1) and its parent material in the other profiles as well, whereas the lower parts of profiles 1–6 and 8 are assumed to be older, which is consistent with the finding of cummingtonite traces in the same stratigraphic position in six out of the seven profiles of Fig. 6.7 (excluding profile 7). The mineral cummingtonite was delivered by a few eruptions of Mt. St. Helens (Mullineaux, 1986), the oldest being luminescence dated at 50 ka (Berger, 1991). Table 6.5 includes this accessory mineral because it is a potential stratigraphic marker mineral (Kleber, 1999a); if so, the age of the parent material of horizon (3) may be assumed to be as old or older than 50 ka. In profile 8, there is a yet undated lake advance documented between the two soils assumed to be correlative to the Cutler Dam (Kleber, 2000), which provides further evidence that the hiatus between the two layers is remarkable.

Thus, indicated by disconformities, its pedogenic overprint, and by intercalated, pedogenetically unaffected substrates, the paleosol between the last two major lake highstands in the northern Great Basin persistently consists of two different parent materials, that is, loess-mixed cover beds, separated by a pronounced episode of soil formation. The cover beds likely are older and younger than the early Wisconsinan, respectively. Definitely, both cover beds have been deposited between the last two major lake highstands. At the time of the formation of the upper part of the paleosol, the lower one was already a buried soil, overlain by a younger deposit.

Utilizing the existing local nomenclature of soils and subaerial layers, Kleber (2000) proposed naming the upper and lower cover bed "upper Wyemaha layer" and "lower Wyemaha layer," respectively, and the upper and lower paleosols "upper Promontory soil" and "lower Promontory soil," respectively.

6.2.2.2 *Surficial Cover Beds*

Cover beds are also frequent in the north-eastern Great Basin outside of former lakes (Kleber, 1993). To correlate buried cover beds with surficial ones, one must be able to trace the former to outside the ancient lakes where they eventually merge or do not merge with the latter. This is difficult to achieve in the north-eastern Great Basin, because deep soils rarely occur in the direct surroundings of former Lake Bonneville. The probable reason is that, during a rapid fall by 100 m of the lake level in the Late Pleistocene (O'Connor, 1993), the so-called Bonneville Flood at around 14.5 ^{14}C-ka BP (Oviatt et al., 1992), erosion truncated most surfaces around the lake (Jewell, 2010).

However, this event was less drastic in Rush Valley, which was separated from Lake Bonneville by a gigantic spit (Burr and Currey, 1988; Jewell, 2007). This is one of the few places where argillic soils occur near the modern surface on the former shores of Lake Bonneville.

They have from top to bottom an ochric (moderately humic) horizon, and an argillic, a calcic, and a compound horizon, all developed in cover beds bounded by disconformities; the underlying alluvium hosts another calcic horizon (Kleber, 2000). The succession, thus, equals the buried pedocomplex described above except for the uppermost horizon. Auger cores, drilled every meter from the location of the profile toward the former lake, revealed that the lower part of the cover-bed succession continues below the marginal lake deposits without significant changes, indicating a connection of lower cover beds and the buried ones, whereas the uppermost cover bed, which hosts the ochric horizon, remains surficial, indicating a lower age than the lake deposits. Contrary to the lower cover beds, this upper cover bed, hosting a soil of low to intermediate maturity, is frequent in this part of the basin. This indicates that this layer is younger than the Bonneville Flood.

The last deep paleolake in the north-eastern Great Basin has left behind several shorelines well visible in the landscape, which have been dated (Oviatt et al., 1992a,b; Patrickson et al., 2010; Sack, 1989). The resulting chronology is depicted in Fig. 6.8. Mapping of the distribution of the upper cover bed revealed a relationship to certain paleolake shorelines. Figure 6.8 depicts this

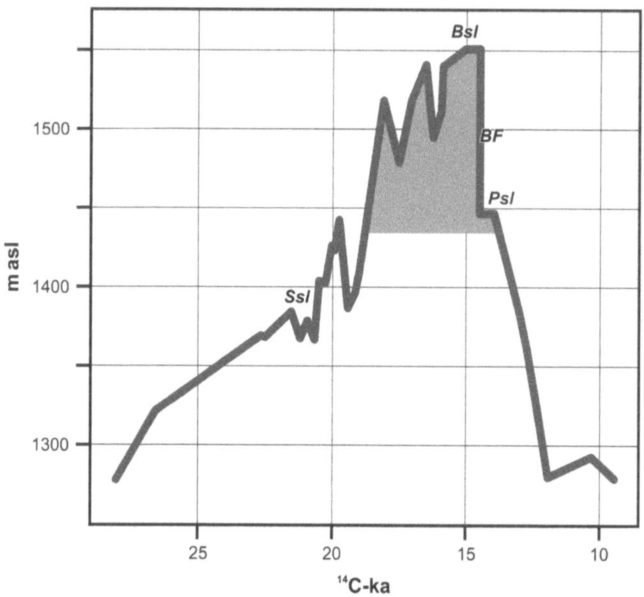

FIGURE 6.8 Levels of Lake Bonneville (*modified from Patrickson et al., 2010*) and distribution of the upper cover bed (shaded). The elevation is given in m above sea level, corrected for isostatic rebound. Ages are ^{14}C-ka BP. The major shorelines in the lake history are: Ssl = Stansbury shoreline, Bsl = Bonneville shoreline, and Psl = Provo shoreline. BF = Bonneville Flood.

distributional pattern: the layer occurs upon shores that the lake had formed before around 14.2 and 13 ka (radiocarbon dates) and which were abandoned at that time. This indicates that the cover bed was last active at around this time. Because the Provo shoreline of Lake Bonneville is essentially the youngest relief unit this layer is related to, it was termed "Provo layer" (Kleber, 1993).

The basic hypothesis underlying the subsequent sections is that processes leaving these distinct traces in the basin were driven by climatic changes. These certainly had similar effects in the areas outside the former lake. In fact, very similar successions of cover beds and soils may be observed throughout the northern Great Basin and rim: deep profiles consist of three members at least, that is, three cover beds with a proprietary soil developed in each of them, which Fig. 6.9 tentatively correlates to the succession in the basin, where there is some age control. The uppermost cover bed, the inferred Provo layer, usually 40–50 cm thick, hosts a soil of intermediate maturity at most; it consists of an ochric (moderately humic) to mollic (rather humic) horizon within a certain elevation range with a weak to moderate imprint by pedogenic carbonate translocated to the underlying horizon. The next two layers, the inferred upper and lower Wyemaha layers, are often thicker. They contain argillic horizons,

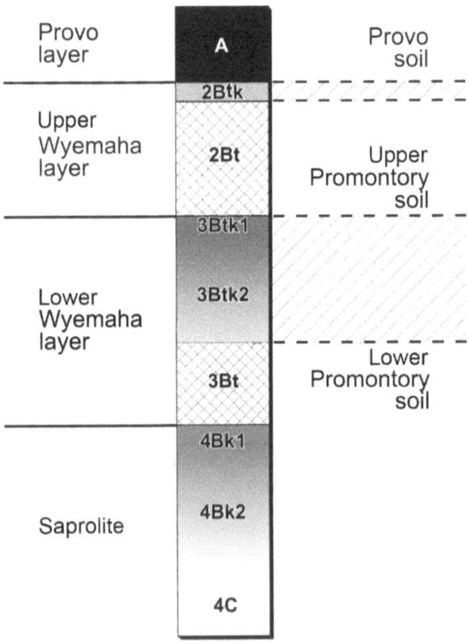

FIGURE 6.9 Idealized soil profile including soil horizons and layers typically found in the Great Basin outside of former lakes. To the right, the inferred pedostratigraphy is shown with those parts hatched where a younger soil is welded onto an older one. To the left, the inferred cover beds are listed. Horizons and their designations may be found in the center column. The profile is not to scale, because particularly its lower parts may vary in thickness, whereas the Provo layer usually is 40–50 cm thick (*modified from Kleber, 1994a*).

succeeded by calcic horizons at low and intermediate elevation (Kleber, 2011). Typically, the calcic horizons affect respective layers and soils beneath, forming compound calcic and argillic horizons (termed "welded ranges" in Fig. 6.9). Furthermore, there is evidence that the chronostratigraphic framework for a threefold layer sequence derived from moraines in the Rocky Mountains is similar to that of the Great Basin (Kleber, 1993, 2001). The threefold layer/soil sequence is interpreted in terms of a sequence stratigraphy (Catuneanu, 2006). This approach is especially encouraged by the remarkable changes which are not reflected in disconformities alone but also in variations in soil formation and other evidences of a considerable hiatus. The entire stratigraphic record is summarized in Table 6.6.

TABLE 6.6 Stratigraphic Framework of the Cover Beds and Soils in and Near the Northern Great Basin (*Reproduced from Kleber, 1999a*)

	Age[a]	Deposit	h (basin)[b]	h (mtns.)[c]	cum.[d]
Provo soil	<13		A-(Bk)I-(C)	Bw-(C)	
Provo layer	14.2–13	Cover bed			+
Last deep-lake cycle (Lake Bonneville)	25–14.2	Lake deposits			
Pinedale glacial advance	20	Till			
Upper Promontory soil	<50		Bt-(Bk)II+-(C)	Bt-(Bk)I-(C)	
Upper Wyemaha layer	50–60	Cover bed			+
Early Wisconsinan lake (Cutler Dam) and glacial advance	75–60	Lake deposits/till			
Lower Promontory soil	<140		Bt-BkII-C	Bt-C	
Lower Wyemaha layer	<140	Cover bed			–
Penultimate deep-lake cycle (Little Valley)/glacial advance (Bull Lake)[e]	≥140	Lake deposits/till			

Relationships to tills are discussed in Kleber (1993, 2001).
[a]Approximate age as discussed in the text.
[b]Characteristic horizon sequence in the northern Great Basin; Roman suffixes indicate typical carbonate morphology (Bachmann and Machette, 1977); (Bk), (C) = only where there is no underlying, older soil horizon. Otherwise Btk and Bt horizons, respectively.
[c]Characteristic horizon sequence in the Rocky Mountains of Wyoming (see Section 6.3.2.1).
[d]Traces of cummingtonite typically observed (+) or absent (–).
[e]Ages *modified from Benson et al. (2004).*

6.2.2.3 *Genesis and Paleoenvironmental Implications of Great Basin Cover Beds*

Similar to Section 2.5, correlation analyses of properties of the cover beds in the Great Basin were conducted by Kleber (1994a). Some results are displayed in Table 6.7.

Possible sources of constituents of a cover bed are the underlying rock (the term regolith is preferred here because massive bedrock was rarely observed), regolith up-slope from the profile, underlying and up-slope occurrences of older cover beds, and eolian (including volcanic) additions. Loess is dominated by a distinct textural range, usually coarse silt, whereas the other sources are usually not.

TABLE 6.7 Correlations Between Layer and Soil Properties (*Modified from Kleber, 1994a*, but with Additional Profiles Included and Profiles on South-Facing Slopes Excluded)

	PL/PS//			UWL/UPS//		LWL/ LPS//
	Regolith	*2C*	*UWL/ UPS*	*Regolith*	*LWL/ LPS*	*Regolith*
Clay	0.66***	0.70***	0.69***	0.50**	0.75***	0.59**
Fine silt	0.54***	0.46*	0.78***	−0.12	0.57**	0.18
Medium silt	0.63***	0.54**	0.68***	0.20	0.26	0.60**
Coarse silt	0.48***	0.62***	0.61***	0.20	0.37	0.41
Fine sand	0.51***	0.54***	0.49**	0.81***	0.64***	0.39
Medium sand	0.56***	0.55**	0.86***	0.61***	0.62***	0.35
Coarse sand	0.49***	0.62***	0.56**	0.60***	0.45*	0.21
Clasts	0.33	0.49*	0.41*	0.12	0.03	0.51*
Carbonate	0.67***	0.89***	0.32*	0.15	0.24	0.86***
Organic carbon	0.24	0.46	−0.01	0.55**	0.52*	0.67**
Thickness			0.21		0.36	
Clasts//coarse silt	−0.05			−0.41*		−0.42*
Clay//coarse silt	0.27*			0.61***		−0.04

A maximum of 85 profiles is included in the calculations. Numbers are correlation coefficients (*r*). Concerning carbonate, the enriched horizons are used for calculation, which are usually developed beneath the layer from which they once evolved. //=is correlated with, PL/PS=inferred Provo layer and soil, UWL/UPS=inferred upper Wyemaha layer and upper Promontory soil, LWL/LPS=inferred lower Wyemaha layer and lower Promontory soil, 2C=the correlations of Provo layer properties with regolith are separated into two columns: the first (headed "regolith") includes all Provo layer–regolith pairs. The second one ("2C") is restricted to cases where the inferred Provo layer is directly overlying a regolith without any intercalated layers. Levels of significance: *=>95%, **=>99%, ***=>99.9%.

If loess admixture plays an important role in a cover bed, particles of that size should be derived from underlying substrates (regolith, older cover beds) to a lesser degree than other textural classes. Coarse silt actually displays this effect as it correlates less with coarse silt in underlying material than all other grain-size classes do (except for clasts). The difference between the Provo and Wyemaha layers in this respect (Table 6.7) is interpreted to result from the influence of loess on the Wyemaha layers being larger than on the Provo layer. The general importance of loess is supported by many heavy-mineral analyses which indicate allochthonous particles in the cover beds (57 analyzed profiles: Kleber, 2001).

There is good evidence that clay is often accompanying airborne silt particles rather than that the clay is exclusively pedogenic (Bronger, 1978; Colman, 1982). The cover-bed clay content correlates significantly with all underlying substrates. Therefore, it is assumed that a significant part of the clay is inherited from underlying material. The correlation of clay and coarse silt should be a measure of the clay that was deposited along with the loess, adhering to silt particles. This is indicated in all but the lower Wyemaha layer and—of course—in the regolith ($r=0.12$). This is considered to result from a noteworthy amount of clay having been deposited along with the loess in some layers (Kleber, 1995). Applegarth and Dahms (2004) also described a statistical correlation between the clay and the coarse-silt content. Similar geogenic differences between different layers are also known from Central European cover beds (see Bargon et al., 1971 and Section 3.3).

It has been proposed that eolian matter preferentially becomes deposited in the interstices between coarse clasts (Blank et al., 1996; Goossens, 1994; McFadden et al., 1987). This would be reflected in a positive correlation between the clast and the coarse-silt content. This is, however, not found in the profiles of this study; instead there is no (Provo layer) or a slightly significant negative correlation (other layers including regolith: $r=-0.30^*$) (note that fines were determined clast-free, so that this relationship is not an analytical artifact). This finding indicates that clasts rather hinder than support eolian admixture during cover-bed dislocation, possibly due to surface sealing or indirectly via vegetation density. The existence of such a relationship implies, however, that a remarkable part of the coarse-silt addition cannot be explained by sole inheritance from any underlying materials (with the possible exception of, again, the Provo layer) but was deposited during or after cover-bed dislocation.

All other textural classes, except for clasts, display some correlation with the underlying substrata. In the Provo layer, correlations are quite similar, regardless whether all regoliths or only regoliths having a direct contact with a Provo layer are considered. This suggests that not the regolith at that distinct site but the up-slope regolith, which is apparently not very different from that at the site on average, is the major source. In summary, underlying substrata largely influence the Provo layer. The upper Wyemaha layer, on the other hand, appears influenced by the underlying lower Wyemaha layer regarding the fine fractions, but from both the lower Wyemaha layer and the regolith concerning sand.

Clasts form the only size fraction that is neither airborne nor correlates with the regolith or underlying cover beds. All correlations indicate that the clast content may be random or controlled by factors not captured in these analyses. Maybe the regolith changes up-slope from the profiles much more rapidly in clast content than in other textural classes (field observations, however, do not support this idea), or the dislocation of coarse material depends on properties not measured, such as relief roughness. Another possible explanation is that clasts once were parts of desert pavements in which clasts become preferentially dislocated (Dietze and Kleber, 2011), so that the clasts might have been transported considerably farther than other textural classes.

The carbonate enrichment of the Provo soil largely correlates with the regolith carbonate, regardless of whether both are in direct contact or not. Thus, this correlation does not reflect carbonate translocation from the Provo soil down into the regolith. It rather indicates that the primary carbonate content of the Provo layer, which was leached from the layer and translocated into the subsoil during pedogenesis, is in a considerable part derived from the (up-slope) regolith instead of being solely eolian, as suggested by most authors (e.g., Birkeland, 1985; Muhs, 1983). In the Bk horizons of the Promontory soils much more carbonate is concentrated (mean values are 13–14% compared to 6% in the Provo soil). There is some correlation of this pedogenic carbonate with the respective underlying substrata, suggesting some inheritance from there, but most of the carbonate cannot be explained this way. Thus, these layers appear influenced by a significant amount of airborne carbonate. Based on observations by Yaalon and Ganor (1973), it is often assumed that permanent carbonate addition through eolian deposition and Ca^{2+} ion fallout onto the soils took place during soil formation (Machette, 1982, Machette, 1985; McFadden and Tinsley, 1985), contributing to the calcic horizons. One may doubt whether the argillic properties could have formed while being permanently disturbed by considerable addition of contradictory (Birkeland, 1999) carbonate (provided exchangeable sodium was absent from the soils, which is denoted by the lack of natric properties as defined by SSS, 2010, even in the basins' Promontory soils). There may or may not have been episodes when such processes took place, besides the major loess pulses (cf. Chadwick and Davis, 1990; Pelletier, 2007) reflected in the cover beds, but there is no clear evidence, since no accretional inclusion of loess material into the layers has been observed in this area (contrary to adjacent typical loess areas: Karlstrom, 1988; McDonald and Busacca, 1990). To assume later (i.e., during most of the time since deposition of the respective cover bed), addition of calcium is not needed to explain the carbonate content of the soils under study. If simply the thickness multiplied by the carbonate content of a calcic horizon is calculated back into the overlying layer, estimating its potential original thickness, and deducting a reasonable primary content of the enriched underlying deposit, a primary carbonate content of the cover beds of up to 35% results. This is a likely range in an area where carbonate-contributing lakes and rocks are in close proximity. Later addition of carbonate must be stressed to explain several generations of calcic horizons in one profile merely as long as the entire soil profile is believed to have formed during one continuous period after all parent materials had been deposited, which does not hold true for the cover beds, which consist of several layers, each bearing a unique soil.

Organic carbon is mainly humus in the Provo soil but is translocated complexed with clay in the Promontory soils. Therefore, there is no correlation between the Provo soil and the Promontory soils or the regoliths, whereas the latter correlate among each other, suggesting translocation into the respective subsoils. Other influences on the organic-carbon content of the Promontory soil than this are not evident from the data.

6.2.2.3.1 Late Wisconsinan Paleoenvironments

There are two major sources of paleoenvironmental information for the Late Pleistocene in the Great Basin: paleobotany (mainly pollen and packrat (*Neotoma* spp.) middens) and geomorphology. The latter are currently limited to a few—usually the coldest—phases during which deposits were formed; thus, a fairly homogeneous picture of Quaternary paleoenvironments could not evolve. Some of this mosaic, however, may be supplemented through indications provided by cover beds and soils.

The full-glacial Late Wisconsinan is characterized by glacier advances from approx. 22 to 14 ka with minimum temperatures and a maximum glacier extent at around 20 ka (Richmond, 1986b). Only during the later part of this time interval, the lakes of the Great Basin stood high (Currey, 1990). Glaciers and lakes apparently reacted in different ways. This points to a change from very cold and dry (Galloway, 1983; Leonard, 1989) to slightly less cold but much moister conditions at around 17 ka (McCoy, 1981). Barnosky (1981) and Spaulding et al. (1983) point out that the seasonality of precipitation was rather different from the present at the expense of summer rainfall. This may have prevented the formation of argillic soils. If, however, there was any soil formation during this phase, not exceeding a depth of some 40 cm, its remnants would have been destroyed by subsequent dislocation of surficial substrates and incorporated into the Provo layer. There is, however, some indication that the Provo soil was preceded by the formation of soils with weakly oxidized horizons (Kleber, 1990).

Accompanied by a rapid lake regression (Oviatt et al., 1992a,b) and by deglaciation (Beget, 1984; Porter et al., 1983), the most intense vegetation shift of the Late Wisconsinan, often even more evident than the later Pleistocene--Holocene transition (Martin and Mehringer, 1965; Wells, 1979), occurred between 14 and 12 ka (Cole, 1990; Davis, 1990; Spaulding, 1990a). Since both glaciers and lakes responded in the same way to this climatic impulse, the change must have been a simultaneous increase and decrease in temperature and precipitation, respectively. The lakes would not have fallen so rapidly down to or below modern levels (Sack, 1989; Scott et al., 1983) without significantly lower precipitation compared to the warmer present conditions. Some paleobotanic data support dry conditions (Madsen and Currey, 1979), whereas other data suggest increased moisture succeeding this shift (Barnosky, 1981; Heusser, 1983; Spaulding, 1990b). Spaulding and Graumlich (1986), Cole (1990), and Betancourt (1990) offered an explanation for this discrepancy by proposing a regime of dominant winter rainfall and summer drought, and displayed evidence for this. This shift is approximately contemporary with the Provo layer. The Provo layer likely formed during its onset while vegetation shifted from full-glacial to late-glacial conditions: the permafrost diminished (see evidence from Wyoming ice-wedges in Section 7.3.4) but the climate still must have been sufficiently cold to permit gelifluction, as indicated by the clast orientation in the cover beds. The short time span between the extinction of the former plant community and the resettlement by a new one should have provided sufficiently barren surfaces to let the layer form (vegetation inertia as proposed by Kleber, 1994a). The solidifying renewed plant cover and/or the warming likely put an end to the dislocation forming the Provo layer that probably was, accordingly, restricted to a short event. This connection between slope activity and vegetation reorganization is based on an assumption regarding the character of the Late Pleistocene vegetation reshape: plant extinction during the transition from full-glacial to late-glacial conditions was not due to competition between former and new species, but to the fact that the plants could no longer survive or rejuvenate—for example, after forest fires—under the climatic conditions at a certain site, while the new vegetation needed some time to resettle that site and to establish a fairly continuous cover. A time lag is thus assumed between forest extinction and resettlement, providing instability to the surfaces. Actually, a vegetation interlude is reported from north of the study area at around 13 ka that was essentially treeless (Mehringer, 1985). Apparently, this lag between extinction and arrival constituted a unique, short-lasting community. Spaulding (1990a) and Cole (1990) display vegetation inertia expressed by an excess of departures over arrivals of species from the southern deserts. In the northern Great Basin, this shift is not readily reflected by the packrat midden record (Betancourt et al., 1990; Thompson, 1990). However, steppe, grassland, or woodland, accompanied by only an insular pattern of subalpine forest, prevailed there at around the time when the Provo layer formed (Betancourt et al., 1990; Mead et al., 1992; Thompson, 1992; Thompson and Mead, 1982). This vegetation could not effectively impede dislocation. The loess influence on this layer may also be well attributed to this phase of lake and glacier retreat, as outlined below.

6.2.2.3.2 Late Illinoian to Mid-Wisconsinan

Aside from the conditions needed for the dislocation of the layers, the environmental causes of loess deposition manifested in the cover beds are similar for all phases concerned: there is a need for barren ground to yield eolian fines and for some surrounding steppe-like vegetation to filter them. As previously shown, the formation of the youngest cover bed was associated with such conditions, indicated by lake retreat and deglaciation. The older cover beds postdate such retreat phases as well. Though they cannot be constrained more precisely, interpretation that rests on these parallelisms will be favored in the following.

The climate of the penultimate glaciation was cold (glacier advances: Richmond, 1986b). Its later part, at least, was also moist (high standing lakes: Scott et al., 1983; paleobotany: Adam, 1988; Baker and Waln, 1985). The subsequent interglacial reached temperatures as high as they are presently, or summers were even somewhat warmer and/or drier (Adam, 1988; Baker, 1983; Baker and Richmond, 1978; Heusser and Heusser, 1990). The vegetation shift at and before its onset, documented in a few continuous pollen records of the western USA (Adam, 1988; Heusser and Heusser, 1990), was drastic. Vegetation inertia (reasoning outlined above) accompanying this rapid climatic shift to interglacial conditions may have provided an environment of relatively scarce overgrowth to form the lower Wyemaha layer. Possibly, dropping lake levels, exposing the beaches to the wind, led to dislocation of eolian fines rather than to a back-and-forth redistribution, which is likely the prevailing mode as long as large deficit areas are absent. Eolian-influx pulses induced by falling lake levels have often been invoked (Chadwick and Davis, 1990; McFadden et al., 1986; Pelletier, 2007; Wells et al., 1985, 1987). The idea that desiccating lakes provided sufficient material to account for all of the loess in the area is, however, inconsistent with the results presented above. Lake-related layers certainly are dominated by particles deflated from the enlarging lake beaches. This, however, does not apply to areas apart from any lake, where other sources must be assumed. The only explanation is the exposition of alluvium to the wind through desiccation or, more likely, incision of creeks; retreating glaciers and thawing permafrost may have caused rivers to cut down due to additional discharge.

The later part of the interglacial ("Eowisconsin") was characterized by oscillations of the temperature (between not as warm as before and not as cold as afterwards) and of effective moisture. This is indicated by pollen spectra (Adam, 1988; Baker and Waln, 1985; Heusser and Heusser, 1990) and by minor glacial advances (Richmond, 1986a). This phase likely provided the environmental conditions needed for the formation of the lower Promontory soil, with Alfisols forming during episodes of increased moisture. Moisture is shown by decalcification and by subsequent deep clay translocation even in the basins. Since stagnic properties (due to stagnation of soil water), if any occur at all, were better expressed in the lower Promontory soil than in the upper one (Kleber, 1992b), either its pedogenesis lasted notably longer, providing an occasion for self-sealing that is characteristic of aging Alfisols, or conditions were moister than during the later argillic pedogenesis.

The early Wisconsinan cold climate is indicated by glacier and lake advances (Colman and Pierce, 1984, 1986; Oviatt et al., 1987; Robison and McCalpin, 1987) and by some pollen data (Adam, 1988; Baker and Waln,

1985). A long series of interstadial–stadial oscillations followed during the middle Wisconsinan, reaching neither the temperature level of the early Wisconsinan nor that of the previous interglacial (Martin and Mehringer, 1965). Within this time span, the upper Wyemaha layer was formed. The only date to provide a minimum age suggestion for this layer is that of 50 ka for the Mt. St. Helens tephra-layer C, heavy-mineral traces of which are inferred to occur in the upper Wyemaha layer (see above). Neither glacier advances have been recorded from this phase (Richmond, 1986a,b, with the only exception of the Yellowstone area: Pierce, 1979) nor high lakes (Currey, 1990). Within this time span, a treeless tundra existed in northern Wyoming (Baker, 1983; Baker and Richmond, 1978). Dry conditions have been reported several times from the southern deserts (Van Devender, 1990a,b). A likely time for the formation of the upper Wyemaha layer is the climatic and accompanying vegetation shift from very cold to warmer and drier at the end of the early Wisconsinan (Adam, 1988; Baker and Waln, 1985), with falling lake levels and possibly down-cutting rivers.

The Upper Promontory soil, as the lower one, needed more moisture to form than is available at present. Moist conditions may have occurred during the mid-Wisconsinan according to Jacobs (1985), who reported enhanced moisture, particularly during summers, in Arizona for the time after approx. 37 ka from pollen evidence.

Cover beds and related soils provide valuable advantages for paleoecological reconstruction. In the ideal case, they give the opportunity to study several phases of geomorphic and pedologic activity within one profile. Through simple statistical methods, it is possible to compare several time-slices throughout a considerably large area regarding some environmental circumstances of their genesis. In the area studied, most phases of the formation of cover beds and soils are different from those phases where detailed information on the ecologic systems, aside from plant life, already exists. Their knowledge contributes to the understanding of environmental dynamics as a reaction to climatic change. The advantages of using cover beds as proxies for paleoenvironmental reconstruction are that they cover time slots with a typical lack of other terrestrial proxies, and that they cover large parts of the landscape, allowing for spatially continuous reconstructions.

6.2.3 Konya Basin of South-Central Turkey

Deep soils are rare in the Konya Basin of South-Central Turkey, probably because of soil erosion throughout much of the Holocene in this area of early human settlement (Mellaart, 1975). Furthermore, deep soils other than soil sediments are absent on surfaces that were formed during and after the last deep-lake cycle of the Konya Basin paleolake (Zech and Cepel, 1977), which lasted until the early Holocene (Erol, 1978). Four of the rare deep soil profiles were used for a study of cover beds and soils on a variety of underlying rock types:

marl, limestone, intermediate andesite, and felsic tephra (Kleber, 1997). Three sites are in mountainous areas of the basin rim with essentially the same rock types cropping out up-slope of the profiles. The fourth is on marl located at the marginal basin bottom, but still more than half a kilometer away from outcrops of differing rock types.

The marl is overlain by a salt-bearing fluviolacustrine gravel, rendering the overlying layers as allochthonous, though they do not significantly differ in texture of the fines fraction from the marl. The heavy-mineral content supports this interpretation, along with differences in gravel content (Kleber, 1997).

Differences in grain size (gravel, sand) and heavy minerals between the horizons are found in all profiles. All have argillic and calcic horizons, and in particular, composite horizons with both properties combined. Typically, the maximum carbonate enrichment, indicated by nodules rather than by carbonate filaments and coatings, occurs in the uppermost parts of these composite horizons. All profiles have similar layer and horizon successions (Fig. 6.10) despite different underlying rock types, indicating similar geomorphic and pedogenic, certainly climate-driven, histories.

Airborne dust as a component of parent materials is indicated by abundant silt in some layers and by the heavy-mineral content (Kleber, 1997). Therefore, the most likely sources of the carbonate in the soils is airborne dust, leached from upper parts of the profiles and precipitated in lower parts. Other possible sources may be excluded for some of the soils: enrichment from ascending or lateral water flow is impossible due to the lack of nearby ground water, except for the marl site; release of Ca^{2+} ions from bedrock by weathering is unlikely to apply to the tephra and andesite sites, as these rocks probably contain insufficient Ca-bearing silicate minerals. The carbonate in the composite horizons must have been deposited after the argillic properties had been formed (see Section 6.2.2.1). The possible exception, translocation in natric horizons, was not observed in these soils, three of which are distant to sodium sources.

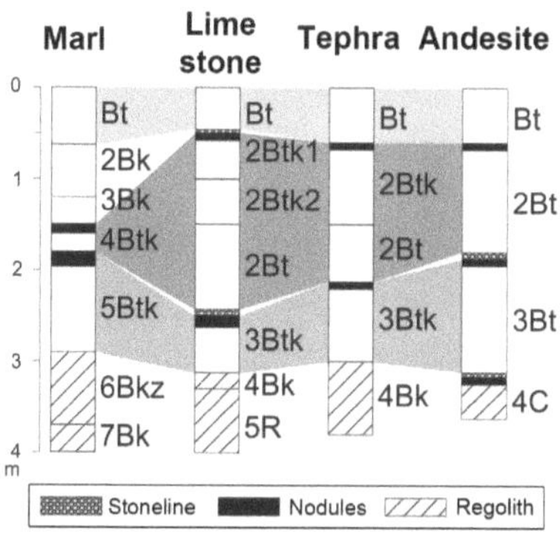

FIGURE 6.10 Correlation of profiles in the Konya area, South-Central Turkey (*reproduced from Kleber, 1997*).

If the carbonate was introduced as a continuous eolian influx, as suggested by Machette (1985) for calcic soils in the south-western USA, the occurrence of several carbonate-enriched horizons at various depths might be explained by repeated climate-driven episodes of varying leaching intensity, each leading to a horizon of carbonate enrichment (see Section 6.2.2.1). Because any intense leaching event would cause the dissolution of older calcic horizons higher in the profiles, the stacked calcic horizons would imply a decreasing trend of leaching intensity with time. This would have happened during the Pleistocene because comparable soils are absent on Holocene surfaces. However, in the lake history of nearby Tuz Gölü, there is no evidence for a comparable climatic trend, as the Pleistocene lake levels are closely spaced vertically (Erol, 1970). Furthermore, calcic horizons occur down to 4 m in the profiles, calling for a hyperthermic soil-temperature regime (McFadden and Tinsley, 1985). The area unlikely has experienced such a climate since the Tertiary (Tzedakis, 2009). The marl profile has a horizon overlain by calcic horizons which is not entirely overprinted by pedogenic carbonate, so that some clay cutans could be recognized if they existed. This fresh, largely unaltered calcareous material identifies the underlying composite horizon with mixed argillic and calcic properties as part of a paleosol, and the overlying layers as younger. Thus, the profile obviously documents at least two different episodes of clay illuviation.

All these soil phenomena may be explained by a multiphase model: three phases of cover-bed formation accompanied by loess admixture were each followed by major pedogenic phases. The latter started with leaching of the carbonate from the loess material, leading to subsoil carbonate enrichment. Decalcification was accompanied and followed by clay translocation. Then, A and E horizons of this soil were probably stripped or incorporated into advancing new deposits, and the remains were buried by the cover bed of the following deposition phase. The subsequent pedogenesis leached this new deposit, and the carbonate was redeposited in the buried soil horizon beneath. All profiles show the most intense carbonate enrichment just in the upper parts of the paleo-argillic horizons because their water permeability may have been decreased by illuvial clay. McFadden and Tinsley (1985) suggested that the permeability for water is the second most important factor after climate to influence the depth of Bk horizons.

6.3 HIGH MOUNTAINS OF THE MID-LATITUDES

6.3.1 European Alps

6.3.1.1 Cover Beds Below Modern Timberline

In the Sisteron Basin, Provence, southern France (Fig. 6.11), up to nine gravel terraces may be mapped along the major rivers. The fifth and sixth merge upstream with moraines of the last glaciation (Kleber, 1997). Thus, the four lower terraces are younger than the Last Glacial Maximum, and the upper ones

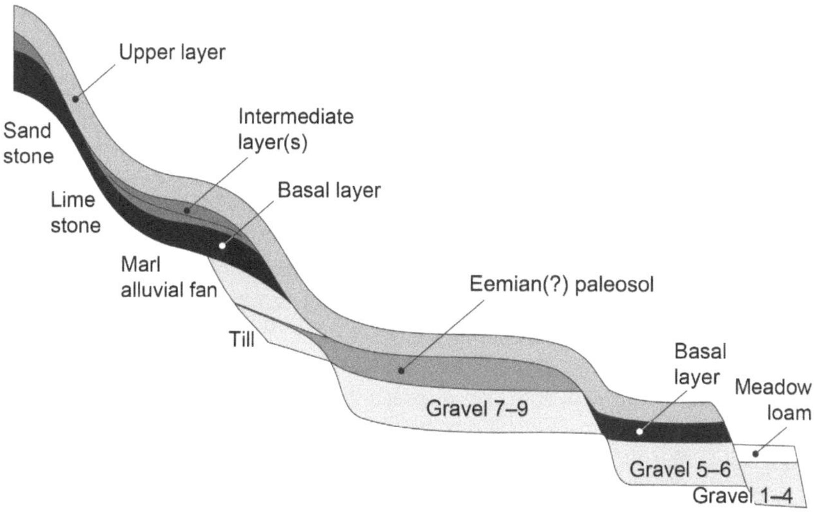

FIGURE 6.11 Distribution of cover beds at different elevations, Sisteron Basin, southern France. This sketch is compiled from various profiles located rather dispersed across the southern basin rim. Therefore, this figure is schematic and not to scale (*reproduced from Kleber, 1997*).

probably correspond to the penultimate and/or earlier glaciations (Kleber, 1997). The basin is surrounded by hogbacks of steeply dipping Mesozoic sedimentary rocks. Summits and steep back slopes are underlain by limestones or, rarely, sandstones. Flattenings and foot slopes are mainly underlain by marls. The highest elevation reached by studied profiles is 1610 m asl. The layers found in most profiles resemble the Central-European cover beds so closely that the terminology of Chapter 2 is applied to them in the following.

On the slopes of the basin rim, at least two layers cover the underlying regolith or bedrock, but three layers occur on some gently inclined upper slopes and on flattenings with slopes of less than some 6°. In all but the deepest layer, local and eolian materials are mixed. The latter is indicated by an increase in coarse silt and by their heavy-mineral content, the former by rock fragments and local grain-size distributions. In the profile almost at the summit of Le Puy, where up-slope influence may essentially be excluded, the upper two layers (Bw and 2Bwg) are significantly richer in heavy minerals than the deeper ones, especially in minerals entirely absent in the regolith. A similarly rich mineral suite is found in the gravel terraces of the Durance Valley, which are potential sources of windblown silt in the area. Rock fragments vary significantly among the cover beds in most profiles with the clasts usually concentrated in the upper layer. The intermediate layers typically contain the largest amount of silt, whereas the basal layers consist of local components only. Typically, the basal layers have high bulk densities, inducing soil-water stagnation even on steeper slopes. Though their down-slope clast orientation is usually not well expressed compared to the upper layer, the basal layers are probably periglacial in origin as well: at higher elevations, they often contain rock fragments, which are entirely weathered and easily smear between the fingers, side by side with fresh fragments of the same rock type and of the same size. This may be explained by dislocation of previously weathered fragments in a frozen condition. Apart from features connected with soil-water stagnation, pedogenesis is restricted to the upper layer and stops abruptly at its lower boundary. Only weak carbonate translocation into deeper parts of the profiles may be observed. In the upper layer, cambic[8] properties are developed. Argillic properties, which might be expected in subsoils rich in loess material, are absent, probably because decalcification as a necessary predecessor was insufficient even in the elevated soils.

On river terraces, three types of soil profiles are present. On the uppermost terraces (7–9), which are probably as old or older than the penultimate glaciation, the most prominent soil feature is a buried horizon with a reddish-brown color, which is rich in clay (about 40%). A similarly mature, clay-rich soil, though less red in color, is exposed on the foot slope of the northern basin rim. It overlies till and is buried by thick fan deposits. These soils result from more intense pedogenesis than the surface soils; thus they are assumed to date from the Eemian, when Mediterranean-type pedogenesis probably reached up into this Alpine foot zone (Kleber, 1997). Fresh rock fragments within these mature horizons indicate that they were not formed entirely *in situ* but were transported after their formation; thus, they are deduced to be soil sediments. On terraces 7–9, this buried soil is overlain by a cover bed, which is brown and contains less weathered terrace material, reflecting reduced weathering intensity. Its heavy-mineral suite is similar to the underlying gravel but different from the Eemian (?) soil due presumably to eolian influx. On the lowest terraces, 1–4, there are no cover beds, and fluvial overbank deposits cover the entire topography, with groundwater, peat, or reworked organic matter dominating the soil properties.

Altogether, the comparison of these cover-bed successions with the Central European ones is hampered by the more southerly position of the area under study, responsible for a type of weathering with its residues admixed to cover beds, which is rarely found in Central Europe, though these residues have many properties in common with the brown loams of Section 2.8.3. At somewhat higher elevations, the cover-bed successions are much closer to the Central European ones. The different pedogenic role of intermediate layers is attributed to the limestone content of the parent materials, hindering clay translocation by frequent release of Ca^{2+} ions, rather than to elevation. Similar observations supporting many resemblances between Central European successions, especially the tri-partite succession, and those in the montane and subalpine altitudinal belts of the Alps below timberline have been reported, especially from the peripheral northern Alps (Artmann and Völkel, 1999). However, the major

8. Soil horizon characterized by brown to reddish color due to oxidization of iron, and often by an increase in clay content.

difference is that these authors could not identify an eolian contribution to the cover beds using clay mineralogy, which finding prevents parallelizing their layer successions to the other successions. Contrary to this source, Stahr (2000) described, from an area close to that studied by Artmann and Völkel (1999), a deposit that resembles all properties of the upper layer (see Chapter 2), including a remarkable admixture of silty material, assumed to be eolian, though he could not exclude derivation of the silt from local rocks through weathering.

Vonlanthen (2000) and Veit et al. (2002) studied cover beds in the Wallis area of the central Swiss Alps throughout an altitudinal range from 1400 to 2200 m asl. They found one ubiquitous cover bed with all properties of the Central-European upper layer, even regarding its high silt content, indicating an eolian contribution. It overlies till of the Daun stade (Older or Oldest Dryas) in places, constricting its age to the Late Glacial because of the remarkably higher elevation reached by forests at the onset of the Holocene (Burga and Perret, 1998). Older cover beds beneath this layer were not identified in these studies, probably because of the glacial overprint that affected the area before the Oldest Dryas. Throughout their study area, several layers overlie this cover bed and occasionally another local one with a patchy distribution and with an uncertain age and genesis. These are attributed to fluvial and gravitational processes during episodes of environmental deterioration during the Holocene (Vonlanthen, 2000). A similar deposit was described by Artmann and Völkel (1999) and interpreted in the same way.

6.3.1.2 Cover Beds Above Timberline

In alpine and subnival relief above timberline, geomorphic processes other than gelifluction, such as rock fall, avalanching, and mud flowage, were and still are widespread and active, competing with the formation of cover beds. Another restriction to the distribution of cover beds is the widespread occurrence of blank rock. Aside from those areas, however, cover beds are ubiquitous in the high alpine landscape (Veit, 1988, 2002; Veit et al., 2002). Eolian deposition took place even in very high parts of the European Alps (Hüttl, 1999; Küfmann, 2003; Veit, 2002) and may be ongoing (Gruber, 1980; Küfmann, 2003). Most of this loess is assumed to have been deposited during the Younger Dryas because it overlies moraines of the Daun stade, which date into the Older or Oldest Dryas (Veit et al., 2002). However, systematic research on the eolian contribution to the composition of cover beds has not yet been carried out.

In the periglacial (subnival) altitudinal belt of the Alps, cover beds are actually forming at present (Emmerich, 1990; Gamper, 1986; Jaesche, 1999; Jaesche et al., 2002, 2003; Krummenacher et al., 1998; Stocker, 1984; Veit et al., 1995): Measured dislocations of up to 1 m a^{-1} of approx. 40-cm thick layers may generate slipping planes where dislocated material overlies substrates retaining an essentially fixed position (Fig. 2.15). The resulting

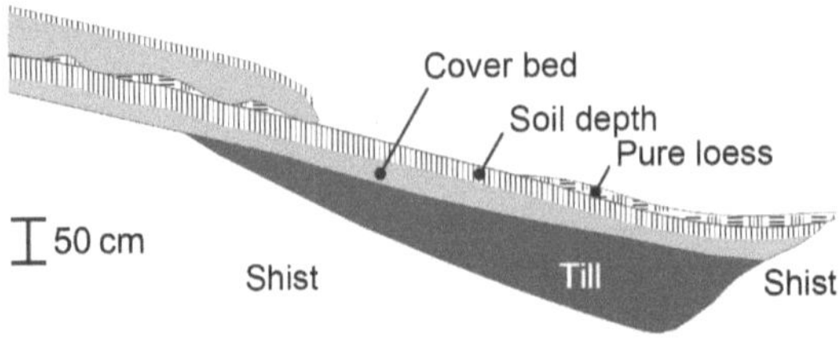

FIGURE 6.12 Overriding of a stable surface by a gelifluction lobe in the Schober Group, Austria. Note the different depths of soils evolved from the younger and the older lobe material (*modified from Veit et al., 2002*).

FIGURE 6.13 Gelifluction lobe overriding an older cover bed. The disconformity between the two substrata is indicated by the grayish line, which represents a soil buried by overriding (*photo A. Kleber*). (For color version of this figure, the reader is referred to the online version of this chapter.)

disconformity is often easily recognized in the field because of paleosols overridden by advancing gelifluction lobes (Figs. 6.12, 6.13). These paleosols' maturity varies remarkably. Some are thin Entisols,[9] but Mollisols,[10] Inceptisols,[11] and several decimeters thick Spodosols[12] occur as well, which all may appear side by side, closely spaced at the surface (Vonlanthen, 2000). Contrary to the much older cover beds of the subdued mountains of Central Europe, buried soils have formed, probably during warm pedogenic episodes during the Holocene, and have persisted to the present. This provides an intriguing

9. Immature soil type characterized by slight to absent weathering and thin to discontinuous humus accumulation.
10. Soil type characterized by thick humus accumulation in the topsoil.
11. Soil type characterized by a cambic (Bw) subsurface horizon with evident oxidization of iron-bearing minerals.
12. Soil type characterized by spodic (Bh, Bs) subsurface horizons, enriched by translocated iron, aluminum, and humus, usually overlain by a depleted E horizon.

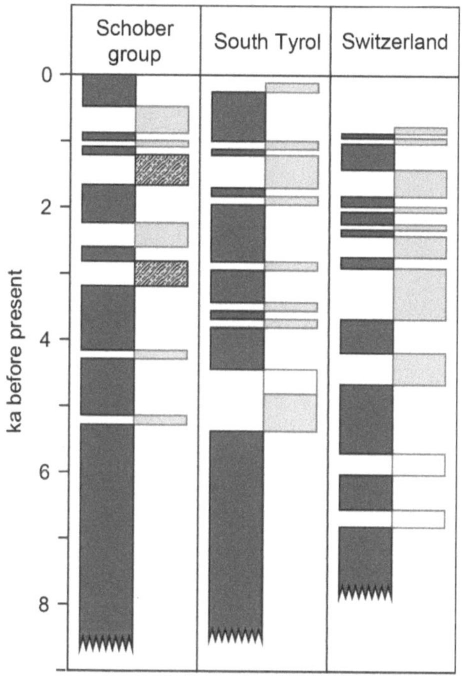

FIGURE 6.14 Phases of pedogenesis (brown) and gelifluction (blue) in three mountain ranges of the European Alps. Data sources: East Tyrol, Austria: Veit (1988); South Tyrol, Italy: Steinmann (1978); various mountain ranges, mainly Albula Pass, Switzerland: Gamper (1986). Zigzag: unknown onset of phase; hatched: activity of rock glaciers in addition to gelifluction lobes indicates remarkable expansion of permafrost; open square: uncertain (*modified from Veit et al., 2002*). (For interpretation of the references to color in this figure legend, the reader is referred to the online version of this chapter.)

opportunity for getting age handles, because soil organic matter may be suitable for radiocarbon dating.

Such dating was pursued by Steinmann (1978), Gamper (1986), Veit (1988, 1993), and Veit and Höfner (1993) in the alpine altitudinal belt, resulting in a puzzling picture of activity of gelifluction interrupted by various soil-forming episodes (Fig. 6.14). Their findings agree regarding rather intense pedogenesis, which lasted until about 5 ka, and in two phases of intense gelifluction at around 3 and 1.5 ka, interrupted by phases when comparably immature soils were formed. Other phases do correspond badly, which may be due to regional differences, different data densities in the three studies, errors induced by the dating method, or overriding is a continuous process acting synchronous to pedogenesis so that the exact timing of a gelifluction lobe to bury some soil may be just stochastic. As a consequence of these processes, there may be several layers of Holocene cover beds overlying one another, that is, layer

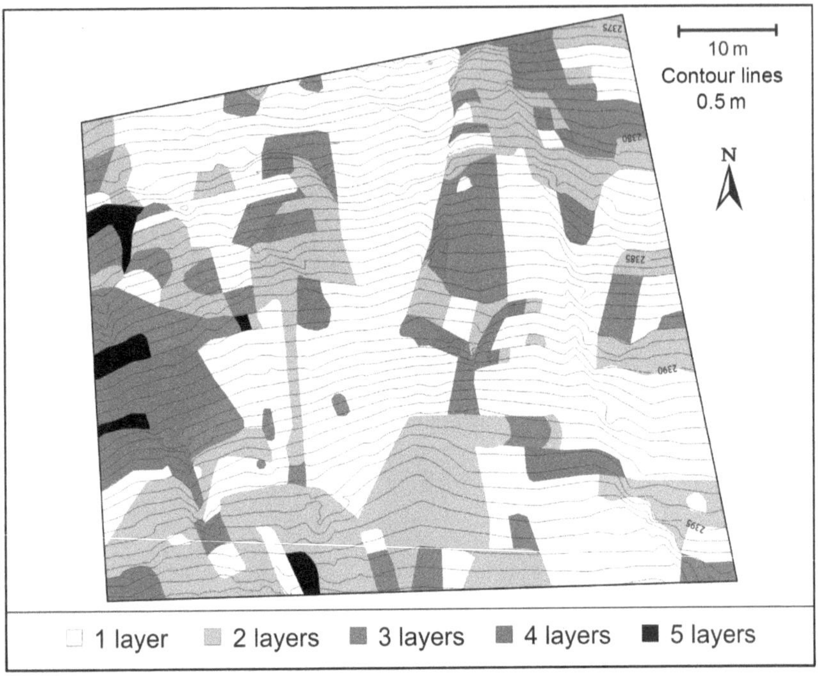

FIGURE 6.15 Number of cover beds mapped at small scale via coring auger at Furkapass, Oberwald, Switzerland (*modified from Vonlanten, 2000*). (For color version of this figure, the reader is referred to the online version of this chapter.)

successions with different numbers of members may occur side by side in a rather irregular mosaic (Fig. 6.15).

6.3.2 Mountains of the Western USA

This section follows a concept similar to Section 6.3.1 with case studies from the North-American Rocky Mountains and the La Sal Mountains, a high-mountain range sitting upon the Colorado Plateau. The first set of studies reaches intermediate heights in these mountain ranges approximately up to timberline, whereas the later one pursues cover beds at high altitudes of the Rocky Mountains.

6.3.2.1 *Cover Beds Below Timberline*

In the northern Great Basin, the adjacent Rocky Mountains, and the La Sal Mountains, Colorado Plateau, there is a recurring record of three major episodes when loess-mixed cover beds were laid down, each with a subsequent pedogenic phase. Their relationships to Great Basin paleolake phenomena constrain their ages (Section 6.2.2.1). An attempt to interpret surface-near cover-bed/

(paleo)soil successions in terms of sequence stratigraphy has been corroborated by findings upon moraines in the Rocky Mountains (Kleber, 1993).

Based upon relative-age criteria (Birkeland et al., 1979), and recently on numerical ages, a chronology of the glacial landforms in the Rocky Mountains has been established: the most prominent moraines are the Pinedale[13] (15–23 ka, from cosmogenic nuclides: Briner, 2009; Marchetti et al., 2005; Thackray, 2008) and the Bull Lake (about 150 ka from cosmogenic nuclides: Licciardi and Pierce, 2008 and from U/Th-series dating: Sharp et al., 2003), which ages approximately agree with datings from the Colorado Plateau (Marchetti et al., 2007). Furthermore, there is evidence of another glacial advance of early Wisconsinan age, though less well preserved and only in extremely rare cases expressed as a moraine relief (Bursik and Gillespie, 1993; Colman and Pierce, 1981; Forman et al., 1993; Gillespie and Molnar, 1995; Phillips et al., 1990; Richmond, 1986a,b; Sharp et al., 2003; see also the review by Pierce, 2003).

Cover beds were studied by Kleber (1993) on Pinedale and Bull Lake moraines in the Medicine Bow Mountains, Wyoming (see for ages McCahon and Munn, 1991; McCallum, 1962; Mears, 2001). The post-Pinedale soil is developed in a 45-cm-thick cover bed with coarse clasts being considerably more frequent than in the underlying till; they are oriented with their long axes parallel to the slope, a feature absent in the till. Also, the amount of coarse silt and of fine sand is larger than in the till. Its soil is immature; it has a pale Bw horizon restricted to the cover bed. The heavy-mineral content of the cover bed and till slightly differ in that the cover bed is mineralogically more diversified with regard to trace minerals that are entirely absent in the till. The post-Bull Lake soil is deeper and more mature. It starts from the top with an upper section containing many clasts oriented parallel to the slope, rich in coarse silt, and mineralogically different from subsequent layers. The soil developed in this upper section has properties close to those of the post-Pinedale soil. Taking the similar clast fabric into account, these layers likely belong to the same group of deposits as the one on the Pinedale moraine; if true, this upper section has an inferred post-Pinedale age. Bt horizons follow down to 240 cm. At 120 cm, they are separated by a major disconformity indicated by changing clast contents and heavy-mineral compositions. The soil horizon that starts abruptly just below this disconformity is notably enriched with carbonate. This indicates that the two soil horizons belong to different soil-forming episodes, welding the upper to the lower horizon. The heavy-mineral content in the various layers supports the stratification. This disconformity also separates layers containing traces of cummingtonite, presumed to be a 50 ka old stratigraphic marker (Section 6.2.2.3). These findings corroborate the sequence-stratigraphic approach based upon interference with the lake history in the basin, because the stratigraphic control

13. See Table 6.1 for an stratigraphic overview.

provided by the moraines supports classifying the upper cover bed as younger than the Last Glacial Maximum, and the subsequent two as younger than the penultimate glaciation.

Yet, there are no own studies focusing on cover beds upon early Wisconsinan moraines. We know of just two studies with detailed soil descriptions from moraines of that age (Colman and Pierce, 1986; Dahms, 2004). Thereafter, early Wisconsinan moraines carry argillic soils, indicating that these formed after the early Wisconsinan; however, argillic soils are mostly thicker and more complexly stratified on the Bull Lake moraines. This supports a classification of the major disconformity separating the lower two cover beds as early Wisconsinan.

There are a few studies to corroborate the inferred geomorphic activity connected with the cover beds (older studies are reviewed in Kleber, 2001). Hall (1999) deduced from absent age relationships in surface horizons, contrary to deeper horizons, truncation, and deposition of fresh material. Schaller et al. (2002) inferred from cosmogenic-^{10}Be concentrations that overriding has occurred of a previously eroded, mixed deposit by a fresh, about 30-cm thick layer.

Pursuing the sequence-stratigraphic approach, Table 6.8 attempts a correlation of elevation with sediment and soil properties separately for the three cover beds and underlying substrates (regolith).

TABLE 6.8 Correlations Between Elevation and Layer/Soil Properties (*Modified from Kleber, 1994a, 2011,* Including Additional Profiles and Excluding Profiles on South-Facing Slopes)

	1st Cover bed		2nd Cover bed		3rd Cover bed		Regolith	
	n	r	n	r	n	r	n	r
Clay	66	0.01	41	0.38*	30	0.47**	70	0.07
Fine silt	66	0.12	41	0.44**	30	0.40*	70	−0.20
Medium silt	66	0.35**	41	0.44**	30	−0.11	70	0.00
Coarse silt	66	0.27	41	0.25	30	0.00	70	0.09
Fine sand	66	−0.31*	41	−0.12	30	0.00	70	−0.09
Medium sand	66	−0.07	41	−0.14	30	−0.11	70	−0.08
Coarse sand	66	−0.06	41	0.07	30	−0.05	70	0.04
Clasts	64	0.32*	39	−0.25	28	−0.24	67	0.12
Thickness	66	0.01	42	0.38*	31	0.11	–	–
Organic carbon	56	0.34*	39	0.23	28	−0.02	66	0.24

n=number of cases; r=correlation coefficient; levels of significance: *=>95%, **=>99%.

Whereas all textural classes are randomly distributed in the regolith, the fine fractions increase with elevation in the two lower cover beds. This is interpreted by frost-induced weathering, which increases with elevation. Since particularly the 3rd layer is often so deeply buried that any influence of modern frost-shattering appears unlikely, it may be presumed that the weathering has taken effect before the layer was buried by the subsequent cover bed. The same holds true for the 2nd layer, since the overlying 1st layer appears almost unaffected by this elevation-dependent effect, indicating that the time during which it was formed and exposed to periglacial conditions was insufficient to develop noticeable weathering results.

Other layer properties are not significantly influenced by elevation. Inheritance of material from underlying older layers plays a greater role for the layers' particle sizes than climatic influences such as elevation do (see Section 6.2.2.3). Coarse silt is enriched in cover beds compared to the regolith—the median coarse-silt content is 13%, 12%, and 16% in the 1st to 3rd layers, but only 9% in the regolith (Kleber, 1994a). From a large number of heavy-mineral analyses (Kleber, 2001), it is known that the cover beds are enriched in allochthonous particles, interpreted as admixed loess. The layers are broadly contemporary to episodes of receding lakes in the Great Basin; one might, therefore, expect the major loess fraction, coarse silt, to decrease significantly with distance from the basin, that is, with elevation. Consequently, the lack of an elevational trend of coarse silt is surprising: local loess sources in the mountains must be assumed to have substituted the decreasing basin-derived loess. This may cast some light on the paleoenvironmental conditions in the Rocky Mountains: barren ground to yield fines ready for deflation must have existed. Because the slopes as a whole apparently acted as loess sinks rather than sources, blow-out from alluvium is the most likely local loess source in the mountains (Kleber, 2001), possibly over several stages of reworking of older loesses (see Van Loon, 2006).

Soils typically form in strong concordance with substrate properties, especially with disconformities. Figure 6.16 compiles the altitudinal ranges of soil horizons separately for the three cover beds and for different study areas. Each area has, aside from individual characteristics, resemblances with the other areas.

Within distinct elevational ranges, the Provo soil contains rather thick mollic and carbonate-enriched horizons, but with the latter often too thin to meet the criteria of calcic horizons. The upper limit of the latter is somewhat higher elevated than the lower limit of the former so that both overlap. Below the mollic in elevation, a slightly developed cambic horizon combined with a carbonate-enriched horizon is developed in areas where the profiles under study reach sufficiently low altitudes. The main factor to determine the upper boundary of mollic horizons is assumed to be soil moisture (particularly during summers), that is, as soon as soil moisture is sufficiently high, forest is able to establish, supporting the formation of cambic rather than mollic horizons. The lower limit

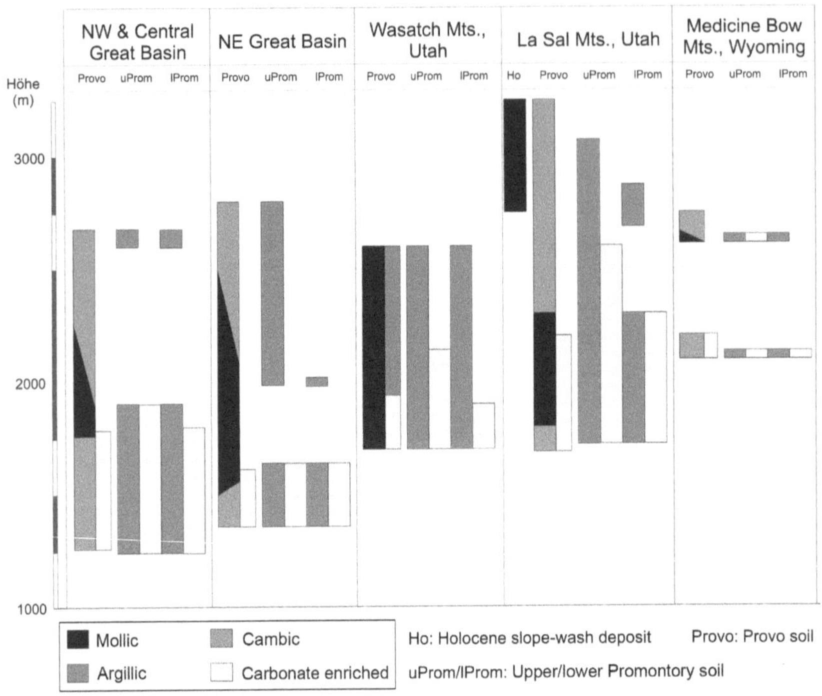

FIGURE 6.16 Comparison of soil horizons depending on altitude in different areas of the western USA (main columns), separately identified as to the layer in which or below of which (regarding carbonate-enriched horizons) they are developed (secondary columns). A slope-wash deposit in the La Sal Mountains is not further mentioned in this book. (For color version of this figure, the reader is referred to the online version of this chapter.)

of the mollic horizons is supposedly connected with soil moisture as well, because there is insufficient organic matter to accumulate in soils when it gets too dry. The temperature also influences this limit, since a low temperature may prevent the mineralization of organic matter.

Both Promontory soils are argillic throughout, for the most part accompanied by calcic horizons. The calcic horizon of the upper Promontory soil is the most elevated horizon of carbonate enrichment; in some profiles, therefore, it is the only carbonate-enriched horizon, since others do not reach that high. This may also be due to soil moisture, which, in case it is too high, leaches carbonate to groundwater rather than letting it precipitate in the subsoil.

All these limits differ in height in each mountain range concerned: in the La Sal Mountains, the upper limit of mollic horizons in the Provo soil is low, whereas their lower limit is elevated above the average. Thus, the altitudinal belt of mollic horizons is thinnest there. However, carbonate-enriched horizons of all soils reach highest in this mountain range. This indicates a steep soil-moisture gradient, which is in accordance with modern climate: the La Sal

Mountains have a steeper gradient of precipitation and of temperature than other areas concerned (Ruffner, 1978).

In the Wasatch Mountains transect, clay translocation from the Provo soil into the underlying substratum is found. This is unique among the areas studied but may be not singular in the Rocky Mountains (Kleber, 2001). Both mollic and argillic properties likely do not form concurrently; rather, these soils indicate a shift to enhanced soil moisture. Furthermore, the lower Promontory soil in this area is often characterized by properties of a seasonally perched water table, pointing at some higher soil moisture compared to other areas. Today, precipitation is relatively high in this area; its annual distribution is characterized by a rather short dry season (cf. station data in Ruffner, 1978).

Unique to the Wasatch Front is a pattern of cambic and mollic horizons in the Provo soil above 2100 m asl. Though this is not combined with stagnic properties, soil moisture may also have been the main factor influencing this pattern, since it provides a major limitation for the development of mollic horizons. A likely explanation of this soil pattern is that precipitation was near the upper limit to form mollic properties; where additional factors such as an impermeable substratum affected the soil moisture balance, the formation of a mollic horizon was impeded. This leads to the conclusion that precipitation at the Wasatch Front was rather high during the formation of the Provo soil above 2100 m asl, permitting local forest growth where the soil moisture was additionally enhanced by edaphic means. The Wasatch Front is the major barrier for winds that have passed over the Great Basin—accordingly, parts of this region receive Utah's precipitation maximum today (Ruffner, 1978).

Similar variations of soils may be observed in the Central Great Basin. The upper limits of all horizons, as far as the studied sites reach sufficiently high, are more elevated than in the east. However, there is much scatter in this area, probably due to locally varying rain-shadow effects.

As previously mentioned, the elevational ranges of humus-rich mollic horizons vary regionally. This lets one expect quite a scattered cloud of data points when plotting organic carbon versus elevation for all sites at once. Yet, Kleber (1995) showed a trend of Provo-soil organic carbon to increase with elevation until around 2000–2500 m asl, to decrease again above this level. This is in agreement with the average limits of the mollic horizons. If, for instance, the data from the La Sal Mountains are computed separately for the two point clouds, there is a positive correlation between the organic-carbon content and elevation up to 2250 m asl ($r = 0.85^{***}$) and a negative one for the more elevated profiles ($r = -0.61^{**}$).

Soils in the western USA often contain one or more carbonate-enriched zones. Most of the literature assumes that the source of pedogenic calcium is continuous eolian influx as calcareous dust (Birkeland, 1990; Capo and Chadwick, 1999; Dahms, 1993; Machette, 1985; Reheis et al., 1992, 1995) onto a geomorphologically inactive, stable landforms (Birkeland et al., 2003). The eolian contribution to the soils is indicated by the silt abundance in topsoils

(Colman, 1982; Dahms, 1993; Kleber, 1993; Muhs and Benedict, 2006; though Munn and Spackman, 1990, came to a divergent conclusion) and by the carbonate that has engulfed originally carbonate-free substrates (McCrea, 1950; McFadden and Tinsley, 1985; Reheis et al., 1995) or substrates which had been leached before they were enriched in carbonate (Kleber, 2000; Machette, 1985; Reheis, 1987; Reheis et al., 1992; see Section 6.2.2.1). Properties of soils in the Great Basin near lake basins are explained by pulses of eolian deposition triggered by lake desiccation rather than by a steady influx (Chadwick and Davis, 1990; Chadwick et al., 1995; McFadden et al., 1998). Furthermore, bifurcating Great Basin paleosols with multiple calcic horizons imply cycles of colluviation and accompanying or subsequent loess admixture (Kleber, 2000).

Recurring episodes of variable leaching intensity interfering with phases of steady influx, each resulting in a proprietary carbonate-enriched horizon, (Birkeland, 1984; McFadden, 1988), are assumed to explain the occurrence of several carbonate-enriched zones (as defined by Kleber, 2011) at various depths. Intense leaching would have dissolved older carbonate-enriched horizons higher in the soils (Pendall et al., 1994), implying a decreasing trend of leaching intensities with time, assumed to be glacial and postglacial in time (Birkeland, 1985; Mayer et al., 1988). If this holds true, one may wonder about the paleoclimatic framework of carbonate horizons at depths of several meters and manifold sequences of such horizons.

Paleoclimatic trends, for example, a moisture increase driven by elevation, must have existed during the proposed leaching episodes similar as they occur at present. Thus, carbonate contents are expected to decrease with elevation (Gile, 1977; Machette, 1985). The depth to the carbonate-enriched horizon of any particular leaching episode should increase accordingly, because of the close relationship of this depth with moisture (Chadwick et al., 1995; Gile, 1977; McFadden and Tinsley, 1985; Retallack, 2005). Here, we test for elevation-related trends in carbonate-enriched horizons to identify the importance of slope deposits for the carbonate enrichment.

Kleber (2011) gave three examples of profiles at different altitudes with several carbonate-enriched horizons (Table 6.9). Profile 1 has a cambic horizon (Bw) overlying a carbonate-enriched argillic horizon (2Btbk). Carbonate enrichment decreases downward (2Btb), to increase again beneath (3Bkb). Within the 4Btkb, it decreases again (field evidence from HCl tests) to reach a third maximum at greater depth (6Bkb). Profile 2 from the western foot zone of Wheeler Peek has three zones of maximum carbonate enrichment in the 2Bk, 4Btkb, and 6Bkb horizons. Profile 3, upon a fluvial terrace (see Section 7.3.2) has carbonate-content peaks in the 2Btbk1, 3Bkb, and 5Bkb horizons.

These soils have argillic properties at least below the first and the second carbonate-enriched horizon, Profiles 2 and 3 also in the deepest horizon, indicating that parts of the soils were leached prior to carbonate enrichment, because carbonate enrichment and clay illuviation usually cannot form in a particular soil horizon simultaneously (see Section 6.2.2.1).

TABLE 6.9 Properties of Example Profiles (*Modified from Kleber, 2011*)

	Depth	Clay	fSi	mSi	cSi	fS	mS	cS	fcl	ccl	pH	Carb.	C_{org}
Horizon	*(cm)*					*(%)*					–	*(g kg^{-1})*	
Profile 1. Pyramid Lake, Nevada, 1250 m asl, on lake gravel													
Bw	−60	9.5	2.6	5.9	12.0	17.5	16.5	36.0	19.2	15	7.5	2.3	3.8
2Btbk	−105	21.4	2.2	3.9	11.8	20.5	20.7	19.6	4.0	7	7.5	13.2	3.8
3Btbk	−135	9.3	2.4	3.1	10.3	22.5	25.1	27.3	9.0	7	7.1	1.8	2.1
3Bkb	−185	5.0	2.3	3.6	11.3	21.5	24.9	31.4	11.0	2	7.9	35.7	2.6
4Btkb	−215	10.6	6.0	15.1	29.2	21.9	7.6	9.5	1.0	7	8.7	135.6	1.8
5Bkb	−225	0.0	0.0	0.0	0.0	0.0	0.0	0.0	0.0	100	n.d.	5*	n.d.
6Bkb	−240	15.4	4.4	10.3	31.7	29.9	5.9	2.4	1.0	0	9.2	37.4	1.3
7Bkb	−330+	5.3	0.7	2.2	9.8	17.5	19.4	45.2	32.6	20	7.9	66.2	2.0
Profile 2. Spring Valley, Nevada, 1770 m asl, on alluvial gravel													
Bw	−40	8.8	4.9	7.4	20.3	15.3	16.9	26.4	25.2	2	7.7	15.7	5.7
2Bk	−110	3.4	1.4	2.1	7.3	8.2	28.9	48.6	50.8	30	7.5	34.4	8.2
3Btb	−140	12.2	4.6	4.4	13.7	22.7	20.1	22.2	22.8	5	7.2	1.6	5.4
3Bkb	−210	8.4	3.6	4.0	8.3	11.2	25.4	39.1	30.9	7	7.6	130.0	4.5
4Btkb	−240	29.6	7.0	5.3	10.8	20.9	16.5	9.9	6.7	3	7.5	135.5	7.6

Continued

TABLE 6.9 Properties of Example Profiles (*Modified from Kleber, 2011*)—Cont'd

	Depth	Clay	fSi	mSi	cSi	fS	mS	cS	fcl	ccl	pH	Carb.	C_{org}
Horizon	*(cm)*					*(%)*					–	*(g kg^{-1})*	
5Btb	−260	20.6	5.2	4.0	11.1	19.7	20.9	18.5	33.4	15	7.6	62.9	4.6
6Bkb	−295	15.6	6.3	8.7	7.4	20.9	21.6	19.4	5.0	5	7.7	498.0	5.3
7Btb	−315	8.3	3.4	2.8	7.8	29.8	32.3	15.7	11.6	5	7.3	2.6	4.9
Profile 3. La Sal Mountains, Utah, 2130 m asl, on older cover beds													
A	−10	14.4	3.7	8.9	26.6	19.6	11.2	15.5	9.1	10	7.7	29.5	18.3
Bw	−40	14.9	3.3	8.9	16.7	23.5	14.4	18.4	8.6	15	7.6	150	14.2
2Btbk1	−65	27.0	5.4	8.6	22.2	23.1	7.3	6.4	0.9	2	7.7	171	5.4
2Btbk2	−130	21.4	5.9	8.0	26.4	24.7	7.5	6.0	0.0	0	8.0	144	2.4
2Bkb	−155	21.4	5.9	8.0	26.4	24.7	7.5	6.0	0.0	2	8.0	191	1.5
3Btb	−180	18.9	5.9	8.4	19.9	21.2	12.7	13.0	4.8	10	8.1	68.8	2.3
4Bkb	−225	24.3	6.5	11.2	4.5	16.4	22.2	14.8	6.1	25	7.7	185.0	9.4
5Btkb	−245	18.8	3.3	6.5	23.3	22.0	10.9	15.3	6.5	5	8.1	68.9	1.6

Horizon = designations according to SSS (2010); depth = measured to lower horizon boundary; + = boundary not reached. Grain-size classes as in Table 6.4; carb. = total carbonate content; C_{org} = organic-carbon content; n.d. = not determined.

TABLE 6.10 Correlations Between Elevation, Soil Carbonate, and Layer Properties

	Provo soil/ cover bed		Upper Promontory soil/cover bed		Lower Promontory soil/cover bed	
	n	r	n	r	n	r
Carb.//elevation	42	−0.30*	31	−0.59***	20	−0.47*
Depth//elevation	42	0.05	31	0.06	20	−0.24
Depth//carb.	42	0.10	31	−0.16	20	0.28
Carb.//ccl	42	−0.17	31	0.59***	20	0.58**
Carb.//clay	42	−0.13	31	−0.49***	20	−0.19
Carb.//PM carb.	34	0.88***	25	0.09	11	0.32
Carb.//coarse silt	42	−0.01	31	0.11	20	−0.27

Note that the carbonate-enriched horizon of a particular soil is usually located just beneath the associated cover bed (*modified from Kleber, 2011*). //=is correlated with; *n*=number of cases; *r*=correlation coefficient; levels of significance: *=>95%, **=>99%, ***=>99.9%; depth=depth from modern surface to upper boundary of carbonate-enriched zone; carb.=carbonate content; ccl=coarse clasts; PM=parent material.

The negative correlation between elevation and carbonate content is significant for all three carbonate-enriched zones up to approx. 2850 m asl (Table 6.10), though to different degrees, probably due to the larger effects of inheritance for the Provo layer (Table 6.7). This indicates the existence of a paleoclimatic elevational trend throughout the studied areas. This result is trifling; it mainly is a test whether the selection of sites was sophisticated enough to establish such a relationship and to demonstrate that correlation analyses allow determining this kind of trend.

The steady-influx/episodic-leaching hypothesis implies strengthened leaching with climatic gradients, in particular with elevation, resulting in greater depths to the carbonate-enriched horizons. However, there is an insignificant correlation between elevation and depth to carbonate enrichment. Rather, the depth to the first carbonate-enriched horizon is essentially not scattered throughout all elevational ranges. Depths to deeper carbonate-enriched horizons vary more, but uncorrelated with elevation. Rather it is controlled by the thickness of the parent material, that is, by the variations in cover-bed thicknesses, because the carbonate content almost always peaks just beneath the basal disconformity of the overlying cover bed.

Because the depths to carbonate-enriched horizons might be related to the available calcium influx (Machette, 1985), which could have concealed elevational trends, depth, and carbonate content correlations were also computed.

For none of the three ubiquitous carbonate-enriched horizons, an indication for such relationships could be found.

The correlations of the carbonate contents with various other soil and relief properties, mainly the particle-size distribution and the slope angle (Table 6.10) revealed that the carbonate contents of the carbonate-enriched horizons of both the Promontory soils are significantly correlated to the particle-size distribution within the respective overlying cover bed (not within the horizon itself). The carbonate-enriched horizon of the upper Promontory soil displays a positive trend between carbonate and gravel content and a negative one between carbonate and clay content in the upper Wyemaha layer. The latter correlation is also significant in the carbonate-enriched horizon of the lower Promontory soil. The Provo soil, in contrast, does not show such correlations; rather there is a tight correlation between its carbonate content and the carbonate content of the deepest exposed underlying material beneath the whole succession, almost independent of whether underlying layers shield the soil from direct contact to this deep substrate (Kleber, 2011). These findings indicate that the carbonate addition to this layer was mainly by inheritance and detrital sedimentation, rather than by direct atmospheric influx, supporting the conclusions already derived above from other correlations.

That pedogenic carbonate and gravel content are correlated in a way is not unexpected. However, model calculations indicate that gravel has a stronger influence on the depth to the carbonate enrichment than on the carbonate content (McFadden and Tinsley, 1985), which is not supported for our soils by the data. Thus, water retention and permeability of the substratum cannot explain the relationship between carbonate and gravel contents. Gravelly deposits are better suited than fine ones for trapping eolian accumulation (Blank et al., 1996; Goossens, 2005a; Mabbutt, 1979; McFadden et al., 1987). This would imply that the traps were at the surface when the eolian dust accumulated, whereas they are buried by at least one more deposit now. However, this conclusion is inconsistent with the correlations between coarse-silt and the clast content (Table 6.7). A possible explanation for this discrepancy is that a considerable fraction of the dust settled on clast-rich surfaces for a while but became redeflated rather than being entrenched, after having lost some of its carbonate through leaching. This is corroborated by the lack of correlation between the carbonate content and the coarse-silt content (Table 6.10), both assumed to largely be airborne components.

The Rocky Mountains and the Colorado Plateau obviously were affected by several drastic environmental shifts, leading to the formation of either loess-influenced cover beds or soils of varying maturity. These phenomena were traced from the Great Basin up to the mountains right below the modern timberline. Some of these shifts, that is, cover-bed formation and argillic pedogenesis, affected this elevational range in its entirety in a similar way. This is interpreted to have resulted in so dominant processes induced by the major shifts that elevational control played just a minor role. The formation of argillic properties in the lower two cover beds is the expression of a rather stable geomorphic setting under a moister soil climate than today. Other phenomena, on the other hand, display some influence of elevation. This is expressed in the Provo soil, the horizons of which are mostly restricted to certain elevational ranges.

One soil property, the carbonate content, is especially influenced by elevation, which finding allows drawing conclusions on cover beds, because carbonate is not only a product of pedogenesis but also was a sediment property before it became leached. The existence of paleoclimatic trends is indicated by decreasing carbonate content with elevation, whereas the depths to carbonate-enriched horizons are uncorrelated. Reworking of up-slope material appears to influence the carbonate content in the uppermost carbonate-enriched horizon, whereas the eolian influx largely determines deeper carbonate-enriched zones. Furthermore, the now buried—but once surficial—clast-rich layers call for discontinuities in the geomorphic evolution.

These findings are consistent with multiple cycles of cover-bed formation (cf. Liu et al., 1994), including the admixture of loess, separated by episodes of soil formation including carbonate-enriched horizons. The latter usually overprinted buried, older soil horizons to form welded soil profiles.

Carbonate-enriched horizons were usually found just beneath disconformities. This indicates that disconformities are strong triggers of carbonate enrichment: Fining of particle sizes and/or increasing bulk density may perch wetting fronts, thus favoring carbonate accumulation (Gile, 1999), whereas abrupt parent-material coarsening may cause carbon dioxide to degas forcing carbonate to precipitate. Thus, any disconformity may induce carbonate accumulation.

Below 3500 m asl, the altitude of the current timberline, on the eastern slope of the Colorado Front Range, the subalpine to montane zones are characterized by abundant stratified slope deposits exhibiting thicknesses exceeding 1 m, and by recurring sediment characteristics within the respective strata complexes. Apart from Leopold et al. (2008a,b), they have not been investigated up to now regarding both their genesis and age. However, glacial, fluvial, and colluvial sediments of the Front Range have been intensively studied in the past due to their significance in the context of soil-geomorphological approaches (for a compendium, see Birkeland et al., 2003).

In general, these slope deposits show characteristics in their lower parts similar to those described from the Niwot Ridge (see Section 6.3.2.2). The pronounced orientation of the clast length axes parallel to the slope inclination and slight grain-size sorting indicate a periglacial origin of the cover beds. Upper layers rich in fine sediment are part of the slope–sediment complex. However, similar to the high alpine sites, they are not ubiquitously distributed, which fact—similar again to the sites above timberline—might be the consequence of high wind speeds. First datings of layers derived from local rocks are assumed to indicate the principal applicability of the OSL method

FIGURE 6.17 Sections Gold Hill 7 and 8, 2710 m asl, 40 ± 3 ka (profile depth 35–45 cm) and 36 ± 3 ka (profile depth 50–60 cm), OSL datings, Lab. No. Risœ 09 54 06 and 09 54 07. Reddish hydromorphological features indicate frequent changes is soil-water contents. The profiles were dug along a road cut between Ward and Gold Hill (location and no. of section: see white arrow on left image). Road cuts allow to follow the slope deposits for hundreds of meters and demonstrate their wide distribution. Left image: the Niwot Ridge (3.744 m asl) is located in the background. (For color version of this figure, the reader is referred to the online version of this chapter.)

in this context (but see Section 2.7.2.3 for a critical evaluation of attempts to date basal layers) and delivered last glacial ages ranging between 40 and 12 ka (Völkel et al., 2010). Figure 6.17 shows two well-developed sections of Pleistocene periglacial slope deposits at 2700 m asl. The sedimentary properties vary broadly between the various layers, both in the fine and in the coarse grain-size range, probably indicating different mechanisms of formation. As the deeper layers are the major aquifers for interflowing water mainly in spring time, characteristic hydromorphological features such as bands of iron oxidation may be observed in many profiles throughout the Front Range (Fig. 6.17).

6.3.2.2 *Cover Beds in the Colorado Front Range Above Timberline*

In contrast to Central Europe, no commonly accepted classification system for Pleistocene periglacial slope deposits exists for North America. A few studies applied the German cover-bed approach to regions of the USA (Kleber, 1993, 1994a, 1997, 1999a,b; Leopold et al., 2008a,b). Leopold et al. (2008a,b) described the overall thicknesses of several periglacial slope deposits on top of the Niwot Ridge (3744 m asl), an area with a modern alpine tundra climate in the Colorado Front Range, using geophysical methods. Furthermore, they described the shape and structure of the various layers using soil pits. The basic idea of their study was to lay out a topo-sequence beginning at the top of the Niwot Ridge at sites with actively forming periglacial slope deposits and to extend it to lower positions in order to compare the results with inactive Pleistocene periglacial slope deposits at intermediate elevation.

Active periglacial layers on the Niwot Ridge are restricted to moist sites, which do not necessarily provide underlying permafrost. Leopold et al. (2010) documented annual ice lenses, which melt until the end of August, rather than permafrost, to be sufficient for active movement at the Niwot Ridge (cf. Section 2.6.3 and Veit, 1993).

The Niwot Ridge, which is assumed to have been unglaciated since the Pleistocene, provides thick layers of active and inactive periglacial slope deposits over large parts, extensively described by Benedict (1970). However, little is known about altitude-related thickness variations, which were first described by Leopold et al. (2008a).

Alpine slopes on the Niwot Ridge are underlain by layers of fine to coarse, blocky deposits of periglacial origin. In general, the depth to the bedrock ranges from 4 to 10 m and may reach a maximum of more than 15 m. Structure, distribution, and thickness of the deepest layer(s) are similar to Central-European basal layers.

Though the uppermost cover bed at the Niwot Ridge is akin in composition to the Central-European upper layer, it shows differences in topographic distribution: it is restricted to lee positions at the Niwot Ridge (see Muhs and Benedict, 2006). High wind speeds of up to 120 m s^{-1} are assumed to be the major cause of this phenomenon. However, the amount of fine sand and silt in the upper layer is large, indicating a considerable eolian admixture (see Dahms, 1993). The long axes of the clasts are not well orientated parallel to the slope inclination, indicating that cryoturbation was active and may still be.

In contrast, the orientation of the longest axes parallel to the slope inclination is well pronounced in the underlying layers. These layers form a mixture of

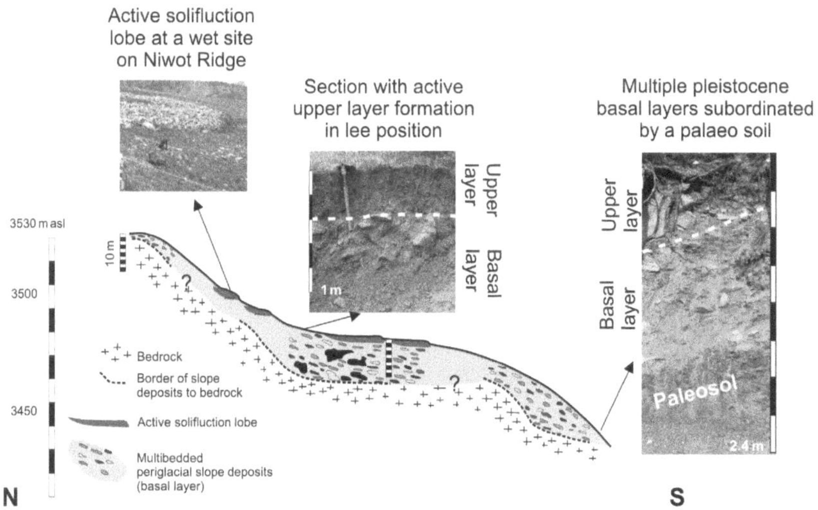

FIGURE 6.18 Transect across the top of the Niwot Ridge, Colorado Front Range, USA (comp. Leopold et al., 2008b). The transect is about 1 km long. Active gelifluction is restricted to moist sites. Upper layers are well developed, but their distribution is restricted to lee positions. Multiple basal layers may be found throughout the transect and the thickness of the layers relies on paleo-relief positions more than simply on their distance from the top.

coarse rock fragments and fine, mainly sandy material. Their thickness depends on their relative slope position, as indicated in Fig. 6.18. Close to the hill top, basal layers start with a thickness of only one meter and form multiple-layer complexes of several meters thick in concave parts of the paleorelief. They function as a main space of living for many ground-digging animals, and these deposits carry an important share of interflowing water, a vital water source in this environment. Where upper layers are absent they also provide the parent material for soil development. Deposits resembling intermediate layers could not be detected at the high altitudes of the Colorado Rocky Mountains.

Figure 6.18 gives an overview of a transect across the Niwot Ridge, together with three images of periglacial slope deposits from open gravel pits.

6.4 CONCLUSIONS

6.4.1 Significance of Cover Beds

Cover beds deposited essentially parallel to the slopes are a common feature of all profiles reported in this chapter from various environments in the mid-latitudes. Based on the areas discussed, we propose that deep soils are commonly layered and that cover beds are the major soil parent material on mid-latitude hillslopes. Similar research in other areas should evaluate the general validity of this claim. This should help increasing understanding soil properties and predicting their distribution.

Soils have not developed from sole bedrock but in cover beds, and their development was decisively influenced by this fact. Soil horizons—be they formed by weathering or translocation of, for instance, clay or carbonate—typically have their boundaries at disconformities; this is easily misinterpreted as being solely pedogenic. Disconformities within soils may be identified by changes in the heavy mineralogy, by grain-size variations (mainly silt, often indicative of incorporated loess material, and/or the coarse fraction that may be enriched contrary to expected weathering trends), and by the bifurcation of layers and/or soil horizons. Since bulk densities often differ among cover beds, they also affect soil-water stagnation, and slope hydrology in general (see Chapter 4). In all studied profiles, some or all cover beds contain admixed loess material. This reduces the influence of bedrock properties on the sediments and their soils; therefore, many soils are entirely decoupled from bedrock influence with the loess material determining the soil properties.

Theories of soil evolution should take the stratified, partially allochthonous nature of parent material into account (Phillips and Lorz, 2008). Modeling pedogenic processes usually requires parent material that is homogeneous or has known consistent trends (Wang and Arnold, 1973). For example, the estimation of pedogenic clay is problematic, unless variation in the geogenic texture can be excluded (Colman, 1982).

However, there are remarkable differences between cover-bed properties in temperate and in semi-arid regions and between study areas above and below timberline. Thus, transferring the principal concept of cover beds to all these sites is successful, but transferring the terminology developed in Central Europe is not generally applicable, because some layer successions reported in this chapter differ remarkably.

6.4.2 Cover Beds in Humid Areas

Typical of study sites in humid temperate areas are layers with high bulk density, which are usually the deepest in the profiles. Overlying layers are usually looser. The admixed loess material of the latter is essential for their sedimentary properties as well as for their pedogenic overprint.

The temperate examples are, thus, characterized by at least two types of layers, usually having distinct, distinguishing properties. These layers may be diachronous or at least have no exactly determinable ages. The explanation of Kleber (1992a, 2004a) is that they could have resulted from many of the climatic oscillations of the Pleistocene; which particular oscillation produced deposits persisting to the present may depend on local factors. Only where buried soils occur may layers also be relatively dated.

6.4.3 Cover Beds in Semi-arid Areas

Different from those in humid areas, cover beds in semi-arid areas are less variable in their sediment properties. The distinction of layers is based on buried soils rather than on sediment properties. Each sequence comprising a cover bed and a (paleo)soil probably represents a major climatic cycle (Kleber, 1994a) rather than any of the numerous minor climatic oscillations assumed to be responsible for cover beds in temperate areas. Therefore, there is evidence that these successions are chronostratigraphically better determined than those in temperate climates, that is, they probably are not diachronous as was suggested for the humid successions. As a result, cover beds may be used for relative dating of underlying deposits or landforms as soon as they themselves can be dated (see Section 7.3).

In semi-arid areas leached carbonate is concentrated in the subsoils. Thus, pedogenic episodes must have been sufficiently moist to allow for topsoil decalcification. The supply of calcium necessary to form the multiple carbonate-enriched horizons in most profiles may be explained by cyclic deposition of fresh eolian material. Synchronous airborne carbonate accretion and pedogenesis, often invoked to explain composite soil profiles (Birkeland, 1984; Yaalon and Ganor, 1973), cannot apply to cover-bed successions, because en-bloc movement dominates during gelifluction so that—contrary to pure loess—not every single part of such a deposit was at or near the surface at any time.

6.4.4 Cover Beds and Elevation

In both studied high-mountain ranges, there appears to be a threshold between two types of cover-bed successions at around the modern timberline. Above this line, processes of cover-bed formation are young or even ongoing. Holocene periglacial processes have formed all (Alps) or parts (Rocky Mountains) of the successions.

Below the timberline, elevation has mostly a gradual rather than a boundary-type effects on the properties of cover beds, and this implies that pedogenic, that is, postdepositional, properties are much more evident than sedimentary, that is, primary, ones.

Chapter 7

Relative Dating with Cover Beds

B. Terhorst, A. Kleber and E. Bibus

7.1 INTRODUCTION

Retrospective analysis with respect to chronological scales serves to understand present geosystems (see Damm and Hagedorn, 2010; Heine, 1987; Leser, 1996; Semmel, 1996a). The reconstruction of former landscapes and ecosystems is especially relevant for evaluating the anthropogenic impact (Terhorst et al., 2009). In this context, periglacial layers as well as soils and paleosols help to indicate the relative ages of land surfaces. Long-term stable surfaces provide sufficient time for pedogenesis, whereas dynamic landscapes with high geomorphodynamic activity prevent the formation of mature soils. Consequently, stable and old surfaces are characterized by mature soils or paleosols. In contrast, active and young surfaces carry initial, less developed soils.

In general, dating of soils is difficult because of bioturbation processes and vertical displacement of particulate or soluble components. Furthermore, organic datable material is rare, but relative ages may also be established by analyzing the weathering intensity and the thicknesses of soils, paleosols, and sediments (Birkeland et al., 2003; Miller et al., 1993). However, the stage of development of (paleo)soils, combined with the occurrence of periglacial layers, may contribute to a reliable differentiation of Holocene and Pleistocene landscape surfaces (Semmel, 1996a). Studies, which successfully applied this approach, have been conducted by Kleber (1999a), Bibus and Terhorst (2001), Terhorst (2007), and Damm and Terhorst (2010).

Section 6.2.2 provided a stratigraphic framework of the cover beds of the Great Basin, southwestern North America, and of the soils associated with them. Section 7.3 aims at applying a sequence-stratigraphic approach to test whether this makes dating of underlying landforms feasible and to identify the potential of cover beds for dating purposes. The types of relief to which this approach is applied are river terraces providing an independent chronosequence as to control the sequence of relative ages derived from cover beds; furthermore, a case of apparent discrepancy between cover-bed stratigraphy and tephrochronology will be discussed, and the dating approach is applied to generations of ice-wedge casts and to till of uncertain age.

Developments in Sedimentology, Vol. 66. http://dx.doi.org/10.1016/B978-0-444-53118-6.00007-6

Soil terminology in Section 7.2 follows World Reference Base for Soil Resources (WRB, 2006), and in Section 7.3, it follows Soil Taxonomy (SSS, 2010). Mass-movement terminology follows Varnes (1978, 1984) in the version advanced by Hungr et al. (2001).

7.2 CASE STUDY AT THE SWABIAN JURASSIC ESCARPMENT

7.2.1 Study Area

The study area comprises parts of the central Swabian Jurassic escarpment in the area of the "Reutlinger Alb" (Fig. 2.1). The overall occurrence of Pleistocene slide masses is characteristic of the slopes of the cuesta scarp (Terhorst, 2007). Numerous slide events indicate intense Holocene geomorphodynamics, permanently causing damage to infrastructure and buildings. The most famous mass movement is the Mössingen landslide of 1983, which damaged 0.6 km^2 of forest area.

The steep slopes of the Swabian Alb are susceptible to landslides, due to geologic, geomorphologic, hydrologic, and climatic conditions. In the past, the area has been the object of several geoscientific studies (see Bibus, 1986; Bleich, 1960; Schädel and Stober, 1988). More recently, GIS assessment related to slope stability was conducted by Neuhäuser and Terhorst (2006) and Terhorst and Kreja (2009).

The escarpment of the Swabian Alb abruptly rises by 300–400 m above its foreland. The geological conditions are determined by intercalations of permeable and impervious bedrock. In general, Oxfordian and Kimmeridgian limestones (Upper Jurassic) overlie marls and clays. Ductile behavior of the Callovian clays (Middle Jurassic) and the superposition with swellable Oxfordian marls favor slope instabilities.

In addition, hydrological conditions cause the general disposition for mass movements, such as the drainage from the shallow limestone karst down to the Oxfordian marls and the Callovian clays as well as the progressive incision of the Rhenish system related to spring erosion in the slope areas (Terhorst, 1997). Triggering factors for landslides are precipitation rates above the average or increased runoff due to intensive snowmelt in winter and spring.

Earlier studies, which were mainly based on field surveys, pedological and sedimentological analyses, as well as on dating methods, gave evidence that landforms and processes related to mass movements may be divided into two main classes (Terhorst, 2001). During glacial times, large complex slump-earth flows occurred. In middle slope areas, rotated blocks 200–300 m long and 20–50 m wide, on average, are present. The foot zones of these complex areas of mass wasting are formed by large tongue-like dams, related to earthflow processes.

The Holocene processes are characterized mainly by smaller earthflows and translational slides. For the major part, their occurrence is related to Pleistocene landslide areas, so they may be classified as reactivation failures according to Brunsden (1999).

7.2.2 Geomorphological Setting in Landslide Areas

The geological structure of the landslide areas is formed by permeable limestones of the Upper Jurassic Oxford Formation (Figs. 7.1 and 7.2), which is underlain by Oxfordian marls and impermeable, ductile Callovian clays (Middle Jurassic).

In both units, large semicircular scarps are present on the upper slopes. Blocks were involved in rotational slips and slid down from the scarps of the upper slopes to the middle slopes. The main scarps comprise thick debris accumulations, partly covering the former surfaces of the upper-rotated blocks. The lower slope consists of large, tongue-shaped ridges resulting from earthflows, which can be seen best in the Lache landslide area (Fig. 7.1). These landslide complexes are best described as a combined form of slump and earthflow. Large slumps combined with earthflows, like the ones presented here, are considered as Pleistocene movements in the study area (Bibus and Terhorst, 2001; Terhorst, 2007). A ^{14}C age from tufa precipitated in the landslide scarp of 21,135 ($\pm$220) years BP (HV 19756) supports this.

The Scheibenbergle landslide area is composed of two rotated blocks, with one lying on top of the other (Fig. 7.2). The uppermost one partially overlaps the lower one. The lower block has been covered by limestone debris after the sliding event. Inside the debris, pure secondary calcites are developed, which yielded a U/Th age of 173,200 ($\pm$7600) years BP (Terhorst, 2007). Thus, the basal slide mass represents a Middle Pleistocene event.

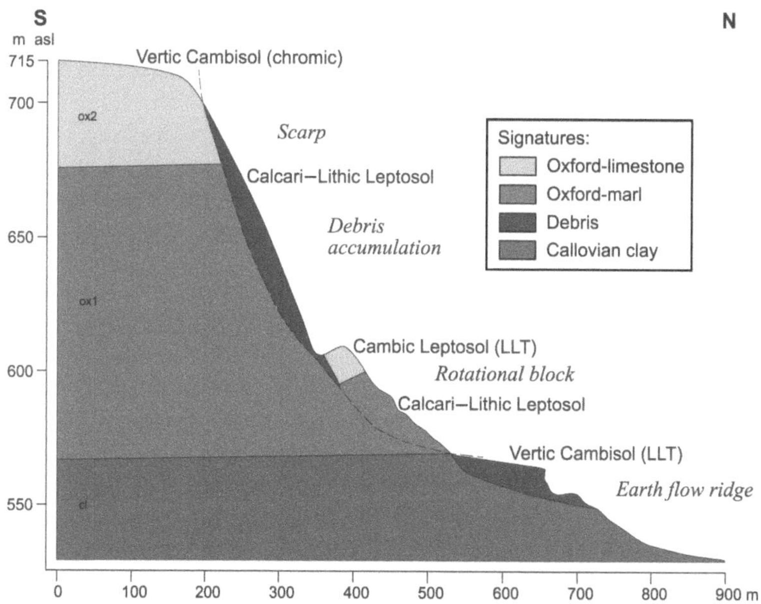

FIGURE 7.1 Landslide area "Lache": landforms, soil groups, and geologic setting. "LLT" corresponds to the minerals of the Lake Laach Tephra, included in the upper layer. (For color version of this figure, the reader is referred to the online version of this chapter.)

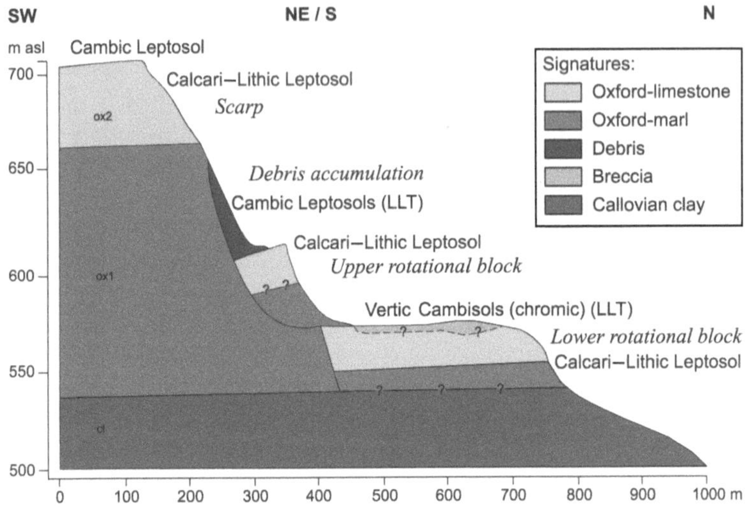

FIGURE 7.2 Landslide area “Scheibenbergle”: landforms, periglacial layer, soil groups, and geology. “LLT” correspond to the minerals of the Lake Laach Tephra, included in the upper layer. (For color version of this figure, the reader is referred to the online version of this chapter.)

7.2.3 Pedological and Sedimentological Setting in Landslide Areas

7.2.3.1 Unstable, Active Slope Areas of the Swabian Alb

In modern as well as in historical times, active slope movements such as soil creep and small earthflows occurred within Pleistocene landslide areas. Concerning the application of the concept of periglacial layers and soils, the main scarp, for example, is a suitable location to provide a minimum age because mass wasting always creates new surfaces. However, slumping usually results in extremely steep slope angles in scarp areas, which are responsible for intense slope activity after the main slope failure. However, there is continuous debris fall coming down and, thus, active detrital cones may form. Consequently, the time for decalcification of parent material and formation of mature soils was insufficient in landslide scarps.

In the same way, the slopes of rotated blocks are disturbed intensely through the rotation, so that they tend to form new slope surfaces. Earthflows consequently occur at the foot of the blocks (Fig. 7.1), formed by a mixture of limestone debris and marls. The areas of the resulting tongue-shaped ridges are important indicators for the relative age of the landslide area, as soils and periglacial layers would have completely been destroyed by flow processes, provided the latter occurred later than the formation of the former and, thus, cannot have been preserved there near the surface. Furthermore, the tongue-shaped ridges provide rather flat landforms, supporting long-term pedogenesis under stable conditions.

In general, young, initial soils are situated on the steep slopes of the main scarps, the permanently creeping slopes of the rotated blocks, and in areas

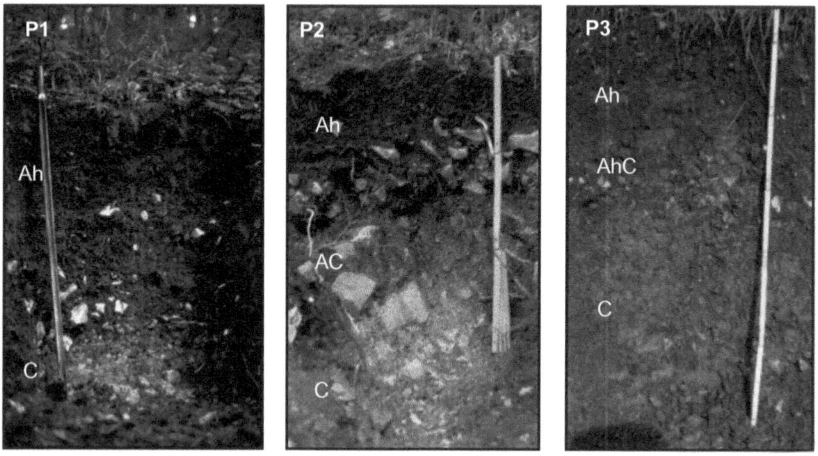

FIGURE 7.3 Different calcareous Leptosols in limestone and marl debris as present in young slide masses. (For color version of this figure, the reader is referred to the online version of this chapter.)

affected by modern and historic mass movements. The corresponding major soil groups are Leptosols[1] and Gleysols,[2] more specifically Calcari–Lithic Leptosols, Rendzic Leptosols, Skeletic Leptosols, and Calcic to Mollic Gleysols. They have formed in limestone and marl debris of slide masses or directly in bedrock. The listed soils tend to have black humic horizons of varying thicknesses, and they are strongly to extremely calcareous (Fig. 7.3).

As a general rule, in active areas and those with historic slope movement, periglacial layers are absent. The composition of slide masses varies extremely, and there is no evidence of periglacial layers. The field results are supported by the heavy-mineral content of soils in the slide masses (Table 7.1). Foreign minerals of the Lake Laach Tephra (LLT) as well as an eolian component are almost absent or strongly mixed, whereas the Jurassic bedrock component dominates the heavy-mineral spectrum.

7.2.3.2 Stable Slope Areas of the Swabian Alb

In contrast to slopes with permanently active geomorphodynamics, well-developed soils are present on stable slope surfaces. These mature soils combined with the occurrence of periglacial layers allow a relative dating and discrimination between Pleistocene and Holocene surfaces. Periglacial

1. Leptosol, shallow or extremely gravelly soil; Calcari–Lithic, consolidated rock occurs at <10 cm depth, free calcium carbonate is present in the soil; Rendzic, humus-rich soil horizon overlying calcaric material; Skeletic, rich in coarse rock fragments; Cambic, with evidence of alteration through weathering.
2. Gleysol, soil affected by groundwater-related reduction and oxidization processes; Calcic, containing secondary carbonate; Mollic, rich in humus; Vertic, clayey, affected by shrinking and swelling.

TABLE 7.1 Heavy-Mineral Association of Characteristic C Horizons in Landslide Areas at the Swabian Jurassic Escarpment (in %)

Sample[a]	Loess minerals	Bedrock minerals	Stable minerals	LLT minerals[b]	Brown amphibole	Titanite	Clinopyroxene
VI/3b, 67	1.1	94.3	0.8	2.7	0.8	0.4	1.5
VI/5, 69	0.0	98.6	1.4	0.0	0.0	0.0	0.0
VI/7, 71	0.0	94.7	0.8	0.8	0.8	0.0	0.0
SBII/Cv, 74	30.4	26.1	21.7	8.7	4.3	4.3	0.0
Mean value	7.9	78.4	6.2	3.1	1.5	1.2	0.4

[a]Sample numbers refer to Terhorst (1997).
[b]The resulting mineral assemblages are grouped as to their provenience, with Lake Laach Tephra minerals as a group and its three typical constituents listed separately in the last three columns.

FIGURE 7.4 Left: Decalcified Vertic Cambisol in Upper Jurassic limestone debris. The upper layer with volcanic minerals is present from top to 30 cm. Center: Vertic Cambisol in Middle Jurassic clays. The upper layer comprises the section down to 15 cm. Right: Vertic Cambisol (chromic) in Upper Jurassic limestone debris. The upper layer corresponds to the upper 15 cm. (For color version of this figure, the reader is referred to the online version of this chapter.)

layers in debris cones and earthflow deposits that overlie slide blocks are particularly useful for these purposes. Furthermore, periglacial layers and young soils on top of earthflow deposits and on the slopes of rotated blocks are most important for relative dating. Both examples of landslide areas (Figs. 7.1 and 7.2) show periglacial layers (upper layer) and soils in stratigraphically relevant geomorphological positions. Figure 7.1 displays Cambic Leptosols (Fig. 7.3) formed on the slope of the slide block and Vertic Cambisols formed on the surface of the earthflow deposits at the foot of the slump-earth flow (Fig. 7.4). The occurrence of the upper layer combined with mature soils on disturbed landslide surfaces indicates that the primary landslide occurred during the Pleistocene.

In the upper part of both the Cambic Leptosols and Vertic Cambisols, a cambic horizon is developed below a thin humus horizon. These cambic (Bw) horizons are characterized by a clay content ranging from 30% to 50% and a noteworthy silt content of 40–60% (Terhorst, 1997). The clay content in the respective underlying soil horizons as well as in the parent materials is, for most profiles, higher, whereas the silt content is lower. This tendency indicates an admixture of exterior material in the Bw horizons.

These findings coincide with the heavy-mineral composition of the Bw horizons. For all studied profiles, the uppermost horizon contains the characteristic volcanic minerals of the Lake Laach Tephra (brown amphibole, clinopyroxene, titanite) with a mean value of 34.5% (Table 7.2). On average, brown amphibole constitutes 17.1%, titanite 6.4%, and clinopyroxene 11.0%. The loess component comprises 10.3% of the transparent heavy minerals, and the remaining part is constituted by the minerals rutile, tourmaline, and zircon as stable components and unstable minerals, that is, muscovite and opaque minerals, all derived from the Jurassic.

TABLE 7.2 Heavy-Mineral Association of Upper Layers, in Landslide Areas at the Swabian Jurassic Escarpment (in %)

Sample	Loess minerals	Bedrock minerals	Stable minerals	LLT minerals	Brown amphibole	Titanite	Clinopyroxene
SIII/1, 21	15.8	31.8	20.3	23.8	12.3	5.5	6.0
GII/1, 33	9.0	10.6	10.0	65.2	35.6	7.9	21.6
LI/1, 39	3.4	58.5	4.4	25.0	5.3	1.2	18.6
LII/1, 40	3.7	65.3	15.7	11.9	8.3	1.5	2.1
LIV/1, 48	4.3	72.6	12.0	7.1	3.1	4.0	0.0
TII/1, 53	16.8	12.7	6.4	62.1	28.4	15.5	18.5
TII/2, 54	11.9	15.0	7.7	57.5	27.6	9.2	19.9
HSTIII/1, 64	6.7	60.3	15.0	16.3	5.0	9.0	2.3
SBII/1, 74	14.5	26.0	4.6	34.1	19.1	5.8	9.3
SBII/2, 75	16.7	21.2	6.1	42.4	26.5	4.6	11.4
Mean value	10.3	37.4	10.2	34.5	17.1	6.4	11.0

See Table 7.1 for explanation.

7.2.4 Model of the Distribution of Periglacial Layers and Soils in Landslide Areas

A model of slope development (Fig. 7.5) describes Pleistocene landslide areas in different stages of Holocene erosion and reworking. Stage 1A represents an area of Pleistocene landslides without any disturbance during the Holocene.

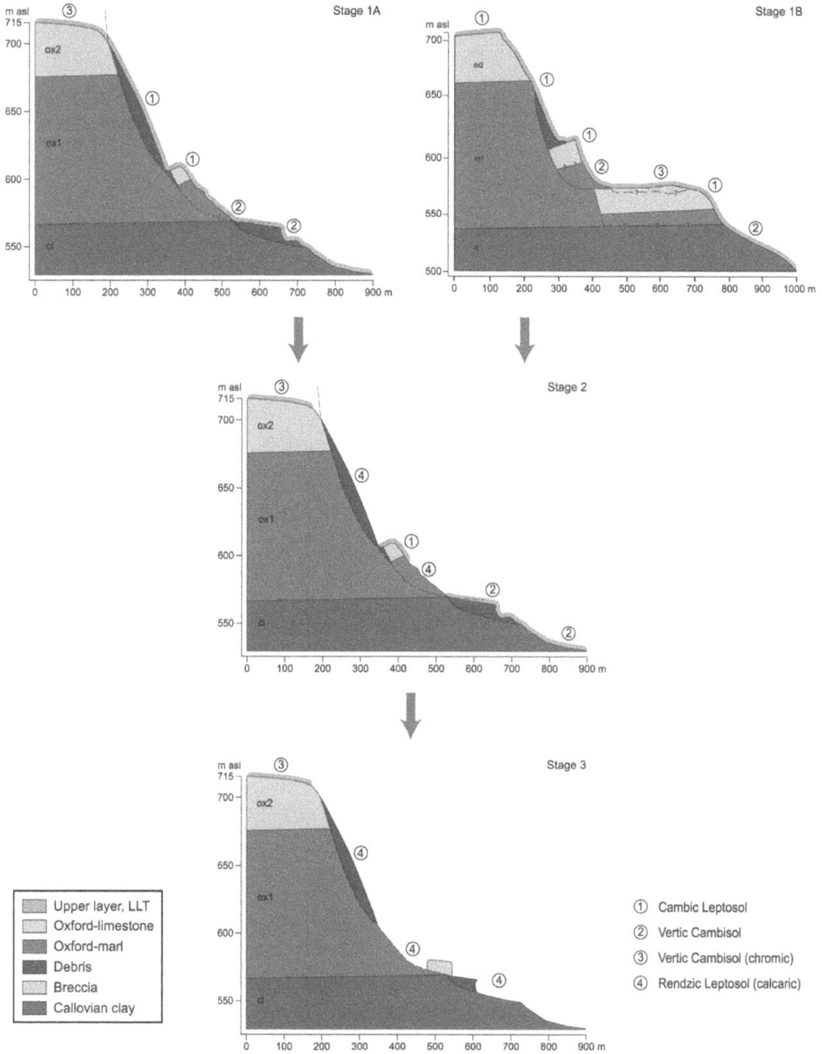

FIGURE 7.5 Slope formation model of characteristic landslide areas at the Swabian Jurassic escarpment. In undisturbed Pleistocene landslide areas (stage 1A and B), the upper layer and well-developed soils cover all landslide elements. Slide masses, which were partly active during the Holocene, are in some parts covered by an upper layer and mature soils (stage 2). Active areas are reflected by initial soils. Stage 3 represents an overall active slope, characterized by young, initial Leptosols. LLT, Lake Laach Tephra. (For color version of this figure, the reader is referred to the online version of this chapter.)

On a slope, the upper layer covers the complete main scarp as well as the related landslide mass. The area is characterized by long-term pedogenesis during the Holocene. Respective soil groups are Vertic Cambisols and Cambic Leptosols, with a silt-bearing, decalcified Bw horizon. Furthermore, on top of Middle Pleistocene sliding blocks, Vertic Cambisols (chromic[3]) may be present (Fig. 7.5, stage 1B). Sliding areas of both stages 1A and 1B (Fig. 7.5) were active during the Pleistocene, before the upper layer was deposited.

Stage 2 (Fig. 7.5) shows the most widespread slope situation of the study area. Slope sections with Pleistocene landslide elements are still present, but Holocene erosional processes in the form of small slides and earthflows are reworking the ancient landforms. Mosaic-like, the residual Pleistocene landforms covered by the upper layer and mature soils alternate with Holocene movements with initial soils, such as Rendzic Leptosols and Calcari–Lithic Leptosols. In these cases, the soil distribution indicates that (1) the main landsliding event took place during the Pleistocene and (2) Holocene processes remove mature soils and sediments in specific, sensitive slope positions.

The final stage is displayed by an overall occurrence of active and historic mass movements (Fig. 7.5, stage 3). Hence, upper layers and mature soils have been completely removed. Without absolute datings, the age of the supposed older landforms remains unclear. Rendzic Leptosols and Calcari–Lithic Leptosol strongly reflect active geomorphodynamics.

7.3 CASE STUDIES IN THE WESTERN USA

A. Kleber

7.3.1 Introduction

Sections 6.2.2 and 6.3.2, especially Table 6.1, provided a stratigraphic framework for the cover beds of the Great Basin and the soils associated with them as related to the established regional chronology of glacier advances and lake highstands. In the following, we aim at applying a sequence-stratigraphic approach in order to find out whether dating of underlying relief by cover beds is feasible. This, thus, intends to identify the potential of cover beds for dating. The types of relief where this approach is applied are river terraces providing an independent chronosequence as to control the sequence of relative ages derived from cover beds, generations of ice-wedge casts, and till of uncertain age.

However, one should bear in mind the common caveats of sequence-stratigraphic approaches: most unconformities are time-transgressive, especially if they connect sequences in inundated basins with subaerial relief (Catuneanu et al., 2010). Thus, there certainly is some chronological inaccuracy inherent to any model of this kind.

---

3. A soil horizon with reddish color.

7.3.2 River Terraces of Castle Valley, Southeast Utah

Several authors used river terraces as frameworks for chronosequence studies. Some traced the terraces back to moraines for age control (Chadwick et al., 1997; Madole et al., 1987; Sharp et al., 2003) or used independently dated terraces (e.g., Busacca, 1987; Chadwick et al., 1997; Hanson et al., 2006; Harden, 1982; McFadden and Hendricks, 1985; Phillips et al., 1997; Sharp et al., 2003). In Castle Valley, no independent age control is available, because the terraces cannot be traced to the few moraines in the catchment area (cf. Shroder and Sewell, 1985, for the reexamination of deposits originally interpreted as tills by Richmond, 1962), and there are no datings other than those reported below. Some workers contributed to pedologic aspects of neighboring valleys but presented age assignments on the oldest terraces only (Biggar et al., 1981; Colman, 1983; Colman et al., 1986; Harden et al., 1985).

Castle Valley is located along the crest of an anticline, collapsed due to subrosive processes connected with salt diapirism (Utah Geological and Mineral Survey, 1987). The valley thus may have a unique history of local tectonics, discouraging the transfer of river-terrace ages from other valleys or from the receiving Colorado River. The terraces are referred to as Qa1, 2, and Qaf3, 4, 5 (abbreviations from Ross, 2006). Qaf5 rises above the valley floor by approx. 100 m in the intermediate section of the valley and by more than 150 m in the upper section, where the profile is located. Qaf4 and Qaf3, approx. 15 and 10 m above the valley floor, respectively, occupy most of the area in the intermediate valley section. Qa2 and Qa1 are restricted to narrow benches along the creeks 3 and 1 m above the valley floor, but Qa2 widens down-valley and occupies most of the valley floor in its lower reach, with the creek being incised by some 10 m into it (a radiocarbon date reported below stems from a soil sediment at the exposed base of Qa2, buried by almost 10 m of calcareous sand). Except for the exemplary profile, the other profiles, discussed in detail in Kleber (1999a), are in the intermediate reach of the valley.

Qa1 and Qa2 are both overlain by 0.5–1 m of meadow loam, the upper half meter of which has lost its bedding through pedogenic alteration on Qa2. However, the meadow loam is still identified as such by lenses of coarse sand and of fine gravel, which shows perpendicular (i.e., fluvial) orientation of clast length axes.

Separated from the meadow loam by a disconformity, a cover bed overlies the alluvial gravel of Qaf3. It follows the slope of a 2-m step in the terrace topography without significantly changing its thickness, and it contains coarse clasts that are oriented parallel to this slope: these two properties are unusual for fluvial deposits. The soil has a low carbonate content in the Bw horizon but is underlain by a Bk horizon of stage I carbonate morphology.[4]

In another profile upon Qaf3, a similar cover bed overlies alluvial gravel, which in turn overlies an argillic horizon. This argillic horizon and the Bw

4. The stages of carbonate morphology after Machette (1985) reach from I (very weak) over III+ (entirely whitened matrix), IV (cemented) to VI (brecciated).

horizon on top are developed in cover beds, enriched in coarse silt, indicating eolian admixture. Two cover beds overlie Qaf4 with a Bw horizon in the upper one and a Btk horizon (clay illuviation and carbonate enrichment of stage I) in the lower one. A calcic horizon (carbonate stage II+) follows in the underlying alluvium. This is virtually the same sequence as in the second profile on Qaf3, except that the intercalated alluvium is missing; thus, the 2Btk horizon, rather than the gravel layer, became the destination of the leachate from the upper layer. Another well-defined argillic horizon is developed in a fine-grained cover bed, overlain by almost 2 m of gravel. It is almost free of lime, whereas the subsequent calcic horizon in the underlying alluvium reaches stage II.

On the slope between Qaf5 and Qaf4 and upon Qaf5, a similar sequence (Bw–2Bt–3Btk, three cover beds) is present, but more cover beds underlie this succession. The basal calcic horizon on Qaf5 is of exceptional intensity (carbonate morphology stage IV).

Figure 7.6 shows the inferred correlations of layers and terrace ages. Qa1 and Qa2, covered by meadow loam with immature soils, are Holocene (radiocarbon date 3675±180 BP on humic acid: Kleber, 1999a); this is consistent with the age suggested by Ross (2006). The uppermost cover bed on Qaf3 (and on higher terraces) may be correlated with the Provo layer (inferred to be Late Glacial), indicating a Pinedale (Last Glacial Maximum) age of Qaf3. Its soil is immature as well but has a horizon of carbonate enrichment. Occasionally, the underlying alluvium overlies a cover bed, which may be correlated with the upper Wyemaha Formation, placing the lower part of the alluvium at least into the early Wisconsinan. Presumably, the same cover bed directly underlies the inferred Provo layer on Qaf4, indicating also an early Wisconsinan age of this terrace. The inferred upper Promontory Soil has an intense calcic horizon in whichever material is beneath. The third cover bed may be correlated

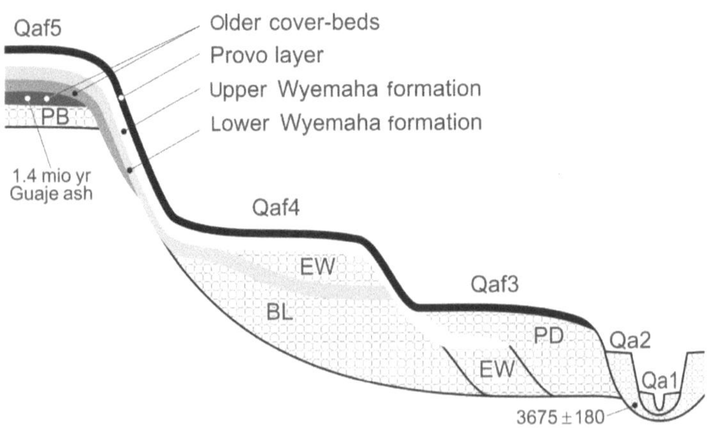

FIGURE 7.6 Schematic cross-section of Castle Valley (*Reproduced from Kleber, 1999a*). PD, Pinedale (Last Glacial Maximum); EW, early Wisconsinan; BL, Bull Lake (penultimate glaciation); PB, pre-Bull Lake. See Table 6.6 for references on the ages of PD, EW, and BL.

with the lower Wyemaha Formation. The inferred lower Promontory Soil has well-defined clay films and a Bk horizon of intermediate intensity. In Qaf4, it is separated from the surficial deposits by intercalated alluvium, indicating a Bull Lake (penultimate glaciation) age of the lower gravel that comprises most of the terrace volume.

Qaf5 is very old. It bears a petrocalcic horizon and is overlain by several cover beds and paleosols, and a tephra layer was found upon it with a chemical composition very close to that of the 1.4 Ma Guaja tephra (see next section). The main river incision occurred between Qaf5 and Qaf4. The soil between Qaf5 and Qaf4 also has a basal calcic horizon older than the inferred lower Promontory Soil but is developed to a lesser degree than that on Qaf5, corroborating the intermediate age of the slope between Qaf5 and Qaf4.

7.3.3 A Case of Misleading Tephrochronology?

Tephrochronology, a relative dating technique based on volcanic tephra layers, is an important tool in geochronology (Froese et al., 2008), particularly where tephra beds are frequent and distinguishable, such as in Iceland (Þórarinsson, 1979), New Zealand (Vucetich and Pullar, 1969), and western USA (Izett, 1981; Sarna-Wojcicki and Davis, 1991).

As soon as tephra layers appear to be pure, that is, not contaminated with other material admixed during reallocation, the danger of reworking of tephra from another location significantly after its primary deposition is rarely considered, because it is supposed that the tephra would have become mixed with the more voluminous local clastic load; rather, it is assumed that reworking could only have occurred shortly after the initial eruption and deposition of the tephra layer (Sarna-Wojcicki and Davis, 1991). Thus, in the case of a discrepancy with independent age estimates, the date provided by a pure tephra layer is usually given preference (e.g., Davis et al., 1986). However, within sediments transported by laminar processes, such as many cover beds are (see book cover photo), mixing is not a necessary consequence of dislocation.

A specific tephra layer, in the following referred to as the "La Sal Mountains (LSM) Tephra," occurs in the northwestern LSM (located 38°36′N, 109°22′30″W, elevation 2130 m a.s.l, 22° slope, in a roadcut of the Manti-La Sal circuit), which had previously been taken for an intense calcic soil horizon due to its bright color (M. Ross, UGMS, pers. comm. 1990). Its chemical composition is very close to that of the approx. 1.45 Ma (Doell et al., 1968) Guaje Tephra, derived from the Jemez Mountains, New Mexico (Table 7.3).

In the field as well as upon microscopy, the tephra layer shows only slight weathering. This preservation is unlikely if the tephra layer had been for 1.45 Ma as close to the surface as the layer is now (O.A. Chadwick, pers. comm. 1993). Thus, the question arises whether material originally overlying the LSM Tephra layer was eroded or whether the tephra layer, despite of its pure appearance, was reworked significantly later than its original deposition. The present

TABLE 7.3 Electron Microprobe Analyses of Glasses from the LSM Tephra and the Closest Matches, the Guaje and the 1.1 Ma Tsankawi Tephras

	LSM Tephra	Guaje	Tsankawi	Tsankawi
SiO_2	77.16	76.64	76.81	77.32
Al_2O_3	12.21	12.49	12.18	11.43
Fe_2O_3	1.49	1.45	1.60	1.64
MgO	0.01	0.00	0.01	0.00
MnO	0.06	0.06	0.05	0.07
CaO	0.27	0.28	0.28	0.27
TiO_2	0.08	0.06	0.05	0.04
Na_2O	4.37	4.49	4.66	4.64
K_2O	4.35	4.52	4.36	4.59
Similarity coefficient	–	0.9740	0.9707	0.9554

Analyst C.E. Meyer, US Geological Survey, Menlo Park, sample NLSM-109. Values are weight-percent oxide, recalculated water free. Particularly, the similar iron content supports the correlation of the LSM Tephra with the Guaje rather than with the Tsankawi Tephra (A.M. Sarna-Wojcicki, USGS, pers. comm. 1993).

contribution discusses evidence for the latter hypothesis, along with consequences for tephrochronology in inclined relief.

For this study, particle sizes were determined by sieving and the pipette method, using $Na_4P_2O_7$ as a dispersant, after removing carbonate by treatment with acetic acid. However, the discussion of the texture will be restricted to the particle-size class of fine sand (0.063–0.2 mm). Carbonate was determined with a gas-volumetric Scheibler apparatus (Page et al., 1982). Light minerals were microscopically counted on a representative fraction of the whole sample. For heavy-mineral analysis, grains of the 100–200 μm fraction, separated by flotation in bromoform, were embedded in resin (refraction index 1.67) for microscopic examination. Approximately, 200 nonopaque grains were identified per sample.

Where the LSM tephra layer was found, the exposure consists of several soil horizons formed from loess-dominated slope deposits during various soil-forming episodes (Fig. 7.7). Various paleosols were distinguished by means of their compound argillic and calcic horizons: the carbonate in these horizons must have been accumulated after the argillic properties had been formed, because carbonate enrichment and clay illuviation cannot occur simultaneously in the same horizon, and clay translocation is possible only during or after the depletion of carbonate (for detailed reasoning, see Section 6.2.2.1 and Kleber, 2000). Furthermore, a much warmer soil-temperature regime than the area is experiencing now would have been needed to form calcic horizons reaching

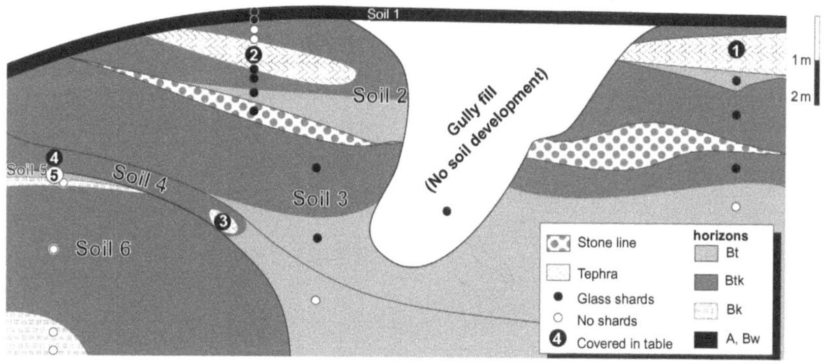

FIGURE 7.7 Sketch of the exposure in the La Sal Mountains, Utah, USA. The horizontal scale is approximate. Circles indicate samplings; filled circles indicate the presence of glass shards. Carbonate concentrations are <10% (Bt horizons), 10–50% (Btk), and >50% (Bk) of the fine earth fraction. There is no visible soil development in the tephra layers and in the gully fill. In the stone lines, Btk horizons are present.

as deep as in this profile, provided the entire parent material was to be seized (McFadden and Tinsley, 1985). Both lines of reasoning lead to the interpretation that the parent material of each soil was deposited after the argillic properties in the respective underlying soil had already been formed, which reveals the latter being part of a paleosol (Kleber, 1993). Accordingly, the soils cover a remarkable time span.

The heavy-mineral composition of the tephra layer is dominated by clinopyroxene (Table 7.4, nr. 1), which is less frequent in loess-dominated slope deposits (nr. 5), where hornblende and titanite dominate instead. Hypersthene is probably unique to the LSM tephra, as it is essentially absent in other deposits within this area (Kleber, 1992b, 2001). Among the light minerals, many glass shards occur. This composition of the LSM Tephra layer is consistent with that of the Guaje tephra (Izett et al., 1972). The grain-size distribution of the tephra layer peaks in the fine-sand fraction, whereas the surrounding slope deposits largely contain coarse silt. Two other samples (nrs. 2 and 3) also contain conspicuous amounts of glass shards, fine sand, and clinopyroxene, as well as some hypersthene. Nr. 2 stems from approximately the same stratigraphic position as nr. 1 but is more contaminated with surrounding loessic material, whereas nr. 3 stems from a small sand lense within the fourth soil. Nr. 4 from the same soil contains some shards, whereas Nr. 5 from the next deeper soil is free of such shards.

Glass shards were found in several layers (Fig. 7.7) down to and including the parent material of the fourth soil, whereas they are absent below. Contrary to other textural classes and minerals (not shown), the tephra constituents glass shards, clinopyroxene, and fine sand correlate significantly with one another (Fig. 7.8): increases in one are accompanied by proportional increases in the others in the entirety of samples. This is interpreted to the effect that these components have one major source in common. Presumably, a layer of the Guaje Tephra has yielded particles from upslope since the time of the formation of

TABLE 7.4 Analytical Characterization of the LSM Tephra (1) and of Selected Samples (2–5)

Number					
Heavy mineral	*(% of Transparent grains)*				
Clinopyroxene	74.1	44.2	47.5	35.7	16.1
Green hornblende	9.9	17.9	18.8	28.6	61.1
Hypersthene	8.8	3.8	4.2	0.0	0.0
Titanite	5.0	30.5	27.0	33.3	22.2
Others	2.2	3.6	2.5	2.4	0.6
Fine sand (% of fine earth)	54.7	36.0	34.9	22.0	16.4
Glass shards (% of light grains)	95	80	80	3	0

See Fig. 7.7 for the position of the samples in the profile. Only values indicative of the La Sal Mountains Tephra are reported. Analyst R. Schill, Geomorphology Lab, Bayreuth University. 1, pure tephra; 2, 3, mixture of cover bed and tephra; 4, cover beds with tephra traces; 5, pure cover bed.

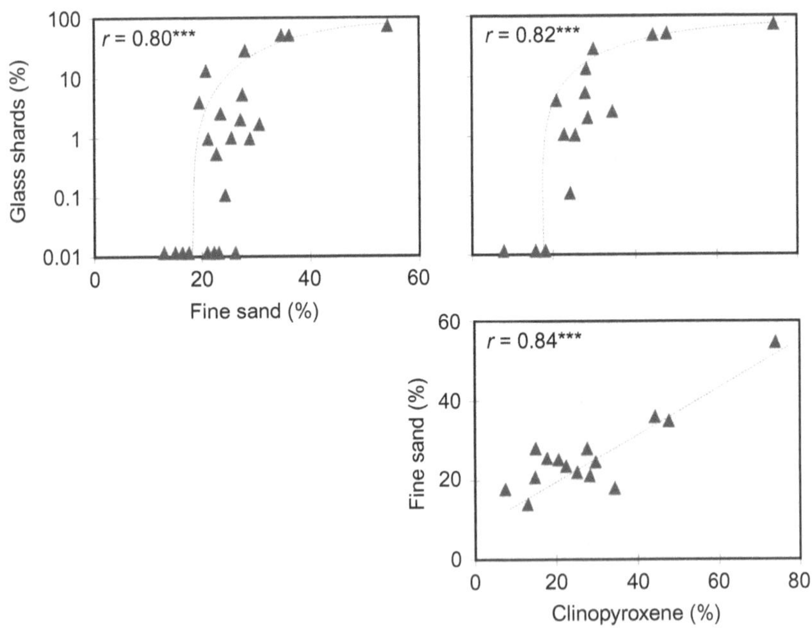

FIGURE 7.8 Correlation of tephra constituents. Top left: glass shards (logarithmic scale) versus fine sand; top right: glass shards (logarithmic scale) versus clinopyroxene; bottom right: fine sand versus clinopyroxene. Linear (note logarithmic scale where applicable) regressions are displayed and *r* values are computed. ***Level of significance >99.9%. (For color version of this figure, the reader is referred to the online version of this chapter.)

FIGURE 7.9 Exposure in the La Sal Mountains, Utah, USA, depicting the right half of the sketch in Fig. 7.7 (*photo Kleber*). (For color version of this figure, the reader is referred to the online version of this chapter.)

the parent material of the fourth soil and became finally eroded and reworked when the parent material of soil 2 was deposited (white layer in the right half of Fig. 7.9). Upslope from the exposure, the loess-rich slope deposits peter out within a few meters, and severe augering yielded no evidence of a first-deposition tephra layer, which is, therefore, assumed to have been entirely eroded.

The tephra constituents down to soil 4 indicate that tephra material was admixed to the parent material of the fourth soil, that is, earlier than the pure tephra layer was formed. This suggests that—particularly on steep slopes—tephra layers may be reworked a long time after their primary deposition, nevertheless retaining an essentially pure character. Laminar transport by gelifluction or transport in a frozen condition due to a periglacial climate could account for this. Thus, age determinations based solely on tephra layers may yield false results due to nonapparent reworking of tephra material. Heavy-mineral analyses may be used to provide evidence of tephra constituents throughout a profile and may allow to unveil the reworking history. Therefore, it is necessary to interpret tephra layers—even if they appear pure and not contaminated with surrounding material—with great care, when applied to other dating methods.

7.3.4 Ice-Wedge Casts in South-Central Wyoming

Fossil ice wedges (in the form of ice-wedge casts) have been reported from southern Wyoming by Mears (1981, 1987). At that time, there were no age handles for these forms. Invited by B. Mears, University of Wyoming in Laramie, sites were visited and analyzed that are believed to represent the various types of wedges characteristic of the area (Mears, 1987).

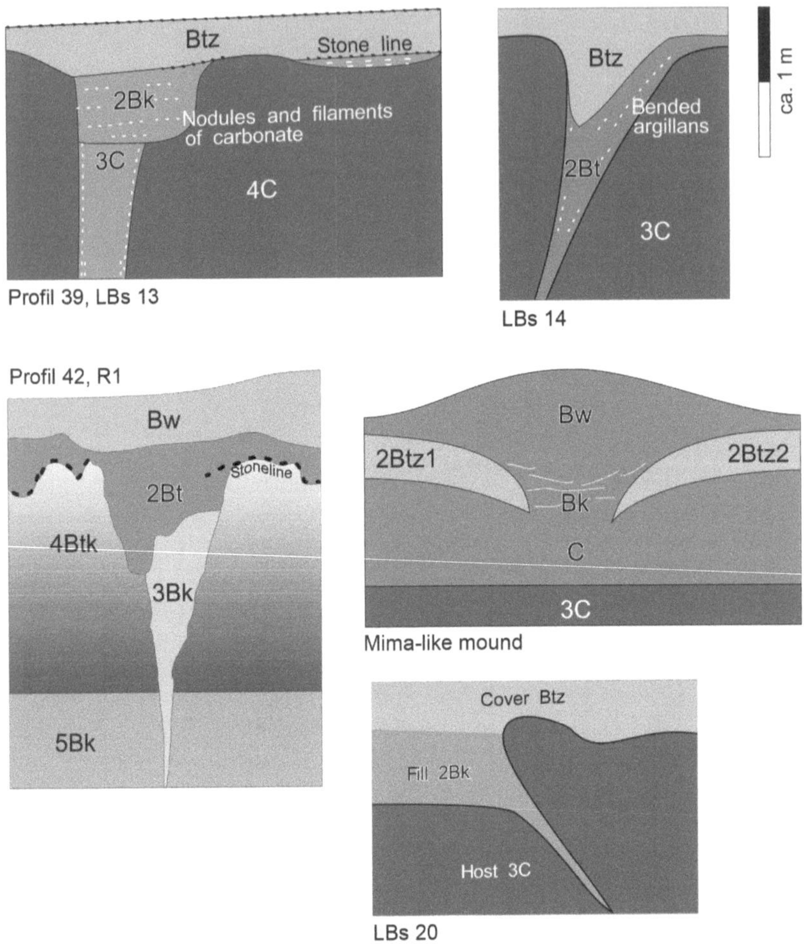

FIGURE 7.10 Sketches of south-central Wyoming ice-wedge casts and their soils (site numbers refer to Mears, 1987; *figure is reproduced from Kleber, 1999a*).

The wedge LBs 13 (Fig. 7.10, site numbers refer to Mears, 1987) is filled with bedded material with carbonate stains. The mantle is rich in salt; carbonate appears redistributed within it rather than leached into the underlying substratum. The soil in the mantle shows occasional clay films, with the illuviation probably induced by exchangeable sodium ions. This is indicated by the salt concentration and by the absence of carbonate leached from the clay-enriched horizon (Kleber, 1999a). The latter process would have preceded clay translocation, if sodium were not involved (Kleber, 1997), because there neither is evidence of later admixture of particulate carbonate to this horizon nor is the Ca-ion concentration in Wyoming rain water (Junge and Werby, 1958) sufficient to explain the soil carbonate. Thus, the cutans probably were rapidly formed in a natric rather than an argillic horizon. Accordingly, they are not indicative of the general environment (Birkeland, 1984). Clay illuviation is absent in the lower parts. The infilling yields traces of cummingtonite (Kleber, 1999a). The underlying shale differs in mineralogy and texture and is highly saline; this rock type presumably is the primary source of the salt in the soils of the area.

In LBs 14, the wedge infilling acted as a host for another wedge, which was in turn filled by mantle material. The lower and upper infillings differ in texture and in their pedogenic properties. The lower infilling (2Bt) is characterized by a fabric parallel to the wedge boundaries, which formed during the origination of the wedge (Mears, 1987). Along this structure, clay has accumulated in a few thin lamellae. It is unclear whether this is from the natric horizon above—there may have been sufficient exchangeable sodium in the 2Bt horizon to keep the clay dispersed—or whether it indicates a separate episode of soil formation. Traces of cummingtonite occur in the mantle but are absent in the lower infilling. The host is essentially the same as in LBs 13.

Wedge R1, located in the former city dump of Rawlins, Wyoming, cuts into alluvium. The mantle consists of two layers, which differ in texture, heavy minerals (reflecting changed sources of material), and in soil development, indicated by argillic properties in the lower layer (2Bt). Both layers contain traces of cummingtonite. Wedge-fill and host material differ from one another and from the overlying cover beds regarding texture and mineralogy. Both are enriched in pedogenic carbonate (the host was not calcareous upon deposition: Munn, 1987), reaching stage II+ in the host and stage I+ in the infilling. This carbonate engulfed a former argillic horizon in the host material (cf. Munn, 1987). Chlorite, probably derived from green hornblende through weathering, primarily accounts for the mineralogical differences between the host layers (4Btk and 5Bk: Kleber, 1999a). This indicates that the lower part of the host is more intensely weathered than the upper one. Therefore, the upper one is assumed to be a cover bed, though eolian admixture cannot be ascertained from the data. Thus, three cover beds may be identified, with the lower two separated by the wedge. In this case, the formation of horizons rich in carbonate was interrupted by the wedge formation, which is consistent with Dillon and Sorenson (2007), who described calcic horizons from southwestern Wyoming decisively disturbed by later periglacial overprint.

In the mima-like mound (see Reider et al., 1996 for site description and a complete data set), a layer with a natric horizon is intersected by a pipe with matrix material from the underlying alluvium, which has migrated upward to form a surficial mound, possibly by bioturbation (cf. Horwath and Johnson, 2006). Grain-size data suggest that pipe and mound materials are mixed underlying alluvium and host material. However, only the host contains cummingtonite (Kleber, 1999a).

The mantle layer common in and near the Laramie Basin is assumed to be equivalent to the Provo layer because it is the uppermost deposit as in the Great Basin; it contains cummingtonite; there is no intense carbonate enrichment beneath, which would be characteristic of the older soils. Clay films, atypical of the Provo soil at such elevation elsewhere, may be attributed to local soil salinity.

The following evolution may be concluded. (1) The oldest wedge is constrained by the inferred lower Promontory Geosol and the inferred upper Wyemaha Formation. The most reasonable age assignment is early Wisconsinan. Based on clast weathering, Mears (1987) also concluded a significantly older age of this type of wedges compared to others. (2) The maximum age of the Laramie Basin wedges described in this study is constrained by the fact that they are located upon an erosional bench of the Laramie River system, considered to be Pinedale in age (Mears, 1991). Furthermore, the age of these wedges is roughly limited by a radiocarbon-dated host of a similar wedge of approx. 25 ka BP (Nissen and Mears, 1990) and by three younger archeological sites (approx. 6000 BP: Mears, 1987). LBs 13 is overlain by the inferred Provo layer but filled mainly with older material. The soil in the older infilling does not resemble any of the major soils described before. Kleber (1990), however, described a local bifurcation of this layer into two sublayers, with a similar soil developed in the lower one; this finding could, however, not be repeated in later studies. Yet, this suggests that the wedge is somewhat older than the Provo layer, perhaps around the Pinedale maximum. (3) The youngest wedges predate the inferred Provo layer only slightly, because infilling and mantle are similar and both contain cummingtonite. This is in good agreement with a radiocarbon date of some 13 ka on a wedge fill (Spackman and Munn, 1984) and by the finding of Late Glacial Clovis artifacts in some fill material (Mayer, 2003). (4) The formation of mima-like mounds postdates the inferred Provo layer. A radiocarbon date of around 7 ka was believed to be too young when a periglacial origin of the mounds was favored (Spackman and Munn, 1984), but if the mounds were formed by a groundwater-vortex mechanism (Reider et al., 1996) or by diapiric folding, this dating might be correct.

7.3.5 Moraines in the Ruby Mountains, Northeast Nevada

Moraines of at least two glacial advances occur on the northwestern slope of the Ruby Mountains. Sharp (1938) assigned an early Wisconsinan age to the older Lamoille till, but Wayne (1984a,b) correlated it with the Bull Lake glaciation mainly based on soil evidence. A fault crosses the moraine, which was active at least once since the disappearance of the last glaciers, that is, after the Pinedale, as it offsets outwash of this age (Sharp, 1939; Wayne, 1984a). Three profiles, two on Lamoille till and one on affiliated outwash, were studied by Kleber (1999a) to test whether the previous age suggestions may be corroborated by analysis of cover beds.

Exposed by the fault scarp, which crosses and offsets the moraine, two cover beds overly fine-bedded alluvium, which is in turn underlain by till. Clay translocation is prominent within the second layer and probably affected parts of the alluvium as lamellae. A horizon of carbonate enrichment follows below the lower boundary of the alluvium. Another profile is located on a flat crest of the same moraine complex but downslope of the fault. Below an A-2Bt sequence as in the first profile, a horizon with compound carbonate-enriched and argillic properties follows above an horizon less enriched with carbonate (3Bt) and, finally, till (4C). The profile on the Lamoille outwash plain is similar to the second profile above, except that the carbonate concentration in the deeper horizons is higher; this is presumably due to calcareous mud beds, which are intercalated in the alluvium.

Consistent with other age suggestions (Birkeland et al., 1991; Chadwick et al., 1995; Wayne, 1984a,b), the till of the second profile and the outwash plain are inferred to be as old as Bull Lake because they carry a threefold sequence of cover beds and soils. However, in the first profile on the fault scarp, the third layer is missing; the carbonate-enriched horizon of the second soil is in till rather than in a paleosol as in the other profiles. Either the lower Wyemaha Formation was stripped off during an early activity of the fault, or Profile 1 resides on an early Wisconsinan till overlying (not exposed) Bull Lake till. Both tills largely differ mineralogically; if they were contemporary, a mixture of their mineral associations would be expected in the outwash material, which is not the case (Kleber, 1999a).

7.4 CONCLUSIONS

The occurrence of the upper layer in landslide areas provides essential evidence for the classification of Pleistocene slide masses. In general, it contains loess as well as the minerals of the Lake Laach Tephra and thus can be used to distinguish Pleistocene and Holocene slope sections. Upper layers superimposed on specific elements of a slide mass ascertain that the landslide is of Pleistocene age. If the upper layer and other datable material are absent, the landslide age remains unclear but is likely to be Holocene.

A problematic case is the genesis of the basal layer in landslide areas. Material mainly derived from surrounding bedrock may also originate from Holocene slope movements. Thus, it cannot be used as a marker bed.

Furthermore, an advanced degree of pedogenesis indicates long-term stable conditions and a lower degree of pedogenesis for geomorphological activity during the Holocene.

As a main result, the present geomorphologic, pedologic, geologic, and structural framework of hillslope systems and their sediments developed over the course of the Quaternary, the regional susceptibility to landslides may be traced back to comparably long-lasting processes and dynamics. As these processes continue today, the consideration of Quaternary landscape formation is an important tool to understand present dynamics of hillslope sediments.

Four attempts to evaluate cover beds for the relative dating of relief in the western USA yielded sound age suggestions when the following principles were applied: (1) Relief without overlying cover beds is assumed to postdate the Late

Glacial Provo layer, that is, to be Late Glacial or Holocene in age; (2) one cover bed or one unit of cover beds with a soil of intermediate maturity indicates an age older than the Provo layer but younger than the upper Wyemaha Formation, that is, a Pinedale (Last Glacial Maximum) age; (3) if there is another underlying cover bed with an argillic horizon, usually succeeded by an intense calcic horizon, the relief is believed to date prior to the upper Wyemaha Formation, that is, into the early Wisconsinan at least; (4) the parent material of the lower of two welded argillic horizons, usually considerably enriched in carbonate derived from the parent material of the upper one, is inferred to correlate with the lower Wyemaha Formation, indicating at least a Bull Lake age of the underlying relief; and (5) additional buried soils below this sequence indicate an even higher age; petrocalcic horizons may be a million or more years old (Bachmann and Machette, 1977).

Heavy-mineral analyses are not frequently used in Quaternary stratigraphy in the USA. As presented here, cummingtonite was found in a variety of sites in the inferred Provo layer and upper Wyemaha layer, but not yet in the inferred lower Wyemaha layer. This is consistent with the eruption history of Mount St. Helens, which yielded this mineral during particular phases (Mullineaux, 1986). If so, cummingtonite may serve as a stratigraphic marker mineral. However, Kleber (1999a) demonstrated that there may be other sources of cummingtonite such as in the Ruby Mountains, Nevada; on the other hand, the mineral does not occur in cover beds in the neighborhood of this source, and bedrock sources would have contributed to all cover beds rather than just to certain ones. Although further research is needed on this matter, the potential of heavy mineralogy as an indicator even of small amounts of admixed tephra is evident, which fact is also known from cover-bed studies in Central European subdued mountains (see Section 2.7.1.4).

The findings of this study appear at odds with the commonly adopted working hypothesis of soil chronofunctions derived from geomorphic chronosequences (e.g., Birkeland et al., 2003; Busacca, 1987; Harden, 1982; McFadden and Hendricks, 1985; Phillips et al., 1997). However, because the formation of a cover bed involves reworking of older material, the degree of weathering and the pedogenic imprint of components even of contemporaneous cover beds may differ markedly, depending on the weathering history of inherited underlying materials.

Several conclusions and caveats drawn from these findings may be of value for employing similar approaches to other areas. (1) A fundamental problem of sequence-stratigraphic approaches are erosion-induced hiatuses; some record may be lost in places, which danger is indicated by the fact that cover beds may vary in thickness within short distances. In many cases, such a hiatus may be recognized by means of soil properties, if these are sufficiently different among the major soil-forming episodes. However, where this is not the case, hiatuses possibly do not become evident. (2) Cover beds are formed in response to climatic conditions. If asynchroneity with climate fluctuations is involved, as in the case of neotectonics or other external influence, additional uncertainty arises. Tectonics may induce deposition out of phase with the sequence of layers elsewhere. However, such "irregular" layers are not supposed to carry the same pedogenic imprints as the "regular" ones. (3) Generally, age suggestions derived from cover beds do not necessarily indicate the age of the underlying deposit but the age of the last major geomorphic activity, when any remnants of older layers or soils were stripped off at the site under study.

This section demonstrates the feasibility of cover-bed studies for stratigraphic purposes, provided isochronous cover beds exist and can be identified in a given profile without the necessity of independent age constraints. The major advantage of using cover beds for relative dating is their wide occurrence (Kleber, 1997) and that the processes of their formation probably affected large areas in a similar manner. Thus, a great number of regions may be appropriate for this kind of approach worldwide. However, up to now, this is untested elsewhere and limited by the lack of independent age control on cover-bed successions.

Chapter 8

Conclusions

A. Kleber and B. Terhorst

8.1 TAKEOUTS OF THIS BOOK

Slopes are the most widespread landform on Earth. Thus, the deposits that cover them are major components of the landscape system. This certainly holds true for the mid-latitude slope deposits, which are in the focus of this book. Sensitive to environmental change, they may be used as important indicators of past and future climate change and anthropogenic intervention.

The subsurface of slopes in the mid-latitudes has not formed from underlying bedrock via weathering alone but consists almost in its entirety of dislocated local materials, partially comprising admixed eolian matter, mainly loess. These so-called cover beds are characterized by their areal extension, that is, they cover slopes almost completely, often without changing their properties remarkably over some distances. These deposits have been studied extensively in Central-European subdued mountains where bi- and tri-partite layer successions are typical, but there is evidence that similar material is ubiquitous in other areas of the mid-latitudes as well. However, there are differences in the particular expression of the phenomenon between humid and semi-arid areas. In the former, layers without eolian components are frequent and the particular layers may rarely be put into a chronostratigraphic context; in the latter, eolian contribution has influenced essentially all layers, and paleosols are ubiquitous to provide age handles, which may be utilized to relatively date landforms or deposits on which those layers have (or have not) been developed.

Slope deposits have remarkable importance in geo-ecological contexts. Without them, soils would have developed very differently, usually to a much less mature stage than they actually have reached. Cover beds control the spatial distribution of soils. Accordingly, they are valuable tools for soil surveying and for the assessment of site quality. Horizon boundaries are usually well adapted to disconformities (i.e., lithological discontinuities) of soil parent materials, and many processes acting to form soils on slopes are decisively influenced by cover-bed properties and successions: tri-partite successions usually tend to form clay-illuviated soils, whereas the soils in the bi-partite cases are more varied, depending especially on the bulk density of deeper and on the loess content

Developments in Sedimentology, Vol. 66. http://dx.doi.org/10.1016/B978-0-444-53118-6.00008-8

of surficial layers. Geochemical properties of soils are also dependent on those of the parent materials, with the loess content again being an important modifier. As a consequence, calculating soil-forming rates must take the indigenous cover-bed properties into account, or Holocene pedogenesis would easily be overestimated.

Furthermore, cover beds influence soil–water paths of slopes, triggering inter- versus groundwater flow through their anisotropic properties. Last but not least, cover beds initiate slope dynamics such as slope failures, as their anisotropic nature may also induce slope instabilities.

8.2 FUTURE RESEARCH DEMANDS ON COVER BEDS

Despite of decades of research on cover beds and all the results obtained by this as outlined in this book, there still remains much demand for further research, which we subdivide in this section into issues of pure and of applied research.

The vast majority of the knowledge on cover beds stems from augering and hand-dug pits. Especially with regard to the thickness of the deeper parts of cover-bed successions, this is often insufficient. A layer change at a depth of, let's say, 1 m seen in an auger core or in a pit is often kept for the boundary between cover beds and underlying saprolized rock. Digging deeper may reveal that the observed substrate change was just the boundary to another cover bed, located deeper in the profile. Areal identification of these deeper layers is, thus, complicated and hampered by the common restriction in depth of the surveying methods most often used. There is a way out of this dilemma: geophysical sounding; but this is more cost- and time-consuming than augering and calls for specially trained researchers. However, there already are some promising attempts to use geo-electrics, ground-penetrating radar, and reflection seismology, which clearly demonstrate that there often is much more to unveil than surficial augering can. This is clearly a methodical approach able to reach new insights into cover-bed distribution patterns.

Despite all similarities of cover-bed properties and successions, every new area under study delivers individual characteristics. There still are mountain ranges in Central Europe that have not yet been investigated in terms of cover beds. There are tremendously more gaps to fill, if the whole mid-latitudes are considered. Accordingly, one may expect by far not the entire bandwidth of phenomena to be known already. This fact lets one assume that every classification needs to be considered as tentative and may call for improvement as new data become available. Even at the present, it is obvious that the commonly used classification of AGB (2005) has its caveats and shortcomings and already needs reconditioning.

Theoretically, cover beds may be valuable archives of Quaternary paleoenvironments, mainly because they are one of the rare types of archives with a widespread, not only local, more or less punctiform, occurrence. Yet, there are two restrictions. One is that several processes interacting in the formation of such layers usually cannot be identified without doubt. As an example, the newest attempt to deduce processes using micromorphological analyses of thin sections brought to the light a number of unexpected findings on the reallocation of soil materials, the modern pedogenesis, and only limited indications of permafrost activity. We are certain that micromorphology will turn out to be an important tool for deciphering paleoenvironment-driven processes of the cover-bed formation and transformation.

As a second restriction, using cover beds as paleoenvironmental archives in Central Europe is decisively hampered by the lack of age handles: what may be the use of understanding a particular paleoenvironment, if one does not know when these conditions actually prevailed, and whether or not they were contemporary to other reconstructed paleoenvironments elsewhere? Thus, it would be extremely important to produce more, and more trustable, numerical datings on cover beds than there are now. However, first attempts to date cover beds were not very promising. For the time being, the only way out of this disappointing situation seems to be single-grain luminescence dating of quite large quantities of grains to perform statistical analyses in order to acquire and separate age populations of (1) eolian grains inherited from older materials, (2) grains admixed during cover-bed dislocation, and (3) grains brought into the substrate after its deposition by bioturbation. Another improvement necessary to achieve this is progress in spatially resolved dosimetry to get a better impression of the exposure to radiation each individual grain may have received. This approach to date loess-bearing deposits may already be very demanding, but in addition it seems, unfortunately, that at present there is no valid method at all to reliably date slope deposits without an eolian contribution. Another approach to the numerical dating of cover beds is by cosmogenic nuclides, which has been successfully tested at least once. However, far more data are necessary to reach valuable conclusions on paleoenvironments archived in cover beds.

A well-established aspect of applied research on cover beds is their influence on pedogenesis. However, there still is demand for additional research, especially concerning inherited soil properties from older soil-forming episodes, which for their part have not been preserved to our days. Micromorphology is the best known technique to approach these issues, in order to better understand soil/landscape relationships.

Another aspect of the application of cover-bed research is slope–water dynamics and, at a more integrating scale, watershed dynamics, with special emphasis on floods and transport of contaminants. First pieces of information show that cover beds actually are important triggers of slope–water paths. However, this knowledge is limited to a few types of environments; especially there is insufficient insight in how far this influence differs upon different bedrock types. Further studies as the ones depicted in this book are needed to overcome this shortcoming. An even more important problem has not yet been addressed in the literature on cover beds at all: the scale transfer from small-scale catchment dynamics to watersheds, especially aiming at improving flood prediction.

Since scale transfer in general is a major challenge for hydrology, this is anything but trivial. The still insufficient knowledge regarding such scale transfer is due to the high spatial variance of physical and hydraulic soil parameters, an incomplete understanding of the processes, as well as to the nonlinear spatio-temporal relationships between superficial and subterranean runoff processes; all these aspects are decisively connected to the layer-induced anisotropy of the subsurface. This calls for a nested approach of future research on catchment modeling: the objective should be to detect structures which control the dominating hydrological processes in catchments at various levels of scale. Only then predictions for yet not examined areas will become possible.

The classical field-measurement approaches as outlined in this book to understand hydrological behavior are much too costly and time-consuming to be applied to a manifold of catchments of different configurations or even at larger scales. Thus, the objective may be met only with a simplified approach, which, in our opinion, is geoelectrical monitoring of subsurface water dynamics, especially at the time of hydrological events. Though less time-consuming, this will still be not sufficiently efficient to allow for the desired scale transfer. Therefore, based on the insights from this type of monitoring and on the knowledge that already exists and that may be derived from geophysical sounding on the spatial distribution of cover beds and their properties, spatial modeling of cover-bed hydrological behavior, that is, of cover-bed properties, will be called for. First attempts on their spatial modeling have already been made, but even these call for intricate preceding field work. Thus, there is an urgent demand to enhance—through the inclusion of process-based modeling—and simplify the models by restriction to known relationships between topography, rocks, and cover beds. Such models would also contribute significantly to the forecasting of slope failures and all other applications that require knowledge of the subsurface.

References

Abdank, H., Zimmerling, B., Steininger, M., Meissner, R., 1999. Präferentielle Fließwege – Ursache für schnellen Stoffeintrag im Festgesteinsbereich der Unterharz-Plateauflächen. Mitteilungen der Deutschen Bodenkundlichen Gesellschaft 91, 731–734.

Abdank, H., Steininger, M., Zimmerling, B., Meissner, R., 2003. Tracerversuche zum Nachweis von Wasser- und Stofftransportpfaden in mehrlagigen Böden. Landnutzung und Landentwicklung 44 (2), 70–76.

Abel, W., 1976. Die Wüstungen des ausgehenden Mittelalters. Gustav Fischer Verlag, Stuttgart.

ACIA (Arctic Climate Impact Assessment), 2004. Impacts of a Warming Arctic. Cambridge University Press, Cambridge.

Adam, D.P., 1988. Pollen zonation and proposed informal climatic units for Clear Lake, California, cores CL-73-4 and CL-73-7. Geol. Soc. Am. Spec. Pap. 214, pp. 81–95.

AGB (Arbeitsgemeinschaft Boden), 1994. Bodenkundliche Kartieranleitung, 4th ed. Schweizerbart, Hannover.

AGB (Arbeitsgemeinschaft Boden), 2005. Bodenkundliche Kartieranleitung, 5th ed. Schweizerbart, Hannover.

Alaily, F., 1972. Entstehung von Decklehmen auf Lias-a-Schichtflächen in Südwestdeutschland und deren Veränderung durch Bodenbildung. PhD-Thesis Univ. Hohenheim.

Aleksandrovsky, A.L., Chichagova, O.A., 1995a. Radiocarbon age of East European forest-steppe Holocene paleosols. Terra Nostra 2/95, 9.

Aleksandrovsky, A.L., Targulian, V.O., Cherinskiy, A.Ye., Chichagova, O.A., 1991. New data on the age and evolution of sod-podzolic soils of the Central Russian Plain. Sov. Soil Sci. 23 (3), 1–5.

Alloway, B.J. (Ed.), 1990. Heavy Metals in Soils. Blackie, Glasgow.

Almond, P.C., Tonkin, P.J., 1999. Pedogenesis by upbilding in an extreme leaching and weathering environment, and slow loess accretion, South Westland, New Zealand. Geoderma 92, 1–36.

Altermann, M., 1990. Zur Petrographie der Lockergesteinsdecken des östlichen Harzes. Mitteilungen der Deutschen Bodenkundlichen Gesellschaft 62, 101–104.

Altermann, M., 1993. Gliederung und Lithologie pleistozäner Lagen im Gebiet Sachsen-Anhalts. Mitteilungen der Deutschen Bodenkundlichen Gesellschaft 72, 819–824.

Altermann, M., Rabitzsch, K., 1976. Quartäre Deckschichten im Raum Elbingerode – Rübeland (Harz). Hercyna N.F. 13 (3), 295–331.

Altermann, M., Mautschke, J., Erbe, C., Pretzschel, M., 1977. Kennzeichnung der quartären Deckschichten im Unterharz. Petermanns Geographische Mitteilungen 121 (2), 95–110.

Altermann, M., Steininger, M., Abdank, H., 1995. Böden, Landnutzung, Natur- und Gewässerschutz im östlichen Harz. Mitteilungen der Deutschen Bodenkundlichen Gesellschaft 77, 155–206.

Altermann, M., Jäger, K.-D., Kopp, D., Kowalkowski, A., Kühn, D., Schwanecke, W., 2008. On characteristics and subdivision of pedospheric differentiations due to previous periglacial conditions. Waldökologie, Landschaftsforschung und Naturschutz 6, 5–42.

Amoozegar, A., 1988. A compact constant-head permeameter for measuring saturated hydraulic conductivity of the vadose zone. Soil Sci. Soc. Am. J. 53, 1356–1361.

Anderson, M.G., Burt, T.P., 1978. The role of topography in controlling throughflow generation. Proc. Earth Surf. 3, 331–334.

Anderson, M.G., Burt, T.P. (Eds.), 1990. Process studies in hillslope hydrology. John Wiley and Sons, Chichester.

Anderson, D.W., Paul, E.A., 1984. Organo-mineral complexes and their study by radiocarbon dating. Soil Sci. Soc. Am. J. 48, 298–301.

André, M.-F., 2002. Rates of postglacial rock weathering on glacially scoured outcrops (Abisko-Riksgränsen area, 68°N). Geogr. Ann. 84 A, 139–150.

Andres, W., Bos, J.A.A., Houben, P., Kalis, A.J., Nolte, S., Rittweger, H., Wunderlich, J., 2001. Environmental change and fluvial activity during the Younger Dryas in central Germany. Quat. Int. 79, 89–100.

Anisimov, O.A., Shiklomanov, N.I., Nelson, F.E., 2002. Variability of seasonal thaw depth in permafrost regions: a stochastic modeling approach. Ecol. Model. 153, 217–227.

Antoine, P., Rousseau, D.-D., Moine, O., Kunesch, S., Hatté, C., Lang, A., Tissoux, H., Zöller, L., 2009. Rapid and cyclic aeolian deposition during the Last Glacial in European loess: a high-resolution record from Nussloch, Germany. Quat. Sci. Rev. 28, 2955–2973.

Apel, K., Fiedler, H.J., 1996. Waldböden auf Muschelkalk am Nordrand des Thüringer Beckens. Chemie der Erde 56, 159–170.

Applegarth, M.T., Dahms, D.E., 2004. Aeolian modification of moraine soils, Whiskey Basin, Wyoming, USA. Earth Surf. Process. Landforms 29, 579–585.

Arnold, R.W., 1968. Pedological significance of lithologic discontinuities. Trans. 9th international Congress Soil Sci. 4, 595–603.

Artmann, S., Völkel, J., 1999. Untersuchungen an periglazialen Deckschichten im Nationalpark Berchtesgaden, Nördliche Kalkalpen. Z. Geomorph. N.F. 43, 463–481.

Astakhov, V.I., 2001. Stratigraphic framework for the Upper Pleistocene of the Russian Arctic: changing paradigms. Glob. Planet. Change 31, 281–293.

Astakhov, V., 2004. Pleistocene ice limits in the Russian northern lowlands. In: Ehlers, J. (Ed.), Quaternary Glaciations: Extent and Chronology Part 1: Europe. Elsevier, San Diego, pp. 309–319.

ASTM (American Society for Testing and Materials), 1985. Classification of soils for engineering purposes. Annual Book of ASTM Standards D 2487-83, 04.08, 04.08. American Society for Testing and Materials 1985, 395–408.

Audorff, V., Beierkuhnlein, C., Heller, K., Kleber, A., 2007. Einzugsgebiete als landschaftsökologische Instrumente – der Frankenwald als Modellregion. Bayreuther Geographische Arbeiten 28, 181–197.

Bachmann, G.O., Machette, M.N., 1977. Calcic soils and calcretes in the southwestern United States. U.S. Geol. Surv. Open-File Rep. 77–794.

Baker, V.R., 1983. Late-Pleistocene fluvial systems. In: Porter, S.C. (Ed.), Late Quaternary Environments of the United States, 1: The Late Pleistocene. University of Minnesota Press, Minneapolis, pp. 115–129.

Baker, R.G., Richmond, G.M., 1978. Geology, Palynology, and climatic significance of two Pre-Pinedale lake sediments in and near Yellowstone National Park. Quat. Res. 10, 226–240.

Baker, R.G., Waln, K., 1985. Quaternary pollen record from the Great Plains and Central United States. In: Bryant Jr., V.M., Holloway, R.G. (Eds.), Pollen records of late-Quaternary North American sediments. American Association of Stratigraphic Palynologists, Dallas, TX, pp. 191–203.

Bargon, E., Fickel, W., Plass, W., Reichmann, H., Semmel, A., Zakosek, H., 1971. Zur Genese und Nomenklatur braunerde- und parabraunerdeähnlicher Böden in Hessen. Notizblatt des hessischen Landesamtes für Bodenforschung 99, 361–372.

Barnosky, C.W., 1981. A record of late Quaternary vegetation from Davis Lake, southern Puget Lowland, Washington. Quat. Res. 16, 221–239.

Barsch, D., Flügel, W.-A., 1988. Niederschlag, Grundwasser, Abfluss. Ergebnisse aus dem hydrologisch-geomorphologischen Versuchsgebiet "Hollmuth". Heidelberger Geographische Arbeiten 66.

Bartels, G., 1967. Stratigraphische und geomorphologische Auswertung von Schuttdecken vor Muschelkalkschichtkämmen und -schichtstufen im niedersächsischen Bergland. Eiszeitalter und Gegenwart 18, 76–81.

Bartels, G., Rohdenburg, H., 1968. Fossile Böden und Eiskeilhorizonte in der Ziegeleigrube Breinum (Niedersächsische Bergland) und ihre Auswertung für die Reliefentwicklung im Jungquartär. Göttinger Bodenkundliche Berichte 6, 109–126.

Bateman, M.D., 2008. Luminescence dating of periglacial sediments and structures. Boreas 37, 547–588.

Bateman, M.D., Frederick, C.D., Jaiswal, M.K., Singhvi, A.K., 2003. Investigations into the potential effects of pedoturbation on luminescence dating. Quat. Sci. Rev. 22, 1169–1176.

Bateman, M.D., Boulter, C.H., Carr, A.S., Frederick, C.D., Pete, D., Wilde, M., 2007. Detecting post-depositional sediment disturbance in sandy deposits using optical luminescence. Quat. Geochronol. 2, 57–64.

Beck, R.K., 1998. Schwermetalle in Waldböden des Schönbuchs: Bestandsaufnahme – ökologische Verhältnisse – Umweltrelevanz. Tübinger Geographische Studien 121.

Beck, R.K., 2002. Schadstoff-Transferpotenziale von Böden in mehrschichtigen periglazialen Lagen im Keuper-Lias-Bergland des Schönbuchs. Berichte zur Deutschen Landeskunde 76 (2/3), 169–185.

Beckmann, T., 1997. Präparation bodenkundlicher Dünnschliffe für mikromorphologische Untersuchungen. Hohenheimer Bodenkundliche Hefte 40, 89–103.

Beget, J.E., 1984. Tephrochronology of late Wisconsin deglaciation and Holocene glacier fluctuations near Glacier Peak, north Cascade Range, Washington. Quat. Res. 21, 304–316.

Behrens, T., 2003. Digitale Reliefanalyse als Basis von Boden-Landschafts-Modellen am Beispiel der Modellierung periglaziärer Lagen im Ostharz. Boden und Landschaft 42.

Beierkuhnlein, C., Kleber, A., 1999. Zur Morphogenese flächiger Quellfluren. Bayreuther Forum Ökologie 71, 27–34.

Bell, F.G., 2007. Basic Environmental and Engineering Geology. Whittles Publishing, Dunbeath/UK.

Bellanger, M., Homand, F., Remy, J.M., 1993. Water behaviour in limestones as a function of pores structure: application to frost resistance of some Lorraine limestones. Eng. Geol. 36, 99–108.

Benedict, J.B., 1970. Downslope soil movement in a Colorado alpine region: rates, processes, and climatic significance. Arctic Alpine Res. 2, 165–226.

Benedict, J.B., 1976. Frost creep and gelifluction features: a review. Quat. Res. 6, 55–76.

Benson, L., 1993. Factors affecting ^{14}C ages of lacustrine carbonates: timing and duration of the last highstand lake in the Lahontan Basin. Quat. Res. 39, 163–174.

Benson, L., Madole, R., Phillips, W., Landis, G., Thomas, T., Kubik, P., 2004. The probable importance of snow and sediment shielding on cosmogenic ages of north-central Colorado Pinedale and pre-Pinedale moraines. Quat. Sci. Rev. 23, 193–206.

Berger, G.W., 1991. The use of glass for dating volcanic ash by thermoluminescence. J. Geophys. Res. B12, 19705–19720.

Berthling, I., Eiken, T., Sollid, J.L., 2000. Continuous measurements of solifluction using carrier-phase differential GPS. Norsk Geografisk Tidsskrift – Norwegian Journal of Geography 54, 182–185.

Bertran, P., Texier, J.-P., 1999. Facies and microfacies of slope deposits. Catena 35, 99–121.
Betancourt, J.L., 1990. Late Quaternary biogeography of the Colorado Plateau. In: Betancourt, J.L., Van Devender, T.R., Martin, P.S. (Eds.), Packrat Middens, The last 40,000 Years of Biotic Change. University of Arizona Press, Tucson, AZ, pp. 259–293.
Bettis III, E.A., Muhs, D.R., Roberts, H.M., Wintle, A.G., 2003. Last glacial loess in the conterminous USA. Quat. Sci. Rev. 22, 1907–1946.
Bibus, E., 1985. Zur jungen Relief- und Bodenentwicklung in der Umgebung von Tübingen. Z. Geomorph. N.F. 50 (Suppl.), 109–124.
Bibus, E., 1986. Die Rutschung am Hirschkopf bei Mössingen (Schwäbischen Alb) – Geowissenschaftliche Rahmenbedingungen – Geoökologische Folgen. Geoökodynamik 7, 333–360.
Bibus, E., Terhorst, B., 2001. Mass movements in South-West Germany. Analyses and results from the Tübingen Work Group of the MABIS Project. Z. Geomorph. N.F. 125 (Suppl.), 53–63.
Bibus, E., Nagel, G., Semmel, A., 1976. Periglaziale Reliefformung im zentralen Spitzbergen. Catena 3 (3/4), 29–44.
Bibus, E., Eberle, J., Kösel, M., Rilling, K., Terhorst, B., 1991. Jungquartäre Reliefformung und ihre Beziehung zur Bodenbildung und Bodenverbreitung im Stromberg und Zabergäu (Bl. Brackenheim). Geologisches Jahrbuch für Baden-Württemberg 33, 7–50.
Bibus, E., Kallinich, J., Terhorst, B., 2001. Dating methods for mass movements studied by the MABIS Project. Z. Geomorph. N.F. 125 (Suppl.), 153–162.
Biggar, N.E., Harden, D.R., Gillam, M.L., 1981. Quaternary deposits in the Paradox Basin. Rocky Mountain Assoc. Geol. Field Conf. 1981, 129–145.
Bíl, M., Müller, I., 2008. The origin of shallow landslides in Moravia (Czech Republic) in the spring of 2006. Geomorphology 99, 246–253.
Birkeland, P., 1974. Pedology, weathering and geomorphological research. Oxford University Press, New York, Oxford.
Birkeland, P.W., 1984. Soils and Geomorphology, 2nd ed. Oxford University Press, New York, Oxford.
Birkeland, P.W., 1985. Quaternary soils of the western United States. In: Boardman, J. (Ed.), Soils and Quaternary Landscape Evolution. John Wiley & Sons, Chichester, pp. 303–324.
Birkeland, P.W., 1990. Soil-geomorphic research – a selective overview. Geomorphology 3, 207–224.
Birkeland, P.W., 1999. Soils and Geomorphology, 3rd ed. Oxford University Press, New York, Oxford.
Birkeland, P.W., Burke, R.M., 1988. Soil catena chronosequences on eastern Sierra Nevada moraines, California, USA. Arctic Alpine Res. 20, 473–484.
Birkeland, P.W., Berry, M.E., Swanson, D.K., 1991. Use of soil catena field data for estimating relative ages of moraines. Geology 19, 281–283.
Birkeland, P.W., Colman, S.M., Burke, R.M., Shroba, R.R., Meierding, T.C., 1979. Nomenclature of alpine glacial deposits, or, what's in a name. Geology 7, 532–536.
Birkeland, P.W., Shroba, R.R., Burns, S.F., Price, A.B., Tonkin, P.J., 2003. Integrating soils and geomorphology in mountains – an example from the Front Range of Colorado. Geomorphology 55, 329–344.
Black, R.F., 1954. Permafrost, a review. Geol. Soc. Am. Bull. 65, 839–855.
Black, R.F., 1976. Periglacial features indicative of permafrost: ice and soil wedges. Quat. Res. 6, 3–26.
Blank, R.R., Young, J.A., Lugaski, T., 1996. Pedogenesis on talus slopes, the Buckskin Range, Nevada, USA. Geoderma 71, 121–142.

Bleich, K.E., 1960. Das Alter des Albtraufs. Jahreshefte des Vereins für Vaterländische Naturkunde in Württemberg 115, 39–92.

Blume, H.-P., 1981. Schwermetallverteilung und -bilanzen typischer Waldböden aus nordischem Geschiebemergel. Zeitschrift für Pflanzenernährung und Bodenkunde 144, 156–163.

Blume, H.-P. (Ed.), 1992. Handbuch des Bodenschutzes. 2nd ed. Ecomed, Landsberg.

Blume, H.-P., Leinweber, P., 2004. Plaggen soils: landscape history, properties, and classification. J. Plant Nutr. Soil Sci. 167, 319–327.

Blume, H.-P., Röper, H.-P., 1977. Der Mineralbestand als bodengenetischer Indikator. Mitteilungen der Deutschen Bodenkundlichen Gesellschaft 25, 797–824.

Blume, H.-P., Schwertmann, U., 1969. Genetic evaluation of profile distribution of Al, Fe, Mn oxides. Soil Sci. Soc. Am. Proc. 33, 438–444.

Boardman, J., 1992. Current erosion in the South Downs: implication for the past. In: Bell, M. and J. Boardman (Eds.), Past and present soil erosion. Oxbow Monograph 22, pp. 9–19.

Bogaard, P.v.D., Schmincke, H.-U., 1985. Laacher See Tephra: a widespread isochronous late Quaternary tephra layer in central and northern Europe. Geol. Soc. Am. Bull. 96, 1554–1571.

Bork, H.-R., Bork, H., Dalchow, C., Faust, B., Piorr, H.-P., Schatz, T., 1998. Landschaftsentwicklung in Mitteleuropa. Wirkungen des Menschen auf Landschaften. Klett-Pertes, Gotha, Stuttgart.

Bos, J.A.A., 2001. Lateglacial and Early Holocene vegetation history of the northern Wetterau and the Amöneburger Basin (Hessen), central-west Germany. Rev. Palaeobot. Palynol. 115, 177–212.

Brady, N.C., Weil, R.R., 2008. The nature and properties of soils, 15th ed. Pearson/Prentice Hall, New Jersey.

Brantley, S.L., White, T.S., White, A.F., Sparks, D., Richter, D., Pregitzer, K., Derry, L., Chorover, J., Chadwick, O.A., April, R., Anderson, S., Amundson, R., 2006. Frontiers in exploration of the critical zone. http://www.czen.org/files/czen/CZEN_Booklet.pdf (retrieved Aug. 31, 2012).

Brantley, S.L., Goldhaber, M.B., Ragnarsdottir, K.V., 2007. Crossing disciplines and scales to understand the critical zone. Elements 3, 307–314.

Brauer, A., Endres, C., Negendank, J.F.W., 1999a. Lateglacial calendar year chronology based on annually laminated sediments from Lake Meerfelder Maar, Germany. Quat. Int. 61, 17–25.

Brauer, A., Endres, C., Günter, C., Litt, T., Stebich, M., Negendank, J.F.W., 1999b. High resolution sediment and vegetation responses to Younger Dryas climate change in varved lake sediments from Meerfelder Maar, Germany. Quat. Sci. Rev. 18, 321–329.

Braukämper, K., 1990. Zur Verbreitung periglazialer Deckschichten in Deutschland. PhD-Thesis, Univ. Bochum.

Breemen van, N., Buurman, P., 2002. Soil Formation, 2nd ed. Kluwer, Dordrecht, Boston, London.

Brewer, R., 1972. Use of macro- and micromorphological data in soils stratigraphy to elucidate surficial geology and soil genesis. J. Geol. Soc. Aust. 19, 331–344.

Bridges, E.M., 1997. Origins, adoption and development of soil horizon designations. Adv. Geoecol. 29, 47–66.

Brimhall, G.H., Lewis, C.J., Ague, J.J., Dietrich, W.E., Hampel, J., Teague, T., Rix, P., 1988. Metal enrichment in bauxites by deposition of chemically mature eolian dust. Nature 333, 819–824.

Briner, J.P., 2009. Moraine pebbles and boulders yield indistinguishable ^{10}Be ages: a case study from Colorado, USA. Quat. Geochronol. 4, 299–305.

Brix, F., 1969. On the stratigraphy and lithology of the Flysch Zone in the surroundings of the Hagenbach Valley (Northern Vienna Woods, Austria). Rocznik Polsk. Tow. Geol. XXXIX (1–3), 455–469.

Brodzikowski, K., van Loon, A.J., 1987. A systematic classification of glacial and periglacial environments, facies and deposits. Earth Sci. Rev. 24, 297–381.

Brodzikowski, K., van Loon, A.J., 1991. Glacigenic sediments. Developments Sedimentology 49, Elsevier, Amsterdam.

Bromhead, E.N., Ibsen, M.-L., 2004. Bedding-controlled coastal landslides in Southeast Britain between Axmouth and the Thames Estuary. Landslides 1, 131–141.

Bronger, A., 1978. Climatic sequences of steppe soils from eastern Europe and the USA with emphasis on the genesis of the "argillic horizon". Catena 5, 33–51.

Brunsden, D., 1999. Some geomorphological considerations for the future development of landslide models. Geomorphology 30, 13–24.

Büdel, J., 1959. Periodische und episodische Solifluktion im Rahmen der klimatischen Solifluktionstypen. Erdkunde 13, 297–314.

Bullmann, H., 2010. Eigenschaften und Genese periglazialer Deckschichten auf Carbonatgesteinen des Muschelkalks in einem Teilgebiet der ostthüringischen Triaslandschaft. PhD-Thesis, Univ. Leipzig.

Bullmann, H., Heinrich, J., 2012. Sensitivity of frost weathering and aeolian deposition during genesis of Late Quaternary periglacial slope covers on calcareous rocks of a Muschelkalk landscape, eastern Thuringian Basin, Germany. http://dx.doi.org/10.1016/j.catena.2012.01.011.

Burga, A., Perret, R., 1998. Vegetations- und Klimageschichte der Schweiz seit dem jüngsten Eiszeitalter. Ott Verlag, Thun.

Burr, T.N., Currey, D.R., 1988. The Stockton Bar. Utah Geological Mineralogical Survey Miscellaneous Publications, 88–1, 66–73.

Bursik, M.I., Gillespie, A.R., 1993. Late Pleistocene glaciation of Mono Basin, California. Quat. Res. 39, 24–35.

Busacca, A.J., 1987. Pedogenesis of a chronosequence in the Sacramento Valley, California, USA, I. Application of a soil development index. Geoderma 41, 123–148.

Bussemer, S., 1994. Geomorphologische und bodenkundliche Untersuchungen an periglaziären Deckserien des mittleren und östlichen Barnim. Berliner Geographische Arbeiten 80.

Bussemer, S., 2002. Periglacial cover beds in the young moraine landscapes of northern Eurasia. Z. Geomorph. N.F. 127 (Suppl.), 81–105.

Bussemer, S., 2003. Periglaziale Deckschichten und Bodenmosaike der Jungmoränenlandschaften von Oberbayern und Nordtirol. Mitteilungen der Geographischen Gesellschaft in München 86, 17–58.

Butler, B.E., 1959. Periodic phenomena in landscapes as a basis for soil studies. Soil Publication 14. CSIRO, Melbourne.

Butler, B.E., 1982. A new system for soil studies. J. Soil Sci. 33, 581–595.

Capo, R.C., Chadwick, O.A., 1999. Sources of strontium and calcium in desert soil and calcrete. Earth Planet. Sci. Lett. 170, 61–72.

Cardinali, M., Galli, M., Guzzetti, F., Reichenbach, P., Borri, G., 1994. Relationships between mass-movements and tectonic setting in the Carpina Basin, northern Umbria. Geografia Fisica e Dinamica Quaternaria 17, 3–17.

Casper, M.C., 2002. Die Identifikation hydrogeologischer Prozesse im Einzugsgebiet des Durreychbaches (Nordschwarzwald). PhD-Thesis, University of Karlsruhe.

Catuneanu, O., 2006. Principles of Sequence Stratigraphy. Elsevier, Amsterdam.

Catuneanu, O., Bhattacharya, J.P., Blum, M.D., Dalrymple, R.W., Eriksson, P.G., Fielding, C.R., Fisher, W.L., Galloway, W.E., Gianolla, P., Gibling, M.R., Giles, K.A., Holbrook, J.M., Jordan, R., Kendall, C.G.St.C., Macurda, B., Martinsen, O.J., Miall, A.D., Nummedal, D., Posamentier, H.W., Pratt, B.R., Shanley, K.W., Steel, R.J., Strasser, A., Tucker, M.E., 2010.

Sequence stratigraphy: common ground after three decades of development. First Break 28 (1), 21–34.

Chadwick, O.A., Davis, J.O., 1990. Soil forming intervals caused by eolian sediment pulses in the Lahontan Basin, northwestern Nevada. Geology 18, 243–246.

Chadwick, O.A., Nettleton, W.D., Staidl, G.J., 1995. Soil polygenesis as a function of Quaternary climate change, northern Great Basin, USA. Geoderma 68, 1–26.

Chadwick, O.A., Hall, R.D., Phillips, F.M., 1997. Chronology of Pleistocene glacial advances in the central Rocky Mountains. Geol. Soc. Am. Bull. 109, 1443–1462.

Chifflard, P., Didszun, J., Zepp, H., 2008. Skalenübergreifende Prozess-Studien zur Abflussbildung in Gebieten mit periglazialen Deckschichten (Sauerland, Deutschland). Grundwasser 13 (1), 27–41.

Chittleborough, D.J., 1992. Formation and pedology of duplex soils. Austr. J. Exp. Agric. 32, 815–825.

Chlopecka, A., Bacon, R.J., Wilson, M.J., Kay, J., 1996. Forms of cadmium, lead and zinc in contaminated soils from southwest Poland. J. Environ. Qual. 25, 69–79.

Clark, P.U., Dyke, A.S., Shakun, J.D., Carlson, A.E., Clark, J., Wohlfarth, B., Mitrovica, J.X., Hostetler, S.W., McCabe, A.M., 2009. The Last Glacial Maximum. Science 325, 710–714.

Clayton, J.S., Ehrlich, W.A., Cann, D.B., Day, J.H., Marshall, I.B., 1977. Soils of Canada. Agric. Can. Publ. 1544, 2 vols, Ottawa.

Clemens, G., Stahr, K., 1994. Present and past soil erosion rates in catchments of the Kraichgau area (SW-Germany). Catena 22, 153–168.

Cole, K.L., 1990. Reconstruction of past desert vegetation along the Colorado River using packrat middens. Palaeogeogr. Palaeoclimatol. Palaeoecol. 76, 349–366.

Colman, S.M., 1982. Clay mineralogy of weathering rinds and possible implications concerning the sources of clay minerals in soils. Geology 10, 370–375.

Colman, S.M., 1983. Influence of the Onion Creek salt diapir on the late Cenozoic history of Fisher Valley, southeastern Utah. Geology 11, 240–243.

Colman, S.M., Pierce, K.L., 1981. Weathering rinds on andesitic and basaltic stones as a Quaternary age indicator, western United States. U.S. Geol. Surv. Prof. Pap. 1210.

Colman, S.M., Pierce, K.L., 1986. Glacial sequence near McCall, Idaho: weathering rinds, soil development, morphology, and other relative age criteria. Quat. Res. 25, 25–42.

Colman, S.M., Pierce, K.L., 1992. Varied records of Early Wiconsinan alpine glaciation in the western United States derived from weathering rind thicknesses. Geol. Soc. Am. Spec. Pap. 270, 269–278.

Colman, S.M., Choquette, A.F., Rosholt, J.N., Miller, G.H., Huntley, D.J., 1986. Dating the upper Cenozoic sediments in Fisher Valley, southeastern Utah. Geol. Soc. Am. Bull. 97, 1422–1431.

Cox, J.W., McFarlane, D.J., 1995. The causes of water logging in shallow soils and their drainage in southwestern Australia. J. Hydrology 167, 175–194.

Cremeens, D.L., Darmody, R.G., George, S.E., 2005. Upper slope landforms and age of bedrock exposures in the St. Francois Mountains, Missouri: a comparison to relict periglacial features in the Appalachian Plateau of West Virginia. Geomorphology 70, 71–84.

Currey, D.R., 1990. Quaternary palaeolakes in the evolution of semidesert basins, with special emphasis on Lake Bonneville and the Great Basin, USA. Palaeogeogr. Palaeoclimatol. Palaeoecol. 76, 189–214.

Currey, D.R., Oviatt, C.G., 1985. Durations, average rates, and probable causes of Lake Bonneville expansion, still-stands, and contractions during the last deep-lake cycle. In: Kay, P.A., Diaz, H.F. (Eds.), Problems of and Prospects for Predicting Great Salt Lake Levels. Proceedings of a NOAA Conference March 1985, University of Salt Lake City, Salt Lake City, UT, pp. 9–24.

Currey, D.R., Atwood, G., Mabey, D.R., 1984. Major levels of Great Salt Lake and Lake Bonneville. Utah Geol. Mineralog. Survey Map 73.

Curtin, D., Steppuhn, H., Selles, F., 1994. Clay dispersion in relation to sodicity, electrolyte concentration, and mechanical effects. Soil Sci. 58, 955–961.

Dahlke, R., 1984. Das Paäozoikum des Winterberges bei Ludwigsstadt/Obfr. (Gräfenthaler Horst; bayerischer Anteil) unter besonderer Berücksichtigung der Tentaculitenschichten (Unter-Devon). Unpubl. Diploma Thesis, Univ. Hamburg.

Dahms, D.E., 1993. Mineralogical evidence for eolian contribution to soils of late Quaternary moraines, Wind River Mountains, Wyoming, USA. Geoderma 59, 175–196.

Dahms, D.E., 2004. Relative and numeric age data for pleistocene glacial deposits and diamictons in and near Sinks Canyon, Wind River Range, Wyoming, U.S.A. Arctic, Antarctic, and Alpine Research 36, 59–77.

Damm, B., 2005. Gravitative Massenbewegungen in Südniedersachsen. Die Altmündener Wand – Analyse und Bewertung eines Rutschungsstandortes. Z. Geomorph. N.F. 138 (Suppl.), 189–209.

Damm, B., 2010. Landslide database: mass movements in subdued mountains of Germany 1137-2009. Göttingen (unpublished).

Damm, B., Hagedorn, J., 2010. Holocene floodplain formation in the southern Cape region, South Africa. Geomorphology 122, 213–222.

Damm, B., Terhorst, B., 2010. A model of slope formation related to landslide activity in the Eastern Prealps, Austria. Geomorphology 122, 338–350.

Damm, B., Terhorst, B., Köttritsch, E., Ottner, F., Mayrhofer, M., 2008. Zum Einfluss bodenphysikalischer und bodenmechanischer Parameter in quartären Deckschichten auf Massenbewegungen im Wienerwald. Abhandlungen der Geologischen Bundesanstalt 62, 33–37.

Damm, B., Varga, K., Heckmann, T., Becht, M., 2009. The impact of bedrock stratification on landslide susceptibility – an example of GIS-based landslide modelling in the Bunter Sandstone areas of northern Hesse and southern Lower Saxony (Germany). Die Erde 140, 175–193.

Damm, B., Terhorst, B., Bork, H.-R. (Eds.), 2010a. Quaternary landscape evolution and morphodynamics of terrestrial ecosystems. Quat. Int. 222, 1–236.

Damm, B., Becht, M., Varga, K., Heckmann, T., 2010b. Relevance of tectonic and structural parameters in Triassic bedrock formations to landslide susceptibility in Quaternary hillslope sediments. Quat. Int. 222, 143–153.

Daniels, R.B., Hammer, R.D., 1992. Soil Geomorphology. Wiley, New York.

Davis, W.M., 1899. The geographical cycle. Geogr. J. 14, 481–504.

Davis, O.K., 1990. Caves as sources of biotic remains in arid western North America. Palaeogeogr. Palaeoclimatol. Palaeoecol. 76, 331–348.

Davis, O.K., Sheppard, J.C., Robertson, S., 1986. Contrasting climatic histories for the Snake River Plain, Idaho, resulting from multiple thermal maxima. Quat. Res. 26, 321–339.

Davydova, N., Servant-Vildary, S., 1996. Late Pleistocene and Holocene history of the lakes in the Kola Peninsula, Karelia and the north-western part of the East European plain. Quat. Sci. Rev. 15, 997–1012.

De Wolf, T., 1988. Stratified slope deposits. In: Clark, M.J. (Ed.), Advances in periglacial Geomorphology. Wiley, Chichester, pp. 91–110.

Derry, L.A., Chadwick, O.A., 2007. Contributions from Earth's atmosphere to soil. Elements 3, 333–338.

Didszun, J., 2004. Experimentelle Untersuchungen zur Skalenabhängigkeit der Abflussbildung. Freiburger Schriften zur Hydrologie 19.

Dietrich, W.E., Montgomery, D.R., 1998. SHALSTAB. A digital terrain model for mapping shallow landslide potential. http://calm.geo.berkeley.edu/geomorph//shalstab/index.htm (accessed Aug. 7, 2010).

Dietze, M., 2006. Untersuchung sedimentpetrographisch-mineralogischer Eigenschaften quartärer Lagen und Modellierung ihrer Verbreitungsmuster im Tharandter Wald (Sachsen). Unpubl. Diploma Thesis, Dept. Geosciences, Dresden University.

Dietze, M., Kleber, A., 2010. Spatial prediction of material and thickness properties of periglacial cover beds. Geoderma 156, 346–356.

Dietze, M., Kleber, A., 2011. Contribution of lateral processes to stone pavement formation in deserts inferred from clast orientation patterns. Geomorphology 139–140, 172–187.

Dietze, M., Roettig, C.B., 2012. Neues Licht auf alten Fragen – zur Entstehung von Steinlagen in aufwachsenden Sedimentprofilen. DEUQUA Annual Meeting. http://www.bayceer.uni-bayreuth.de/deuqua2012/en/top/bayconf/beitrag_detail.php?id_obj=10526 (retrieved Sept. 25, 2012).

Dietze, M., Muhs, S., Dietze, E., 2010. Ambiguities of relative age indicators on abandoned surfaces of arid environments. Z. Geomorph. N.F. 55 (Suppl. Iss. 3), 49–75.

Dietze, M., Groth, J., Kleber, A., 2013. Alignment of stone-pavement clasts by unconcentrated overland flow – implications of numerical and physical modelling. Earth Surface Processes and Landforms in the press, http://dx.doi.org/doi:10.1002/esp.3365.

Dillon, J.S., Sorenson, C.J., 2007. Relict cryopedogenic features in soils with secondary carbonate horizons, western Wyoming, USA. Permafrost Periglac. Proc. 18, 285–299.

DIN 1055-2, 1976. Lastannahmen für Bauten. Bodenkenngrößen, Wichte, Reibungswinkel, Kohäsion, Wandreibungswinkel. Deutscher Normenausschuss, Berlin.

DIN 4084, 1981. Baugrund, Gelände und Böschungsbruchberechnungen. Deutscher Normenausschuss, Berlin.

DIN TB 113, 2002. Erkundung und Untersuchung des Baugrundes. DIN-Taschenbuch 113. Beuth-Verlag, Berlin, Wien, Zürich.

Dotterweich, M., 2008. The history of soil erosion and fluvial deposits in small catchments of Central Europe: deciphering the long term interaction between human and the environment – a review. Geomorphology 101, 192–208.

Dreimanis, A., 1989. Tills: their genetic terminology and classification. In: Goldthwait, R.P., Matsch, C.L. (Eds.), Genetic Classification of Glacigenic Deposits. Balkema, Rotterdam, pp. 17–83.

Durn, G., 2003. Terra rossa in the Mediterranean region: parent materials, composition and origin. Geol. Croat. 56, 83–100.

Eckmeier, E., Gerlach, R., Gehrt, E., Schmidt, M.W.I., 2007. Pedogenesis of chernozems in Central Europe – a review. Geoderma 139, 288–299.

Ehwald, E., 1970. Zur Systematik der Böden der DDR unter Berücksichtigung rezenter und reliktischer Merkmale. In: Anonymus, (Ed.), Beiträge zur Bodensystematik unter besonderer Berücksichtigung reliktischer und rezenter Merkmale, Tagungsber. DADL, Berlin, pp. 9–32, 102.

Eiberweiser, M., 1995. Untersuchungen zur Schwermetall-Tiefenverteilung in Böden und periglazialen Deckschichten des ostbayerischen Kristallins und seiner Randgebiete. Unpubl. PhD-Thesis, Univ. Regensburg.

Eiberweiser, M., Völkel, J., 1993. Schwermetallverteilung als Indikator für Schichtwechsel in Böden des ostbayerischen Grundgebirges und seiner Randgebiete. Mitteilungen der Deutschen Bodenkundlichen Gesellschaft 72, 327–330.

Einsele, G., Agster, G., Elgner, M., 1986. Niederschlag-Bodenwasser-Abflußbeziehungen bei Hochwasserereignissen im Keuper-Lias-Bergland des Schönbuchs. In: Deutsche Forschungsgemeinschaft, (Ed.), Das landschaftsökologische Forschungsprojekt Naturpark Schönbuch. VCH-Verlag, Weinheim, pp. 209–234.

Eitel, B., 1994. Kalkreiche Decksedimente und Kalkkrustengenese in Namibia: Entstehung und geomorphologische Bedeutung. Selbstverlag, Stuttgart.

Eitel, B., 2003. Geomorphogenese zwischen Rhein und Neckar: Stand der Forschung und offene Fragen. GeoArchaeoRhein 4, 127–152.

Emmerich, K.-H., 1990. Das Meßfeld für Bodenbewegungen. Nachrichtenblatt Sektion Eichstätt Deutscher Alpenverein 40 (1), 14–23.

Emmerich, K.-H., 1994. Podsole im Buntsandstein-Odenwald. Geologisches Jahrbuch Hessen 122, 173–184.

Enters, D., Kirilova, E., Lotter, A.F., Lücke, A., Parplies, J., Jahns, S., Kuhn, G., Zolitschka, B., 2010. Climate change and human impact at Sacrower See (NE Germany) during the past 13,000 years: a geochemical record. J. Paleolimnol. 43, 719–737.

Ernstberger, H., Meuser, A., Sokollek, V., Wohlrab, B., 1992. Hydrologische Brachlandforschung im Mittelgebirge – Zehn Jahre Untersuchungen in Mittelhessen. DVWK-Schriften 101.

Erol, O., 1970. Les hauts niveaux pléistocenes du Tuz Gölü (lac Salê) en Anatolie centrale (Turquie). Annales du Géographie 79, 39–50.

Erol, O., 1978. The Quaternary history of the lake basins of central and southern Anatolia. In: Brice, W. (Ed.), The Environmental History of the Near and Middle East Since the Last Ice Age. London Academic Press, London, pp. 111–139.

Fang, X.-M., Li, J.-J., van der Voo, R., 1999. Rock magnetic and grain size evidence for intensified Asian atmospheric circulation since 800,000 years B.P. related to Tibetan uplift. Earth Planet. Sci. Lett. 165, 129–144.

Felix-Henningsen, P., 1990. Die mesozoisch-tertiäre Verwitterungsdecke (MTV) im Rheinischen Schiefergebirge. Relief, Boden, Paläoklima 6.

Felix-Henningsen, P., 1991. Bodenentwicklung in periglazialen Deckschichten des Osthunsrücks, Rheinisches Schiefergebirge. Mitteilungen der Deutschen Bodenkundlichen Gesellschaft 66, 779–782.

Felix-Henningsen, P., Spies, E.D., Zakosek, H., 1991. Genese und Stratigraphie periglazialer Deckschichten auf der Hochfläche des Ost-Hunsrücks (Rheinisches Schiefergebirge). Eiszeitalter und Gegenwart 41, 56–69.

Fiedler, H.J., Hunger, W., 1970. Geologische Grundlagen der Bodenkunde und Standortslehre. Steinkopff, Dresden, Leipzig.

Fieldes, M., Perrott, K.W., 1966. The nature of allophane in soils. New Zealand J. Soil Sci. 9, 623–629.

Filipinski, M., Grupe, M., 1990. Verbreitungsmuster lithogener, pedogener und anthropogener Schwermetalle in Böden. Zeitschrift für Pflanzenernährung und Bodenkunde 153, 69–73.

Finke, P.A., Bouma, J., Hoosbeek, M.R. (Eds.), 1998. Soil and Water Quality at Different Scales. Kluwer, Dordrecht.

Fitzpatrick, E.A., 1956. An indurated soil horizon formed by permafrost. J. Soil Sci. 7, 248–257.

Flügel, W.-A., 1993. Hierarchical structured hydrological process studies to regionalize interflow in a loess covered catchment near Heidelberg. IAHS Publ. 212, 215–223.

Forman, S.L., Smith, P.R., Tullis, J.A., McDaniel, P.A., 1993. Timing of late Quaternary glaciations in the western United States based on the age of loess on the eastern Snake River Plain, Idaho. Quat. Res. 40, 30–37.

Frank, C., Terhorst, B., Damm, B., Thiel, C., Frechen, M., Peticzka, R., 2011. Pleistocene loess deposits and mollusk assemblages in the eastern Pre-Alps. E&G Quat. Sci. J. 60, 126–136.

Frechen, J., Rosauer, E.A., 1959. Aufbau und Gliederung des Würm-Löss-Profils von Kärlich im Neuwieder Becken. Fortschritte in der Geologie des Rheinlands und Westfalen 4, 267–282.

French, C.A.I., 1992. Alluviated fen-edge prehistoric landscapes in Cambridgeshire, England. Archeologia del Paesaggio, Firenze, 709–731.

French, H., 2007. The periglacial environment, 3rd ed. Wiley, Chichester.

Fried, G., 1984. Gestein, Relief und Boden im Buntsandstein-Odenwald. Frankfurter Geowissenschaftliche Arbeiten Ser. D 4.

Friedrich, K., 1996. Digitale Reliefgliederungsverfahren zur Ableitung bodenkundlich relevanter Flächeneinheiten. Frankfurter Geowissenschaftliche Arbeiten D21.

Fritz, P., 1976. Gesteinsbedingte Standorts- und Formendifferenzierung rezenter Periglazialerscheinungen in den Ostalpen. Mitteilungen der Österreichischen Geographischen Gesellschaft 118, 237–272.

Froese, D.G., Slate, J., Knott, J.R., Lowe, D.J., 2008. Global tephra studies: John Westgate and Andrei Sarna-Wojcicki commemorative volume. Quat. Int. 178, 1–320.

Fröhlich, H., Opp, Ch., Scholten, T., 2005. Periglacial layer or anthropogenic layer? – A small scale analysis of their spatial distribution under forest. Z. Geomorph. N.F. 139, 203–219.

Frühauf, M., 1990. Neue Befunde zur Lithologie, Gliederung und Genese periglazialer Lockermaterialdecken. Petermanns Geographaphische Mitteilungen 134, 249–250.

Frühauf, M., 1991. Neue Befunde zur Lithologie, Gliederung und Genese der periglazialen Lockermaterialdecken im Harz: Erfassung und Bewertung postallerödzeitlicher decksedimentbildender Prozesse. Petermanns Geographische Mitteilungen 135, 49–60.

Frühauf, M., 1996. Periglaziale Lockermaterialdecken in ostdeutschen Mittelgebirgen – Forschungsstand und Versuch einer kritischen Bestandsaufnahme. In: Mäusbacher, R., Schulte, A. (Eds.), Beiträge zur Physiogeographie – Festschrift für Dietrich Barsch. Selbstverlag des Geographischen Instituts der Universität, Heidelberg, pp. 133–146.

Fuchs, M., Lang, A., 2009. Luminescence dating of hillslope deposits – a review. Geomorphology 109, 17–26.

Fuchs, M., Fischer, M., Revermann, R., 2010. Colluvial and alluvial sediment archives temporally resolved by OSL dating: implications for reconstructing soil erosion. Quat. Geochronol. 5, 269–273.

Fürst, C., Lorz, C., Makeschin, F., 2007. Development of formerly deposition influenced forest systems considering Dübener Heide as an example and consequences for process-oriented forest management planning. For. Ecol. Manage. 248, 6–16.

Galloway, R.W., 1983. Full-glacial southwestern United States: mild and wet or cold and dry? Quat. Res. 19, 236–248.

Gamper, M., 1986. Mikroklima und Solifluktion: Resultate von Messungen im schweizerischen Nationalpark in den Jahren 1975-1985. Göttinger Geographische Abhandlungen 84, 31–44.

GBA (Geologische Bundesanstalt), 2006, Mass Movements – soil creeps, landslides and rock fall, digital database. Geological Survey of Austria. http://geomap.geolba.ac.at (accessed July, 2006).

Gerasimenko, N., 2006. Upper Pleistocene loess–palaeosol and vegetational successions in the Middle Dnieper Area, Ukraine. Quat. Int. 149, 55–66.

Gerber, R., Salat, C., Junge, A., Felix-Henningsen, P., 2007. GPR-based detection of Pleistocene periglacial slope deposits at a shallow-depth test site. Geoderma 139, 346–356.

Gerber, R., Felix-Henningsen, P., Behrens, T., Scholten, T., 2010. Applicability of ground penetrating radar as a tool for nondestructive soil-depth mapping on Pleistocene periglacial slope deposits. J. Plant Nutr. Soil Sci. 173, 173–184.

Gerrard, J., 1990. Soil variations on hillslopes in humid temperate climates. Geomorphology 3, 225–244.

Gerrard, J., 1995. Soil Geomorphology. Chapman and Hall, London 269 S.

Gerth, J., 1985. Untersuchungen zur Adsorption von Nickel, Zink und Cadmium durch Bodentonfraktionen unterschiedlichen Stoffbestandes und verschiedene Bodenkomponenten. PhD-Dissertation, Univ. Kiel.

Giessübel, J., 1977. Nutzungsbedingte Änderungen im Naturraum dargestellt an Beispielen aus dem Rhein-Main-Gebiet und Nordhessen. Rhein-Mainische Forschungen, 85.

Gile, L.H., 1977. Holocene soils and soil-geomorphic relations in a semiarid region of southern New Mexico. Quat. Res. 7, 112–132.

Gile, L.H., 1999. Eolian and associated pedogenic features of the Jornada Basin floor, southern New Mexico. Soil Sci. Soc. Am. J. 63, 151–163.

Gile, L.H., Peterson, F.F., Grossman, R.B., 1966. Morphological and genetic sequences of carbonate accumulation in desert soils. Soil Sci. 101, 347–360.

Gillespie, A., Molnar, P., 1995. Asynchronous maximum advances of mountain and continental glaciers. Rev. Geophys. 33, 311–364.

Göbel, P., 1977. Vorläufige Ergebnisse der Messung gravitativer Bodenbewegungen auf bewaldeten Hängen im Taunus. Catena 3, 387–398.

Goddard, M.A., Mikhailova, E.A., Post, C.J., Schlautman, M.A., Galbraith, J.M., 2009. Continental United States atmospheric wet calcium deposition and soil inorganic carbon stocks. Soil Sci. Soc. Am. J. 73, 989–994.

Godfrey-Smith, D.I., Huntley, D.J., Chen, W.H., 1988. Optical dating studies of quartz and feldspar sediment extracts. Quat. Sci. Rev. 7, 373–380.

Godsey, H.S., Currey, D.R., Chan, M.A., 2005. New evidence for an extended occupation of the Provo shoreline and implications for regional climate change, Pleistocene Lake Bonneville, Utah, USA. Quat. Res. 63, 212–223.

Goldie, H.S., 2005. Erratic judgements: re-evaluating solutional erosion rates of limestones using erratic-pedestal sites, including Norber, Yorkshire. Area 37, 433–442.

Goossens, D., 1988. The effect of surface curvature on the deposition of loess – a physical model. Catena 15, 179–194.

Goossens, D., 1994. Effect of rock fragments on eolian deposition of atmospheric dust. Catena 23, 167–189.

Goossens, D., 2001. The aeolian dust accumulation curve. Earth Surf. Process. Landforms 26, 1213–1219.

Goossens, D., 2005a. Effect of rock fragment embedding on aeolian deposition of dust on stone-covered surfaces. Earth Surf.Process. Landforms 30, 443–460.

Goossens, D., 2005b. Quantification of the dry aeolian deposition of dust on horizontal surfaces: an experimental comparison of theory and measurements. Sedimentology 52, 859–873.

Goossens, D., 2006. Aeolian deposition of dust over hills: the effect of dust grain size on the deposition pattern. Earth Surf. Process. Landforms 31, 762–776.

Goossens, D., Offer, Z.Y., 1997. Aeolian dust erosion on different types of hills in a rocky desert: wind tunnel simulations and field measurements. J. Arid Environ. 37, 209–229.

Goossens, D., Offer, Z.Y., London, G., 2000. Wind tunnel and field calibration of five aeolian sand traps. Geomorphology 35, 233–252.

Goss, D.W., Smith, S.J., Stewart, B.A., 1973. Movement of added clay through calcareous materials. Geoderma 9, 97–103.

Götzinger, G., 1943. Neue Beobachtungen über Bodenbewegungen in der Flyschzone. Mitteilungen der Österreichischen Geographischen Gesellschaft 86, 87–104.

Govi, M., Mortara, G., Sorzana, P.F., 1985. Hydrological events and landslides. Geologia Applicata e Idrogeologia 20, 359–375.

Graham, C.B., van Verseveld, W., Barnard, H.R., McDonnell, J.J., 2010. Estimating the deep seepage component of the hillslope and catchment water balance within a measurement uncertainty framework. Hydrol. Process. 24, 3631–3647. http://dx.doi.org/10.1002/hyp. 7788.

Grimley, D.A., 2000. Glacial and nonglacial sediment contributions to Wisconsin Episode loess in the central United States. Geol. Soc. Am. Bull. 112, 1475–1495.

Gruber, F., 1980. Die Verstaubung der Hochgebirgsböden im Glocknergebiet. Veröffentlichungen des Österreichischen MAB-Hochgebirgsprogramms Hohe Tauern 3.

Gudehus, G., Goldschneider, M., Lippomann, R., 1985. Ingenieurgeologische und bodenmechanische Untersuchungen an Kriechhängen. In: Heitfeld, K.-H. (Ed.), Ingenieurgeologische Probleme im Grenzbereich zwischen Locker- und Festgesteinen. Springer, Berlin, New York, pp. 316–335.

Hadorn, P., Thew, N., Coope, G.R., Lemdahl, G., Hajdas, I., Bonani, G., 2002. A late-glacial and early Holocene environment and climate history for the Neuchâtel region (CH). Collection Annales Littéraires, Série Environment, Sociétés et Archéologie 3, 75–90.

Haff, P., Werner, B.T., 1996. Dynamical processes on desert pavements and the healing of surficial disturbances. Quat. Res. 45, 38–46.

Hall, R.D., 1999. Effects of climate change on soils in glacial deposits, Wind River Basin, Wyoming. Quat. Res. 51, 248–261.

Hanson, P.R., Mason, J.A., Goble, R.J., 2006. Goble fluvial terrace formation along Wyoming's Laramie range as a response to increased late Pleistocene flood magnitudes. Geomorphology 76, 12–25.

Harden, J.W., 1982. A quantitative index of soil development from field descriptions: examples from a chronosequence in central California. Geoderma 28, 1–12.

Harden, D.R., Biggar, N.E., Gillam, M.L., 1985. Quaternary deposits and soils in and around Spanish Valley, Utah. Geol. Soc. Am. Spec. Pap., 203: 43–64.

Harrach, T., 1974. Kritische Bemerkungen zur Genese und Systematik der "Verwitterungsböden" Mitteilungen der Deutschen Bodenkundlichen Gesellschaft 18, 320–326.

Harris, C., 1998. The micromorphology of paraglacial and periglacial slope deposits: a case study from Morfa Bychan, west Wales, UK. J. Quat. Sci. 13, 73–84.

Harris, C., Davies, M.C.R., 2000. Gelifluction: observations from large-scale laboratory simulations. Arctic Antarct. Alpine Res. 32, 202–207.

Harris, C., Rea, B., Davies, M.C.R., Hamza, O., 2003. Scaled physical modelling in the geotechnical centrifuge. http://www.earth.cf.ac.uk/research/geoenvironment/pace/CH31/GeotechnicalCentrifuge Modelling.htm (accessed June. 14, 2003).

Harris, C., Kern-Luetschg, M., Murton, J., Font, M., Davies, M., Smith, F., 2008a. Solifluction processes on permafrost and non-permafrost slopes: results of a large-scale laboratory simulation. Permafrost Periglac. Proc. 19, 359–378.

Harris, C., Kern-Luetschg, M., Smith, F., Isaksen, K., 2008b. Solifluction processes in an area of seasonal ground freezing, Dovrefjell, Norway. Permafrost Periglac. Proc. 19, 31–47.

Harris, C., Smith, J.S., Davies, M.C.R., Rea, B., 2008c. An investigation of periglacial slope stability in relation to soil properties based on physical modelling in the geotechnical centrifuge. Geomorphology 93, 437–459.

Harrison, S., 2002. Lithological variability of Quaternary slope deposits in the Cheviot Hills, UK. Proc. Geol. Assoc. 113, 121–138.

Häusler, W., Niederbudde, E.-A., 1992. Morphologie und Alter von Böden der Südlichen Frankenalb und der nichtkarbonatische Mineralbestand der Kalksteine. J. Plant Nutr. Soil Sci. 155, 391–399.

Heimsath, A.M., Chappell, J., Spooner, N.A., Questiaux, D.G., 2002. Creeping soil. Geology 30, 111–114.

Heine, K., 1987. Anthropogenic sedimentological changes during the Holocene in Southern Africa. Striae 26, 41–50.

Heinrich, J., 1991. Ursachen von Sturmschäden in Buchenbeständen (*Fagus sylvatica*) aus bodengeographischer Sicht. Allgemeine Forst- und Jagdzeitung 8, 145–149.

Heller, K., 2012, Einfluss periglazialer Deckschichten auf die oberflächennahen Fließwege am Hang – eine Prozessstudie im Osterzgebirge, Sachsen. PhD-Thesis, Dresden, http://nbn-resolving.de/urn:nbn:de:bsz:14-qucosa-98437.

Hempel, L., 1955. Studien über Verwitterung und Formenbildung im Muschelkalkgestein. Ein Beitrag zur klimatischen Morphologie. Göttinger Geographische Abhandlungen 18.

Hesse, P.P., McTainsh, G.H., 2003. Australian dust deposits: modern processes and the Quaternary record. Quat. Sci. Rev. 22, 2007–2035.

Heusser, C.J., 1983. Vegetational history of the northwestern United States including Alaska. In: Porter, S.C. (Ed.), Late Quaternary Environments of the United States, 1: The Late Pleistocene. University of Minnesota Press, Minneapolis, pp. 239–258.

Heusser, C.J., Heusser, L.E., 1990. Long continental pollen sequence from Washington State (USA): correlation of upper levels with marine pollen-oxygen isotope stratigraphy through substage 5e. Palaeogeogr. Palaeoclimatol. Palaeoecol. 79, 63–71.

Hindel, R., Fleige, H., 1990. Geogene Schwermetallgehalte in Böden der Bundesrepublik Deutschland. Verein Deutscher Ingenieure Berichte 837, 55–74.

Hindel, R., Gehrt, E., Kantor, W., Weidner, E., 2004. Spurenelementgehalte in Böden Deutschlands: Geowissenschaftliche Grundlagen und Daten. Bodenschutz 1520, 1–74.

Hoek, W.Z., 2001. Vegetation response to the 14.7 and 11.5 ka cal. BP climate transitions: is vegetation lagging climate? Glob. Planet. Change 30, 103–115.

Hoffmann, W., 1970. Beitrag zur Kenntnis der Bundtsandsteinböden Thüringens. Archiv Forstwissenschaften 19, 1007–1025.

Hoffmann, W., 1977. Böden auf periglazialen Deckschichten der Kreidesedimente des sächsischen Erzgebirges. Internat. Soil Sci. Soc. Working Group Forest Soils Proc., vol. 1, pp. 102–116.

Hoffmann, C., Funk, R., Li, Y., Sommer, M., 2008. Effects of grazing and topography on dust flux and deposition in the Xilingele grassland, Inner Mongolia. J. Arid Environ. 72, 792–807.

Hooke, R.L., 2000. On the history of humans as geomorphic agents. Geology 28, 843–846.

Horwath, J.L., Johnson, D.L., 2006. Mima-type mounds in Southwest Missouri: expressions of point-centered and locally thickened biomantles. Geomorphology 77, 308–319.

Houben, P., 2011. Sediment fluxes and budgets during the Holocene: quantification and modelling. Terra Nostra 2011 (1), 89–90.

Huber, J., Völkel, J., 2009. Untersuchungen zum Wasserfluss und zur stofflichen Differenzierung in der Critical Zone – Beispiele aus dem Bayerischen Wald und der Colorado Front Range, USA. Presentation on German Geographers' Day, 19–26 September 2009, Vienna.

Huggett, R.J., Huggett, R.J., 2011. Fundamentals of Geomorphology, 3rd ed. Routledge, London.

Hughes, M.W., Schmidt, J., Almond, P.C., 2009. Automatic landform stratification and environmental correlation for modelling loess landscapes in North Otago, South Island, New Zealand. Geoderma 149, 92–100.

Huijzer, A.S., Isarin, R.F.B., 1997. The reconstruction of past climates using multi-proxy evidence: an example of the Weichselian Pleniglacial in northwestern and central Europe. Quat. Sci. Rev. 16, 513–533.

Hülle, D., Hilgers, A., Kühn, P., Radtke, U., 2009. The potential of optically stimulated luminescence for dating periglacial slope deposits – a case study from the Taunus area, Germany. Geomorphology 109, 66–78.

Hungr, O., Evans, S.G., Bovis, M.J., Hutchinson, J.N., 2001. A review of the classification of landslides of the flow type. Environ. Eng. Geosci. 7, 221–238.

Huntley, D.J., Lamothe, M., 2001. Ubiquity of anomalous fading in K-feldspars, and the measurement and correction for it in optical dating. Canadian J. Earth Sci. 38, 1093–1106.

Hüttl, C., 1999. Steuerungsfaktoren und Quantifizierung der chemischen verwitterung auf dem Zugspitzplatt (Wettersteingebirge, Deutschland). Münchner Geographische Abhandlungen B 30, 1–171.

Iijima, Y., Fedorov, A.N., Park, H., Suzuki, K., Yabuki, H., Maximov, T.C., Ohata, T., 2010. Abrupt increases in soil temperatures following increased precipitation in a permafrost region, central Lena River basin, Russia. Permafrost Periglac. Proc. 21, 30–41.

Ilyuchev, B.A., Targulian, V.O., 1981. Eolian deposition of silt in the forest zone of the Russian Plain. Doklady AN SSSR 256 (4), 1–17.

Imeson, A.C., Kwaad, F.J.P.M., Mücher, H.J., 1980. Hillslope processes and deposits in forested areas in Luxembourg. In: Cullingford, R.A., Davidson, D.A., Lewin, J. (Eds.), Timescales in Geomorphology. Wiley, New York, pp. 31–43.

Isarin, R.F.B., 1997. Permafrost distribution and temperatures in Europe during the Younger Dryas. Permafrost Periglac. Proc. 8, 313–333.

Isarin, R.F.B., Renssen, H., 1999. Reconstructing and modelling Late Weichselian climates: the Younger Dryas in Europe as a case study. Earth Sci. Rev. 48, 1–38.

Isarin, R.F.B., Renssen, H., Koster, E.A., 1997. Surface wind climate during the Younger Dryas in Europe as inferred from aeolian records and model simulations. Palaeogeogr. Palaeoclimatol. Palaeoecol. 134, 127–148.

Ivy-Ochs, S., Kerschner, H., Reuther, A., Preusser, F., Heine, K., Maisch, M., Kubik, P.W., Schlüchter, C., 2008. Chronology of the last glacial cycle in the European Alps. J. Quat. Sci. 23, 559–573.

Izett, G.A., 1981. Volcanic Tephra beds: recorders of Upper Cenozoic silicic pyroclastic volcanism in the Western United States. J. Geophys. Res. 86, 10200–10222.

Izett, G.A., Wilcox, R.E., Borchardt, G.A., 1972. Correlation of a volcanic Tephra bed in Pleistocene deposits near Mount Blanco, Texas, with the Guaje Pumice bed of the Jemez Mountains, New Mexico. Quat. Res. 2, 554–578.

Jacobs, B.F., 1985. A middle Wisconsin pollen record from Hay Lake, Arizona. Quat. Res. 24, 121–130.

Jacobs, P.M., Mason, J.A., 2007. Late Quaternary climate change, loess sedimentation, and soil profile development in the central Great Plains: a pedosedimentary model. GSA Bull. 119 (3/4), 462–475.

Jacobs, P.M., Mason, J.A., Hanson, P.R., 2012. Loess mantle spatial variability and soil horizonation, southern Wisconsin, USA. Quat. Int. 265, 43–53.

Jaesche, P., 1999. Bodenfrost und Gelifluktionsdynamik in einem alpinen Periglazialgebiet (Hohe Tauern, Osttirol). Bayreuther Geowissenschaftliche Arbeiten 20.

Jaesche, P., Huwe, B., Stingl, H., Veit, H., 2002. Temporal variability of alpine gelifluction – a modelling approach. Geographica Helvetica 57, 157–169.

Jaesche, P., Veit, H., Huwe, B., 2003. Snow cover and soil moisture controls on gelifluction in an area of seasonal frost, Eastern Alps. Permafrost Periglac. Proc. 14, 399–410.

Jäger, K.-D., Altermann, M., 1991. Zur Stratigraphie periglaziärer Decken im thüringischen Bergland – Stratigraphische Auswertung. Archäologisches Korrespondezblatt Mainz 21, 332–333.

Jenny, H., 1980. The soil resource, origin and behavior. Springer, New York.

Jewell, P.W., 2007. Morphology and paleoclimatic significance of Pleistocene Lake Bonneville spits. Quat. Res. 68, 421–430.

Jewell, P.W., 2010. River incision, circulation, and wind regime of Pleistocene Lake Bonneville, USA. Palaeogeogr. Palaeoclimatol. Palaeoecol. 293, 41–50.

Johnson, D.L., 1990. Biomantle evolution and the redistribution of earth materials and artifacts. Soil Sci. 149, 84–102.

Johnson, D.L., Lin, H., 2006. The biomantle-critical zone model. American Geophysical Union, Fall Meeting: H11G-06.

Johnson, D.L., Watson-Stegner, D., 1987. Evolution model of pedogenesis. Soil Sci. 143 (5), 349–366.

Johnson, D.L., Watson-Stegner, D., Johnson, D.N., Schaetzl, R.J., 1987. Prosisotropic and proanisotropic processes of pedoturbation. Soil Sci. 143 (4), 278–292.

Johnson, D.L., Domier, J.E.J., Johnson, D.N., 2005. Reflections on the nature of soil and its biomantle. Ann. Assoc. Am. Geogr. 95, 11–31.

Jost, A., Lunt, D., Kageyama, M., Abe-Ouchi, A., Peyron, O., Valdes, P.J., Ramstein, G., 2005. High-resolution simulations of the last glacial maximum climate over Europe: a solution to discrepancies with continental palaeoclimatic reconstructions? Clim. Dyn. 24, 577–590.

Junge, C.E., Werby, R.T., 1958. The concentration of chloride, sodium, potassium, calcium, and sulfate in rain water over the United States. J. Meteorol. 15, 417–425.

Juvigné, E., Wintle, A., 1988. A new chronostratigraphy of the late Weichselian loess units in Middle Europe based on thermoluminescence dating. Eiszeitalter und Gegenwart 38, 94–105.

Kaiser, M., 1996. Bestimmung von Sulfat-Reduktionsraten mit der $^{35}SO_4$-Methode in säurebelasteten Quelleinzugsgebieten des Frankenwaldes und des Fichtelgebirges (Nordostbayern) – Methodenentwicklung und Standortsvergleich. Unpubl. Diploma Thesis, Univ. Bayreuth.

Kalm, V., 2010. Ice-flow pattern and extent of the last Scandinavian Ice Sheet southeast of the Baltic Sea. Quat. Sci. Rev. 44, 51–59. http://dx.doi.org/10.1016/j.quascirev.2010.01.019.

Karlstrom, E.T., 1988. Multiple paleosols in Pre-wisconsinan drift, north-western Montana and southwestern Alberta. Catena 15, 147–178.

Karte, J., 1983. Grezes litees as a special type of periglacial slope sediments in the German highlands. Polarforschung 53 (2), 67–74.

Kaufman, A., Broecker, W.S., 1965. Comparison of the ^{230}Th and ^{14}C ages for carbonate materials from Lakes Lahontan and Bonneville. J. Geophys. Res. 35, 1155–1183.

Kaufman, D.S., Forman, S.L., Bright, J., 2001. Age of the Cutler Dam Alloformation (Late Pleistocene), Bonneville Basin, Utah. Quat. Res. 56, 322–334.

Kaupenjohann, M., Mai-Peter, C., 1998. Stoffliche Aspekte des Bodenschutzes. Forum Geoökologie 9, 13–18.

Kaupenjohann, M., Wilcke, W., 1995. Heavy metal release from a serpentine soil using a pH-stat technique. Soil Sci. Soc. Am. J. 59, 94–99.

Kerzum, P.P., Rusakov, A.V., Matinyan, N., 1991. Geomorphic position of paleosols and aspects of Holocene soil evolution in the Central Russian Plain. Soviet Soil Sci. 23/1, 1–10.

Kirkby, M.J., 1985. Hillslope hydrology. In: Anderson, M.G., Burt, T.P. (Eds.), Hydrological Forecasting. John Wiley and Sons, Chichester, pp. 37–74.

Kleber, A., 1990. Upper Quaternary sediments and soils in the Great Salt Lake area, USA. Z. Geomorph. N.F. 34, 271–281.

Kleber, A., 1991. Die Gliederung der Schuttdecken am Beispiel einiger oberfränkischer Bodenprofile. Bayreuther Bodenkundl. Ber. 17, 83–105.

Kleber, A., 1992a. Periglacial slope deposits and their pedogenic implications in Germany. Palaeogeogr. Palaeoclimatol. Palaeoecol. 99, 361–372.

Kleber, A., 1992b. Deckschichten und Böden in den nordwestlichen La Sal Mts., Utah, USA. Bonner Geogr. Arb. 85, 114–129.

Kleber, A., 1993. A stratigraphy of slope deposits and soils in the northeastern Great Basin and its vicinity. Z. Geomorph. N.F. 92 (Suppl.), 173–188.

Kleber, A., 1994a. On the paleoecology of the northern Great Basin and adjacent Rocky Mountains. Z. Geomorph. N.F. 38, 421–434.

Kleber, A., 1994b. Traces of Laacher See Tephra in cover beds near Chagny, Bourgogne, France. Rev. Géomorphologie dynamique 45, 71–76.

Kleber, A., 1995. Die Höhenabhängigkeit von Deckschichten und Böden vom nordöstlichen Great Basin in die angrenzenden Rocky Mountains, USA. Mitteilungen der Österreichischen Geographischen Gesellschaft 137, 223–244.

Kleber, A., 1997. Cover-beds as soil parent materials in mid-latitude regions. Catena 30, 197–213.

Kleber, A., 1998. Loess-bearing soil parent materials, Northern Great Basin, USA. Washington State Univ. Coll. Agricult. Home Econ. Misc. Publ. 190, 155–158.

Kleber, A., 1999a. Cover beds as relative-dating tools – examples from the western USA. Z. Geomorph. N.F. 43, 51–59.

Kleber, A., 1999b. Zur Übertragbarkeit des deutschen Deckschichtenkonzepts. Petermanns Geographische Mitteilungen 143 (5–6), 363–372.

Kleber, A., 2000. Compound soil horizons with mixed calcic and argillic properties – examples from the northern Great Basin, USA. Catena 41, 111–131.

Kleber, A., 2001. Jungquartäre Entwicklung von Hängen, Böden und Klima im Westen der USA (Nördliches Great Basin und Randgebirge). Bayreuther Geowiss. Arb. 21.

Kleber, A., 2004a. The age of the Central European upper layer (Hauptlage) – a synthesis deduced from analogues. Z. Geomorp. N.F. 48, 491–499.

Kleber, A., 2004b. Lateraler Wasserfluss in Hangsedimenten unter Wald. In: Lorz, C., Haase, D. (Eds.), Stoff- und Wasserhaushalt in Einzugsgebieten. Beiträge zur EU-Wasserrahmenrichtlinie. Springer, Berlin, pp. 7–22.

Kleber, A., 2011. Influence of elevation on carbonate contents in stratified soils, northern Great Basin and adjacent mountains, USA. Open J. Soil Sci. 1 (2), 62–70.

Kleber, A., 2013. Heavy-mineral analysis as a tool in tephrochronology. With an example from the La Sal Mountains, Utah, U.S.A. Geologos, in the press.

Kleber, A., Gusev, V.V., 1998. Soil parent materials in the Moshaysk-district, Russia. Catena 34, 61–74.

Kleber, A., Mailänder, R., 2000. Heavy Metal sources in bedrock and cover-beds of the Steigerwald mountains. Z. Geomorph. N.F. 123 (Suppl.), 125–133.

Kleber, A., Schellenberger, A., 1998. Slope hydrology triggered by cover-beds. With an example from the Frankenwald Mountains, Northeastern Bavaria. Z. Geomorph. N.F. 42, 469–482.

Kleber, A., Schellenberger, A., 1999. Hydrogeologische Verhältnisse eines Quelleinzugsge biets im Frankenwald. Bayreuther Forum Ökologie 71, 17–25.

Kleber, A., Stingl, H., 2000. Zur Flußgeschichte des Trebgasttals nördlich von Bayreuth. Eine zweiphasige Talverlegung im Rotmainsystem. Bamberger Geographische Schriften Sonderfolge 6, 191–208.

Kleber, A., Mailänder, R., Zech, W., 1998a. Stratigraphic approach to alteration in mineral soils – the heavy metal example. Soil Sci. Soc. Am. J. 62, 1647–1750.

Kleber, A., Lindemann, J., Schellenberger, A., Beierkuhnlein, C., Kaupenjohann, M., Peiffer, S., 1998b. Slope deposits and water paths in a spring catchment, Frankenwald, Bavaria, Germany. Nutr. Cycl. Agroecosyst. 50, 119–126.

Klute, A. (Ed.), 1986. Methods of soil analysis I. Physical and mineralogical methods. 2nd ed. American Society of Agronomy and Soil, Madison, WI.

Knoblich, K., 1967. Mechanische Gesetzmäßigkeiten beim Auftreten von Hangrutschungen. Z. Geomorph. N.F. 11 (Suppl.), 286–299.

Kopp, D., 1970. Kryogene Perstruktion und ihre Beziehung zur Bodenbildung im Moränengebiet. In: Richter, H., Haase, G., Lieberoth, I., Ruske, R. (Eds.), Periglazial – Löß – Paläolithikum im Jungpleistozän der Deutschen Demokratischen Republik. Gotha, Leipzig, pp. 213–268.

Körner, W., 1996. Abflussbildung, Interflow und Stoffbilanzierung im Schönbuch Waldgebiet. Tübinger Geowissenschaftliche Arbeiten C27.

Kösel, M., 1996. Der Einfluß von Relief und periglazialen Deckschichten auf die Bodenausbildung im mittleren Rheingletschergebiet von Oberschwaben. Tübinger Geowissenschaftliche Arbeiten D1.

Kösel, M., 1997. Relief, periglaziale Deckschichten, fossile und rezente Böden in der rißzeitlichen Altmoränenlandschaft und auf der äußeren Jungendmoräne. Tübinger Geowissenschaftliche Arbeiten D3, 14–44.

Koster, E.A., 2005. Recent advances in luminescence dating of late Pleistocene (cold-climate) aeolian sand and loess deposits in Western Europe. Permafrost Periglac. Proc. 16, 131–143.

Kovacik, M., Ondrasic, R., 1991. The effect of Quaternary uplift on the stability of slopes in the central part of the Skorusinske Vrchy Mountains in the West Carpathians. Geol. Soc. London, Engineering Geol. Spec. Publ., doi: http://dx.doi.org/10.1144/GSL.ENG.1991.007.01.66.

Krauter, E., Feuerbach, J., Lauterbach, M., 2004. Risikoabschätzung von Hangbewegungen und Schutzkonzepte. Interpraevent 2004/IX, 113–124.

Krummenacher, B., Budmiger, K., Mihajlovic, D., Blank, B. 1998. Periglaziale Prozesse und Formen im Furggentälti, Gemmipass. Analysen zur räumlich-zeitlichen Entwicklung der Permafrostverbreitung und der Solifluktion basierend auf der Entwicklung und Anwendung moderner Arbeitsmethoden. Mitteilungen des Eidgenössischen Instituts für Schnee- und Lawinenforschung, Davos, 56, 1–245.

Kubiena, W.L., 1953. Bestimmungsbuch und Systematik der Böden Europas. Illustriertes Hilfsbuch zur leichten Diagnose und Einordnung der wichtigsten europäischen Bodenbildungen unter Berücksichtigung ihrer gebräuchlichsten Synonyme. Enke Verlag, Stuttgart.

Kubiena, W.L., 1986. Grundzüge der Geopedologie und der Formenwandel der Böden. Österreichischer Agrarverlag, Wien.

Küfmann, C., 2003. Soil types and eolian dust in high-mountainous karst of the Northern Calcareous Alps (Zugspitzplatt, Wetterstein Mountains, Germany). Catena 53, 211–227.

Kühn, P., 2003. Micromorphology and Late Glacial/Holocene genesis of Luvisols in Mecklenburg-Vorpommern (NE-Germany). Catena 54, 537–555.

Kühn, P., Billwitz, K., Blauriegel, A., Kühn, D., Eckelmann, W., 2006. Distribution and genesis of Fahlerden (Albeluvisols) in Germany. J. Plant Nutr. Soil Sci. 169, 420–433.

Kuntze, H., Roeschmann, G., Schwerdtfeger, G., 1994. Bodenkunde, 5th ed. Ulmer Verlag, Stuttgart.

Kurenkova, E.L., Gribchenko, Y., Grekhova, L.V., Ludmila, V., 1995. Paleolithic archeology and the paleogeographical study of loess-paleosol sequences in eastern Europe. GeoJournal 362, 229–235.

Kutílek, M., Nielsen, D.R., 1994. Soil Hydrology. Catena, Cremlingen-Destedt.

Kwaad, F.J.P.M., Mücher, H.J., 1979. The evolution of soils and slope deposits in the Luxembourg Ardennes near Wiltz. Geoderma 22, 173–192.

Lang, A., Hönscheidt, S., 1999. Age and source of colluvial sediments at Vaihingen-Enz, Germany. Catena 38, 89–107.

Langbein, R., Stepansky, M., 1996. Beiträge zur Geochemie des Muschelkalks in Thüringen. Geowissenschaftliche Mitteilungen von Thüringen Beiheft 5, 111–129.

Lao, Y., Benson, L., 1988. Uranium-series age estimates and palaeoclimatic significance of Pleistocene tufas from the Lahontan Basin, California and Nevada. Quat. Res. 30, 165–176.

Latałowa, M., Borówka, R.K., 2006. The Allerød/Younger Dryas transition in Wolin Island, northwest Poland, as reflected by pollen, macrofossils, and chemical content of an organic layer separating two aeolian series. Veget. Hist. Archaeobot. 15, 321–331.

Leonard, E.M., 1989. Climatic change in the Colorado Rocky Mountains: estimates based on modern climate at late Pleistocene equilibrium lines. Arctic Alpine Res. 21, 245–255.

Leopold, M., 2003. Multivariate Analyse von Geoarchiven zur Rekonstruktion eisenzeitlicher Landnutzung im Umfeld der spätlatènezeitlichen Viereckschanze von Poign, Lkr. Regensburg. Regensburger Beiträge zur Bodenkunde, Landschaftsökologie und Quartärforschung 2. http://www.opus-bayern.de/uni-regensburg/volltexte/2003/233

Leopold, M., Völkel, J., 2002. Zu Fragen der Eignung kolluvialer Sedimente als Paläoklimaindikatoren. Zentralblatt für Geologie und Paläontologie T.1 2001 3/4, 1–18.

Leopold, M., Völkel, J., 2003. GPR images of periglacial slope deposits beneath peat bogs in Middle European Highlands, Germany. In: Bristow, C., Jol, H. (Eds.), Ground Penetrating Radar in Sediments. Royal Geological Society, London, pp. 177–185.

Leopold, M., Völkel, J., 2007. Colluvium: definition, differentiation, and possible suitability for reconstructing Holocene climate data. Quat. Int. 162–163, 133–140.

Leopold, M., Dethier, D., Völkel, J., Raab, T., Rickert, T.C., Caine, N., 2008a. Using geophysical methods to study the shallow subsurface of a sensitive alpine environment, Niwot Ridge, Colorado Front Range, USA. Arctic Antarct. Alpine Res. 40, 519–530.

Leopold, M., Völkel, J., Dethier, D., Raab, T., 2008b. Shape, thickness and distribution of periglacial slope deposits at Niwot Ridge, Rocky Mountains Front Range, Colorado, USA. Zeitschrift für Geomorphologie N. F. 52 (Suppl.-Iss. 2), 77–94.

Leopold, M., Völkel, J., Dethier, D., Williams, M., Caine, N., 2010. Mountain permafrost – a valid archive to study climate change? Examples from the Rocky Mountains Front Range of Colorado, USA. Nova Acta Leopoldina 112, 281–289.

Leser, H., 1996. Pedologisch-sedimentologische Untersuchungen als geomorphologische Methode. Forsch. Fortschr. 40, 296–300.

Licciardi, J.M., Pierce, K.L., 2008. Cosmogenic exposure-age chronologies of Pinedale and Bull Lake glaciations in greater Yellowstone and the Teton Range, USA. Quat. Sci. Rev. 27, 814–831.

Lin, H., 2003. Hydropedology: bridging disciplines, scales, and data. Vadose Zone J. 2, 1–11.

Lin, H.S., 2009. Earth's critical zone and hydropedology. Hydrol. Earth Syst. Sci. Discuss. 6, 3417–3481.

Lindemann, J., 1997. Quantifizierung biogeochemischer Eisen- und Sulfat-Umsetzungen in einem Quellmoor und deren Beitrag zur Säureneutralisierung in einem Einzugsgebiet des Frankenwaldes. Bayreuther Forum Ökologie 51.

Litt, T., Stebich, M., 1999. Bio- and chronostratigraphy of the Lateglacial in the Eifel region, Germany. Quat. Int. 61, 5–16.

Litt, T., Schmincke, H.-U., Kromer, B., 2003. Environmental response to climate and volcanic events in central Europe during the Weichselian Lateglacial. Quat. Sci. Rev. 22, 7–32.

Little, E.C., Lian, O.B., Velichko, A.A., Morozova, T.D., Nechaev, V.P., Dlussky, K.G., Rutter, N.W., 2002. Quaternary stratigraphy and optical dating of loess from the East European Plain (Russia). Quat. Sci. Rev. 21, 1745–1762.

Liu, B., Phillips, F.M., Elmore, D., Sharma, P., 1994. Depth dependence of soil carbonate accumulation based on cosmogenic 36Cl dating. Geology 22, 1071–1074.
Löffler, J., 2002. Altitudinal changes of ecosystem dynamics in the central Norwegian High Mountains. Die Erde 133, 227–258.
Lorz, C., 1996. Zur Problematik des geologischen Aufbaus der Bodendecke und der Verteilung von geogenen Schwermetallen. Geoökodynamik XVII, 25–44.
Lorz, C., 2008. Ein substratorientiertes Boden-Evolutions-Konzept für geschichtete Bodenprofile. Genese und Eigenschaften von lithologisch diskontinuierlichen Böden. Relief, Boden, Paläoklima 23.
Lorz, C., Phillips, J., 2006. Pedo-Ecological consequences of lithological discontinuities in soils – examples from Central Europe. J. Plant Nutr. Soil Sci. 169, 573–581.
Lorz, C., Heller, K., Kleber, A., 2011. The regolith continuum – a key compartment for processes and functions of landscapes. Z. Geomorph. N.F. 55 (Suppl. Iss. 3), 277–292.
Lyell, C., 1832. Principles of Geology. John Murray, London.
Mabbutt, J.A., 1979. Pavements and patterned ground in the Australian stony deserts. Stuttgarter Geographische Studien 93, 107–123.
Machann, R., 1972. Wüstungen im Steigerwald. Mainfränkische Studien 5, 1–202.
Machette, M.N., 1982. Guidebook to the Late Cenozoic geology of the Beaver basin, south-central Utah. Open-file report 82-850, U.S. Geological Survey, Denver, CO.
Machette, M.N., 1985. Calcic soils and calcretes of the southwestern United States. Geol. Soc. Am. Spec. Pap. 203, 1–21.
Madole, R.F., Bradley, W.C., Loewenherz, D.S., Ritter, D.F., Rutter, N.W., Thorn, C.E., 1987. Rocky Mountains. Geol. Soc. Am. Centenn. Spec. 2, 211–257.
Madsen, D.B., Currey, D.R., 1979. Late Quaternary glacial and vegetation changes, Little Cottonwood Canyon area, Wasatch Mountains, Utah. Quat. Res. 12, 254–270.
Mahr, A., 1998. Lockerbraunerden und periglaziale Hangsedimente im Bayerischen Wald. Untersuchungen zu Paläoumwelt und Geomorphodynamik im Spätglazial und ihrem Einfluß auf die Pedogenese. Regensburger Geographische Schriften 30.
Mahr, A., Völkel, J., 2000. Genese von Lockerbraunerden im Bayerischen Wald. J. Plant Nutr. Soil Sci. 162, 471–476.
Mailänder, R., 2002. Stratified soils on the Central Swiss Plateau. Unpubl. PhD-Thesis, University of Berne.
Mailänder, R., Veit, H., 2001. Periglacial cover-beds on the Swiss Plateau: indicators of soil, climate and landscape evolution during the Late Quaternary. Catena 45, 251–272.
Marchetti, D.W., Cerling, T.E., Lips, E.W., 2005. A glacial chronology for the Fish Creek drainage of Boulder Mountain, Utah, USA. Quat. Res. 64, 263–271.
Marchetti, D.W., Cerling, T.E., Dohrenwend, J.C., Gallin, W., 2007. Ages and significance of glacial and mass movement deposits on the west side of Boulder Mountain, Utah, USA. Palaeogeogr. Palaeoclimatol. Palaeoecol. 252, 503–513.
Margielewski, W., 2006. Structural control and types of movements of rock mass in anisotropic rocks: case studies in the Polish Flysch Carpathians. Geomorphol. 77, 47–68.
Marsh, P., Endrizzi, S., Quinton, W.L., Thorne, R., Dall'Amico, M., 2009. Modelling the spatial variability of the depth of ground thaw in a continuous permafrost, arctic tundra landscape. American Geophysical Union, Fall Meeting 2009, abstract #C51A-0456. http://adsabs.harvard.edu/abs/2009AGUFM.C51A0456M (accessed Aug. 7, 2010).
Martin, P.S., Mehringer, P.J., 1965. Pleistocene pollen analysis and biogeography of the Sothwest. In: Wright Jr., H.E., Frey, D.G. (Eds.), The Quaternary of the United States. Princeton University of Press, Princeton, NJ, pp. 433–451.

Mason, J.A., Knox, J.C., 1997. Age of colluvium indicates accelerated late Wisconsinan hillslope erosion in the Upper Mississippi Valley. Geology 25, 267–270.

Mason, J.A., Nater, E.A., Zanner, C.W., Bell, J.C., 1999. A new model of topographic effects on the distribution of loess. Geomorphology 28, 223–236.

Matsumoto, H., Ishikawa, M., 2002. Gelifluction within a solifluction lobe in the Kärkevagge Valley, Swedish Lapland. Geogr. Ann. Ser. A 84, 261–266.

Matsuoka, N., 2001. Gelifluction rates, processes and landforms: a global review. Earth Sci. Rev. 55, 107–134.

Matsuoka, N., Murton, J., 2008. Frost weathering: recent advances and future directions. Permafrost Periglac. Proc. 19, 195–210.

Maus, H., Stahr, K., 1977. Auftreten und Verbreitung von Lößlehmbeimengungen in periglazialen Schuttdecken des Schwarzwaldwestabfalls. Catena 3, 369–386.

Mayer, J.H., 2003. Paleoindian geoarchaeology and paleoenvironments of the western Killpecker dunes, Wyoming, USA. Geoarchaeology 18, 35–69.

Mayer, L., McFadden, L.D., Harden, J.W., 1988. Distribution of calcium carbonate in desert soils: a model. Geology 16, 303–306.

McCahon, T.J., Munn, L.C., 1991. Soils developed in Late Pleistocene till, Medicine Bow Mountains, Wyoming. Soil Sci. 152, 472–484.

McCallum, M.E., 1962. Glaciation of Libby Creek Canyon, east flank of Medicine Bow Mountains, southeastern Wyoming. Univ. Wyoming Contrib. Geol. 1, 21–29.

McCalpin, J.P., 1986. Thermoluminescence (TL) dating in seismic hazard evaluations: an example from the Bonneville Basin, Utah. Proc. Sympos. Eng. Geol. Soils Eng. 22, 156–176.

McCoy, W.D., 1981. Quaternary aminostratigraphy of the Bonneville and Lahontan Basins, western U.S., with palaeoclimatic implications. PhD-Dissertation, University of Colorado, Boulder.

McCrea, J.M., 1950. On the isotopic chemistry of carbonates and a paleotemperature scale. J. Chem. Phys. 18, 849–858.http://dx.doi.org/10.1063/1.1747785.

McDonald, E.V., Busacca, A.J., 1990. Interaction between aggrading geomorphic surfaces and the formation of a Late Pleistocene paleosol in the Palouse loess of eastern Washington State. Geomorphology 3, 449–470.

McDonnell, J.J., 2003. Where does water go when it rains? Moving beyond the variable source area concept of rainfall-runoff response. Hydrol. Proc. 17, 1869–1875.

McDonnell, J.J., Sivapalan, M., Vaché, K., Dunn, S., Grant, G., Haggerty, R., Hinz, C., Hooper, R., Kirchner, J., Roderick, M.L., Selker, J., Weiler, M., 2007. Moving beyond heterogeneity and process complexity: a new vision for watershed hydrology. Water Resourc. Res. 43, W07301. http://dx.doi.org/10.1029/2006WR005467.

McFadden, L.D., 1988, Climatic influences and processes of soil development in Quaternary deposits of southern California. Geol. Soc. Am. Spec. Pap. 216, pp. 153–177.

McFadden, L.D., Hendricks, D.M., 1985. Changes in the content and composition of pedogenic iron oxyhydroxides in a chronosequence of soils in southern California. Quat. Res. 23, 189–204.

McFadden, L.D., Tinsley, J.C., 1985. Rate and depth of pedogenic-carbonate accumulation in soils: formulation and testing of a compartment model. Geological Soc. Am. Spec. Pap. 203, 23–41.

McFadden, L.D., McDonald, E., Wells, S., Anderson, K., Quade, J., Forman, S., 1998. The vesicular layer and carbonate collars of desert soils and pavements: formation, age and relation to climate change. Geomorphology 24, 101–145.

McFadden, L.D., Wells, S.G., Dohrenwend, J.C., 1986. Influences of Quaternary climatic changes on processes of soil development on desert loess deposits of the Cima Volcanic Field, California. Catena 13, 361–389.

McFadden, L.D., Wells, S.G., Jercinovich, M.J., 1987. Influences of eolian and pedogenic processes on the origin and evolution of desert pavements. Geology 15, 504–508.

McFarlane, D., Williamson, D.R., 2002. An overview of water logging and salinity in southwestern Australia as related to the "Ucarro" experimental catchment. Agric. Water Manage. 53, 5–29.

Mead, J.I., Bell, C.J., Murray, L.K., 1992. Mictomys borealis (northern bog lemming) and the Wisconsin paleoecology of the east-central Great Basin. Quat. Res. 37, 229–238.

Mears Jr., B., 1981. Periglacial wedges and the late Pleistocene environment of Wyoming's intermontane basins. Quat. Res. 15, 171–198.

Mears Jr., B., 1987. Late Pleistocene periglacial wedge sites in Wyoming. Geol. Surv. Wyoming Mem. 3.

Mears Jr., B., 1991. Laramie Basin. The Geology of North America K-2, 418–427.

Mears Jr., B., 1997. Pleistocene gelifluction and rock deformation on slopes in southern Wyoming. Permafrost Periglac. Process. 8, 251–255.

Mears Jr., B., 2001. Glacial records in the Medicine Bow Mountains and Sierra Madre of Southern Wyoming and adjacent Colorado, with a traveler's guide to their sites. Wyoming State Geological Survey Public Information Circular 41.

Mehringer Jr., P.J., 1985. Late-Quaternary pollen record from the interior Pacific Northwest and northern Great Basin. In: Bryant Jr., V.M., Holloway, R.G. (Eds.), Pollen records of late-Quaternary North American sediments. American Association of Stratigraphic Palynologists, Dallas, TX, pp. 167–189.

Mellaart, L., 1975. The Neolithic of the Near East. Thames & Hudson, London.

Mensching, H., 1960. Periglazial-Morphologie und quartäre Entwicklungsgeschichte der Hohen Rhön und ihres östlichen Vorlandes. Würzburger geographische Arbeiten 7.

Metz, B., 1997. Glaziale Formen und Formungsprozesse im Schwarzwald. Freiburger Geographische Hefte 36, 51–74.

Meyer, H.-H., Kottmeier, C., 1989. Die atmosphärische Zirkulation in Europa im Hochglazial der Weichselkaltzeit – abgeleitet von Paläowind-Indikatoren und Modellsimulationen. Eiszeitalter und Gegenwart 39, 10–18.

Meyers, N.L., McSweeney, K., 1995. Influence of treethrow on soil properties in Northern Wisconsin. Soil Sci. Soc. Am. J. 59, 871–876.

Miller, D.C., Birkeland, P.W., Rodbell, D.T., 1993. Evidence for Holocene stability of steep slopes, Northern Peruvian Andes, based on Soils and Radiocarbon dates. Catena 20, 1–12.

Mitchell, J.K., 1988. Fundamentals of Soil behaviour, 2nd ed. Wiley, New York.

Miyazaki, T., 1993. Water flow in soils. Marcel Dekker, New York.

Moldenhauer, K.-M., 1993. Quantitative Untersuchungen zu aktuellen fluvial-morphodynamischen Prozessen in bewaldeten Kleineinzugsgebieten von Odenwald und Taunus. Frankfurter Geowissenschaftliche Arbeiten D15.

Molloy, L.F., 1988. Soils in the New Zealand Landscape. Mallinson Rendel, Wellington, NZ.

Montrasio, L., Valentino, R., Losi, G.L., 2009. Rainfall-induced shallow landslides: a model for the triggering mechanism of some case studies in Northern Italy. Landslides 6, 241–251.

Morrison, R.B., 1964. Lake Lahontan, Geology of southern Carson Desert, Nevada. U.S. Geol. Surv. Prof. Pap. 401.

Morrison, R.B., 1965a. Quaternary geology of the great basin. In: Wright Jr., H.E., Frey, D.G. (Eds.), The Quaternary of the United States. Princeton University Press, Princeton, NJ, pp. 265–285.

Morrison, R.B., 1965b. Lake Bonneville, Quaternary stratigraphy of eastern Jordan Valley, south of Salt Lake City, Utah. U.S. Geol. Surv. Prof. Pap. 477.

Morrison, R.B., 1965c. New evidence on Lake Bonneville stratigraphy and history from southern Promontory Point, Utah. U.S. Geol. Surv. Prof. Pap. 525-C, 110–119.

Morrison, R.B., 1991. Quaternary stratigraphic, hydrologic, and climatic history of the Great Basin, with emphasis on Lakes Lahontan, Bonneville, and Tecopy. In: Morrison, R.B. (Ed.), Quaternary Nonglacial Geology: Conterminous U.S. Geol. Soc. Am., Boulder, pp. 283–320.

Mückenhausen, E., 1979. Die Paläoböden der Eifel in Abhängigkeit von der Geomorphologie. Z. Geomorph. N.F. 33 (Suppl.), 16–24.

Muhs, D.R., 1983. Airborne dust fall on the California Channel Islands, USA. J. Arid Environ. 6, 223–238.

Muhs, D.R., Benedict, J.B., 2006. Eolian additions to late Quaternary alpine soils, Indian Peaks wilderness area, Colorado Front Range. Arctic Antarct. Alpine Res. 38, 120–130.

Muhs, D.R., Budahn, J.R., 2009. Geochemical evidence for african dust and volcanic ash inputs to Terra Rossa soils on carbonate reef terraces, northern Jamaica, West Indies.

Muhs, D.R., Budahn, J.R., Prospero, J.M., Carey, S.N., 2007. Geochemical evidence for African dust inputs to soils of western Atlantic islands: Barbados, the Bahamas, and Florida. J. Geophys. Res. 112, F02009. http://dx.doi.org/10.1029/2005JF000445.

Muhs, D.R., Bettis III, E.A., Aleinikoff, J.N., McGeehin, J.P., Beann, J., Skipp, G., Marshall, B.D., Roberts, H.M., Johnson, W.C., Benton, R., 2008. Origin and paleoclimatic significance of late Quaternary loess in Nebraska: evidence from stratigraphy, chronology, sedimentology, and geochemistry. Geol. Soc. Am. Bull. 120, 1378–1407.

Müller, F., 1996. Sulfatreduktion in einem Quellmoor des Frankenwaldes – Optimierung der $^{35}SO_4$-Methode und Bestimmung räumlicher und saisonaler Unterschiede der Sulfatreduktionsraten. Unpubl. Diploma Thesis, Univ. Bayreuth.

Müller, S., 2011. Neue Erkenntnisse zu periglaziären Lagen und Pedogenese im hessischen Spessart. Dissertation, Univ. Frankfurt a. Main. http://publikationen.ub.uni-frankfurt.de/frontdoor/index/index/docId/19257 (accessed Sept. 25, 2012).

Müller, S., Thiemeyer, H., 2009. Mikromorphologische Untersuchungen geschichteter Böden im Hessischen Spessart. Berichte der Deutschen Bodenkundlichen Gesellschaft. http://eprints.dbges.de/116/1/spessart.pdf (accessed June 3, 2010).

Müller, S., Thiemeyer, H., 2010. Potentials of reconstructing the formation and transformation of slope deposits by the use of soil micromorphology. Geophys. Res. Abstr. 12, 2620.

Mullineaux, D.R., 1986a. Summary of the pre-1980 tephra-fall deposits erupted from Mount St. Helens, Washington State, USA. Bull. Vulcanol. 48, 17–26.

Munn, L.C., 1987. Soil genesis associated with periglacial ice wedge casts, southcentral Wyoming. Soil Sci. Soc. Am. J. 51, 1000–1004.

Munn, L.C., Spackman, L.K., 1990. Origin of silt-enriched alpine surface mantles in Indian Basin, Wyoming. Soil Sci. Soc. Am. J. 54, 1670–1677.

Muscheler, R., Kromer, B., Björck, S., Svensson, A., Friedrich, M., Kaiser, K.F., Southon, J., 2008. Tree rings and ice cores reveal ^{14}C calibration uncertainties during the Younger Dryas. Nat. Geosci. 1, 263–267.

NACSN (North American Commission on Stratigraphic Nomenclature), 1983. North American stratigraphic code. Am. Assoc. Petrol. Geol. Bull. 67, 841–875.

Neuhäuser, B., Terhorst, B., 2006. Landslide susceptibility assessment using weights-of-evidence applied on a study site at the Jurassic escarpment of the Swabian Alb (SW-Germany). Geomorphology 86, 12–24.

Neymark, L.A., Paces, J.B., Marshall, B.D., Peterman, Z.E., Whelan, J.F., 2005. Geochemical and C, O, Sr, and U-series isotopic evidence for the meteoric origin of calcrete at Solitario Wash, Crater Flat, Nevada, USA. Environ. Geol. 48, 450–465.

Nicolsky, D.J., Romanovsky, V.E., Panteleev, G.G., 2009. Estimation of soil thermal properties using in-situ temperature measurements in the active layer and permafrost. Cold Regions Sci. Tech. 55, 120–129.

Niessen, F., Lister, G., Giovanoli, F., 1992. Dust transport and palaeoclimate during the Oldest Dryas in Central Europe – implications from varves (Lake Constance). Clim. Dyn. 8, 71–81.
Nikiforoff, C.C., 1949. Weathering and soil evolution. Soil Sci. 67, 219–230.
Niller, H.-P., 1998. Prähistorische Landschaften im Lößgebiet um Regensburg – Kolluvien, Auenlehme und Böden als Archive der Paläoumwelt. Regensburger Geographische Schriften 31.
Nissen, T.C., Mears Jr., B., 1990. Late Pleistocene ice-wedge casts and sand-wedge relics in the Wyoming basins, USA. Permafrost Periglacial Proc. 1, 201–219.
Northcote, K.H., 1979. A Factual Key for the Recognition of Australian Soils, 4th ed. Rellim Tech. Pub, Adelaide.
Novsky, V.A., 1975. Pleystotsen Yaroslavskogo povolzh'ya. Nauka, Moscow.
Nowinski, N.S., Taneva, L., Trumbore, S.E., Welker, J.M., 2010. Decomposition of old organic matter as a result of deeper active layers in a snow depth manipulation experiment. Oecologia 163, 785–792.
Nugteren, G., Vandenberghe, J., 2004. Spatial climatic variability on the Central Loess Plateau (China) as recorded by grain size for the last 250 kyr. Glob. Planet. Change 41, 185–206.
O'Connor, J.E., 1993. Hydrology, hydraulics, and geomorphology of the Bonneville Flood. Geol. Soc. Am. Spec. Pap. 274.
Oelke, C., Zhang, T., Serreze, M.C., Armstrong, R.L., 2003. Regional-scale modeling of soil freeze/thaw over the Arctic drainage basin. J. Geophys. Res. 108/D, 4314–4333.
Offer, Z.Y., Goossens, D., 1995. Wind tunnel experiments and field measurements of aeolian dust deposition on conical hills. Geomorphology 14, 43–56.
Ollier, C., Pain, C., 1996. Regolith, soils and landforms. Wiley, Chichester.
Opp, C., 1998. Geographische Beiträge zur Analyse von Bodendegradation und ihrer Diagnose in der Landschaft. Leipziger Geowissenschaften 8, 1–187.
Oviatt, C.G., 1997. Lake Bonneville fluctuations and global climate change. Geology 25, 155–158.
Oviatt, C.G., McCoy, W.D., 1992. Early Wisconsin lakes and glaciers in the Great Basin, USA. Geol. Soc. Amr. Spec. Pap. 270, 279–287.
Oviatt, C.G., McCoy, W.D., Reider, R.G., 1987. Evidence for a shallow early or middle Wisconsin-age lake in the Bonneville-Basin, Utah. Quat. Res. 27, 248–262.
Oviatt, C.G., Currey, D.R., Sack, D., 1992. Radiocarbon chronology of Lake Bonneville, eastern Great Basin, USA. Palaeogeogr. Palaeclimatol. Palaeoecol. 99, 225–241.
Ozouf, J.-C., Coutard, J.-P., Lautridou, J.-P., 1995. Grèzes, grèzes litées: historique des definitions. Permafrost Periglac. Proc. 6, 85–87.
Page, A., Miller, R., Keeney, D. (Eds.), 1982. Methods of soil analysis: II. Chemical and microbiological properties. 2nd ed. American Society of Agronomy and Soil, Madison, WI.
Panin, P.G., 2007. Specificity of the morphology of interglacial and interstadial paleosol complexes of the middle and late Pleistocene in the center of the East European Plain. Eurasian Soil Sci. 40, 126–139.
Pankhurst, C.E., Pierret, A., Hawke, B.G., Kirby, J.M., 2002. Microbiological and chemical properties of soil associated with macropores at different depths in a red-duplex soil in NSW Australia. Plant Soil 238, 11–20.
Paton, T.R., Humphreys, G.S., Mitchell, P.B., 1995. Soils – A new global view. UCL-Press, London.
Patrickson, S.A., Sack, Dorothy, Brunelle, A.R., Moser, K.A., 2010. Late Pleistocene to early Holocene lake level and paleoclimate insights from Stansbury Island, Bonneville basin, Utah. Quat. Res. 73, 237–246.
Pawelec, H., 2006. Origin and palaeoclimatic significance of the Pleistocene slope covers in the Cracow Upland, southern Poland. Geomorphology 74, 50–69.

Payton, R.W., 1993. Fragipan formation in argillic brown earths (Fragiudalfs) of the Milfield Plain, north-east England. II. Post Devensian developmental processes and the origin of fragipan consistence. J. Soil Sci. 44, 703–723.

Pelletier, J.D., 2007. Cantor set model of eolian dust deposits on desert alluvial fan terraces. Geology 35, 439–442.

Pelletier, J.D., Cline, M., DeLong, S.B., 2007. Desert pavement dynamics: numerical modeling and field-based calibration. Earth Surf. Process. Landforms 32, 1913–1927.

Pendall, E.G., Harden, J.W., Trumbore, S.E., Chadwick, O.A., 1994. Isotopic approach to soil carbonate dynamics and implications for paleoclimatic interpretations. Quat. Res. 42, 60–71.

Pesci, M., Richter, G., 1996. Löss: Herkunft–Gliederung–Landschaften. Z. Geomorph. N.F. 98 (Suppl.).

Peters, C., 2000. Untersuchungen zum Preferential Flow in steinreichen Böden des Buntsandstein-Odenwaldes. Unpubl. Diploma Thesis, University of Frankfurt.

Pfeffer, K.-H., 2004. Zur Bodengeographie der Schwäbischen Alb – eine Bilanz aus Literatur- und Geländebefunden. Tübinger Geowissenschaftliche Arbeiten D10, 73–93.

Pfeiffer, M., 2002. Periglaziale Lagen: ihre Zuordnung zum Relief und ihre sedimentologisch-pedologische Differenzierung sowie ihre Auswirkungen auf Standortqualitäten anhand ausgewählter Beispiele aus dem Steigerwald. PhD-Thesis, Ludwig-Maximilian University, München. http://edoc.ub.uni-muenchen.de/1468/ (accessed Aug. 17, 2010).

Phillips, J.D., 1993. Progressive and regressive pedogenesis and complex soil evolution. Quat. Res. 40, 169–176.

Phillips, J.D., 2001a. Inherited vs. acquired complexity in East Texas weathering profiles. Geomorphology 40, 1–14.

Phillips, J.D., 2001b. Contingency and generalization in pedology, as exemplified by texture-contrast soils. Geoderma 102 (3–4), 347–370.

Phillips, J.D., 2004. Geogenesis, pedogenesis, and multiple causality in the formation of texture-contrast soils. Catena 58 (2), 275–295.

Phillips, J.D., Lorz, C., 2008. Origins and implications of soil layering. Earth Sci. Rev. 89, 144–155.

Phillips, F.M., Zreda, M.G., Smith, S.S., Elmore, D., Kubik, P.W., Sharma, P., 1990. Cosmogenic chlorine-36 chronology for glacial deposits at Bloody Canyon, eastern Sierra Nevada. Science 248, 1529–1532.

Phillips, F.M., Zreda, M.G., Gosse, J.G., Klein, J., Evenson, E.B., Hall, R.D., Chadwick, O.A., Sharma, P., 1997. Cosmogenic ^{36}Cl and ^{10}Be ages of Quaternary glacial and fluvial deposits of the Wind River Mountains, Wyoming. Geol. Soc. Am. Bull. 109, 1453–1463.

Pierce, K.L., 1979. History and dynamics of glaciation in the northern Yellowstone National Park area. U.S. Geol. Surv. Prof. Pap. 729.

Pierce, K.L., 2003. Pleistocene glaciations of the Rocky Mountains. Dev. Quat. Sci. 1, 63–76.

Pissart, A., Juvigné, E., 1980. Genèse et âge d'une trace de butte périglaciaire (pingo ou paöse) de la Konnerzvenn (Hautes-Fagnes, Belgique). Ann. Soc. Géol. Belg. 103, 73–86.

Plass, W., 1968. Braunerden und Parabraunerden in Nordhessen. Zeitschrift für Pflanzenernährung, Düngung und Bodenkunde 114, 12–26.

Plöchinger, B., Prey, S., 1993. Der Wienerwald. Sammlung Geologischer Führer 59, 1–168.

Poetsch, T., 1975. Untersuchungen von bodenbildenden Deckschichten unter besonderer Berücksichtigung ihrer vulkanischen Komponente. Gießener Geologische Schriften 4.

Poisel, R., Eppensteiner, W., 1986. Control of a large natural slope in a suburb of Vienna. Proceedings of the Conference Hong Kong (Institution of Mining and Metallurgy, London. IMM N. American Publications Center, Brookfield, VT), pp. 335–340.

Pomeroya, J.W., Maleb, D.H., 1992. Steady-state suspension of snow. J. Hydrol. 136, 275–301.

Porter, S.C., Pierce, K.L., Hamilton, T.D., 1983. Late Wisconsin mountain glaciation in the western United States. In: Porter, S.C. (Ed.), Late Quaternary Environments of the United States, 1: The Late Pleistocene. University of Minnesota Press, Minneapolis, pp. 71–111.

Preusser, F., 2003. IRSL dating of K-rich feldspars using the SAR protocol: comparison with independent age control. Ancient TL 21, 17–23.

Prick, A., 1997. Critical degree of saturation as a treshold moisture level in frost weathering of limestones. Permafrost Periglac. Proc. 8, 91–99.

Priehäuser, G., 1958. Über den Aufbau und die Oberflächenformen der Ablagerungen aus dem Firneis der letzten Kaltzeit (Endwürm) im Bayerischen Wald. Geologische Blätter für Nordost-Bayern 8, 152–157.

Pye, K., 1987. Aeolian dust and dust deposits, 2nd ed. Academic Press, London, Orlando.

Qiang, M., Lang, L., Wang, Z., 2010. Do fine-grained components of loess indicate westerlies: insights from observations of dust storm deposits at Lenghu (Qaidam Basin, China). J. Arid Environ. 74, 1232–1239.

Raab, T., 1999. Würmzeitliche Vergletscherung des Bayerischen Waldes im Arbergebiet. Regensburger Geographische Schriften 32.

Raab, T., 2005. Erfassung und Bewertung von Landschaftswandel in (prä-)historischen Montangebieten am Beispiel Ostbayerns. Regensburger Beiträge zur Bodenkunde, Landschaftsökologie und Quartärforschung 7. http://www.opus-bayern.de/uni-regensburg/volltexte/2005/581/.

Raab, T., Völkel, J., 2002. Verbreitung und Altersstellung polygenetischer Hangsediment-Komplexe am Kleinen Arbersee im Hinteren Bayerischen Wald. Berichte zur deutschen Landeskunde 76 (2/3), 131–149.

Raab, T., Völkel, J., 2003. Late Pleistocene glaciation of the Kleiner Arbersee area in the Bavarian Forest, South Germany. Quat. Sci. Rev. 22, 581–593.

Raab, T., Völkel, J., 2005. Soil geomorphological studies on the prehistoric to historic landscape change in the former mining area at the Vils River (Bavaria, Germany). Z. Geomorph. N.F. 139 (Suppl.), 129–145.

Raab, T., Völkel, J., Leopold, M., 2007. Character, age, and ecological significance of Pleistocene periglacial slope deposits in Germany. Phys. Geogr. 28, 1–23.

Rasmussen, C., Troch, P.A., Chorover, J., Brooks, P., Pelletier, J., Huxman, T.E., 2010. An open system framework for integrating critical zone structure and function. Biogeochemistry 102, 15–29. http://dx.doi.org/10.1007/s10533-010-9476-8.

Raukas, A., 1991. Eemian interglacial record in the northwestern European part of the Soviet Union. Quat. Int. 10–12, 183–189.

Rees-Jones, J., 1995. Optical dating of young sediments using fine-grain-quartz. Ancient TL 13 (2), 9–15.

Reheis, M.C., 1987. Climatic implications of alternating clay and carbonate formation in semiarid soils of south-central Montana. Quat. Res. 27, 270–282.

Reheis, M.C., 2003. Dust deposition in Nevada, California, and Utah, 1984-2002. U.S. Geol. Surv. Open-file Report 03-138.

Reheis, M.C., Budahn, J.R., Lamothe, P.J., 2002. Geochemical evidence for diversity of dust sources in the southwestern United States. Geochim. Cosmochim. Acta 66, 1569–1587.

Reheis, M.C., Sowers, J.M., Taylor, E.M., McFadden, L.D., Harden, J.W., 1992. Morphology and genesis of carbonate soils on the Kyle Canyon Fan, Nevada, USA. Geoderma 52, 303–342.

Reheis, M.C., Goodmacher, J.C., Harden, J.W., McFadden, L.D., Rockwell, T.K., Shroba, R., Sowers, J.M., Taylor, E.M., 1995. Quaternary soils and dust deposition in southern Nevada and California. Geol. Soc. Am. Bull. 107, 1003–1022.

Rehfuess, K.E., 1990. Waldböden. Entwicklung, Eigenschaften und Nutzung, 2nd ed. Paul Parey Verlag, Hamburg, Berlin.

Reider, R.G., Huss, J.M., Miller, T.W., 1996. A groundwater vortex hypothesis for the mima-like mounds, Laramie Basin, Wyoming. Geomorphology 16, 295–317.

Rennert, R., 1991. Geoökologische Untersuchungen zur Bodengefrornis an der Untergrenze des alpinen Permafrostes unter Einsatz von Hammerschlagseismik, Geoelektrik und Bodentemperaturmessungen. Unpubl. Diploma Thesis, Univ. of Bayreuth.

Renssen, H., Vandenberghe, J., 2003. Investigation of the relationship between permafrost distribution in NW-Europe and extensive winter sea-ice cover in the North Atlantic Ocean during the cold phases of the Last Glaciation. Quat. Sci. Rev. 22, 209–233.

Renssen, H., Isarin, R.F.B., Vandenberghe, J., 2002. Thermal gradients in Europe during the last glacial-interglacial transition. Geologie en Mijnbouw 81, 113–122.

Renssen, H., Kasse, C., Vandenberghe, J., Lorenz, S.J., 2007. Weichselian Late Pleniglacial surface winds over northwest and central Europe: a model–data comparison. J. Quat. Sci. 22, 281–293.

Reszler, Ch., Komma, J., Blöschl, G., Gutknecht, D., 2006. Ein Ansatz zur Identifikation flächendetaillierter Abflussmodelle für die Hochwasservorhersage. Hydrologie und Wasserbewirtschaftung 50 (5), 220–232.

Retallack, G.J., 2005. Pedogenic carbonate proxies for amount and seasonality of precipitation in paleosols. Geology 33, 333–336.

Rezzoug, A., Schumann, A., Chifflard, P., Zepp, H., 2005. Field measurement of soil moisture dynamics and numerical simulation using the kinematic wave approximation. Adv. Water Resour. 28, 917–926.

RGZM (Römisch Germanisches Zentralmuseum Mainz), 1972. Führer zu vor- und frühgeschichtlichen Denkmälern Bd. 21. Hochtaunus, Bad Homburg, Usingen, Königstein, Hofheim, 2nd ed. Verlag Philipp von Zabern, Darmstadt.

Richmond, G.M., 1962. Quaternary stratigraphy of the La Sal Mountains, Utah. U.S. Geol. Surv. Prof. Pap. 324.

Richmond, G.M., 1986a. Stratigraphy and correlation of glacial deposits of the Rocky Mountains, the Colorado Plateau and the ranges of the Great Basin. Quat. Sci. Rev. 5, 99–127.

Richmond, G.M., 1986b. Stratigraphy and correlation of glacial deposits of the Rocky Mountains, the Colorado Plateau, and the ranges of the Great Basin. In: Sibrava, V., Bowen, D.Q., Richmond, G.M. (Eds.), Quaternary Glaciations of the Northern Hemisphere. Pergamon Press, Oxford, United Kingdom, pp. 99–127.

Richter, H., Ruske, R., Schwanecke, W., 1970. Die periglaziäre Fazies im lößfreien Hügelland und im Mittelgebirge. Petermanns Geographische Mitteilungen Erg.-H. 274, 57–97.

Ridefelt, H., Åkerman, J., Beylich, A.A., Boelhouwers, J., Kolstrup, E., Nyberg, R., 2009. 56 years of solifluction measurements in the Abisko Mountains, Northern Sweden – analysis of temporal and spatial variations of slow soil surface movement. Geogr. Ann. 91A, 215–232.

Robison, R.M., McCalpin, J.P., 1987. Surficial geology of Hansel Valley, Box Elder County, Utah. Utah Geol. Assoc. Publ. 16, 335–349.

Rohdenburg, H., 1965. Untersuchungen zur pleistozänen Formung am Beispiel der Wetabdachung des Göttinger Waldes. Giessener Geographische Schriften 7.

Rohdenburg, H., 1968. Jungpleistozäne Hangformung in Mitteleuropa – Beiträge zur Kenntnis, Deutung und Bedeutung ihrer räumlichen und zeitlichen Differenzierung. Göttinger Bodenkundliche Berichte 6, 3–107.

Rohdenburg, H., 1978. Zur Problematik der spätglazialen und holozänen Bodenbildung in Mitteleuropa. In: Nagl, H. (Ed.), Festschrift für Julius Fink. Hirt, Wien, pp. 467–471.

Rohdenburg, H., 1989. Landschaftsökologie – Geomorphologie. Catena, Cremlingen.

Rohdenburg, H., Meyer, B., 1963. Rezente Mikroformung in Kalkgebieten durch inneren Abtrag und die Rolle der periglazialen Gesteinsverwitterung. Eine geomorphologische Auswertung von Bilanzanalysen an Bodenprofilen. Z. Geomorph. N.F. 7, 120–146.

Rohdenburg, H., Meyer, B., 1968. Zur Datierung und Bodengeschichte mitteleuropäischer Oberflächenböden, Schwarzerde, Parabraunerde, Kalksteinbraunlehm. Spätglazial oder Holozän? Göttinger Bodenkundliche Berichte 6, 127–212.

Rosenberg, F., 1988. Geochemie, Erläuterungen zur Geologischen Karte von Hessen 1: 25 000 Blatt Nr. 5621 Wenings, 186–209.

Ross, M., 2006. Preliminary geologic map of the Warner Lake Quadrangle, Grand County, Utah. Utah Geol. Surv. Open-File Rep. 497.

Rousseau, D.-D., Antoine, P., Kunesch, S., Hatte, C., Rossignol, J., Packman, S., Lang, A., Gauthier, C., 2007. Evidence of cyclic dust deposition in the US Great Plains during the last deglaciation from the high-resolution analysis of the Peoria Loess in the Eustis sequence (Nebraska, USA). Earth Planet. Sci. Lett. 262, 159–174.

Ruffner, J.A., 1978. Climates of the States, vol. 2. – Gale Research, Detroit.

Ruppert, H., 1994. Einfluß periglazialer Deckschichten auf die natürlichen Schwermetallgehalte von Böden. In: Matschullat, J., Müller, G. (Eds.), Geowissenschaft und Umwelt. Springer, Berlin, pp. 21–33.

Ruppert, H., Schmidt, F., 1987. Natürliche Grundgehalte und anthropogene Anreicherung von Schwermetallen in Böden Bayerns. GLA Fachberichte 2, 1–95.

Rusakov, A.V., Korkka, M.A., 2004. The Bryansk fossil soil of the extraglacial zone of the Valday glaciation as an indicator of landscape and soil forming processes in the center of the Russian Plain. Revista Mexicana de Ciencias Geológicas 21, 94–100.

Rusakov, A.V., Korkka, M.A., Kerzum, P.P., Simakova, A.N., 2007. Paleosols in the moraine-mantle loam sequence of northeastern Europe: the memory of pedogenesis rates and evolution of the environment during OIS3. Catena 71, 456–466.

Russow, F., Heinrich, J., 2001. Jungholozäne Überprägungen von quartären Deckschichten und Böden des Mittelgebirgsraumes durch biomechanische Prozesse in Mitteleuropa. Geoöko 22, 37–58.

Sabel, K.-J., 1989. Zur Renaissance der Gliederung periglazialer Deckschichten in der deutschen Bodenkunde. Frankfurter Geowiss. Arb. D 10, 9–16.

Sack, D., 1989. Reconstructing the chronology of Lake Bonneville: an historical review. Binghampton Symp. Geomorphol. Internat. Ser. 19, 223–256.

Šály, R., Ciesarik, M., Mihálik, A., 1976. Die "Terra calcis" – Böden der slowakischen Wälder – ihre Entstehung, Eigenschaften und Bedeutung. Zborník Vedeckých Prác, Lesníckej Fakulty VŠld vo Zvolene XVIII-2, 33–63.

Sarna-Wojcicki, A.M., Davis, J.O., 1991. Quaternary tephrochronology. In: Morrison, R.B. (Ed.), Quaternary Nonglacial Geology: Conterminous U.S. Geol. Soc. Am., Boulder, pp. 93–116.

Sauer, D., 2002. Genese, Verbreitung und Eigenschaften periglaziärer Lagen im Rheinischen Schiefergebirge – anhand von Beispielen aus Westerwald, Hunsrück und Eifel. Boden und Landschaft 36.

Sauer, D., 2004. Bildungsprozesse periglaziärer Lagen in den deutschen Mittelgebirgen und ihre zeitliche Einstufung. Tübinger Geowissenschaftliche Arbeiten D10, 19–38.

Sauer, D., Felix-Henningsen, P., 2004. Application of ground penetrating radar for the determination of the thickness of Pleistocene periglacial slope deposits (PPSD) on hard rock. J. Plant Nutr. Soil Sci. 167, 752–760.

Sauer, D., Felix-Henningsen, P., 2006. Saprolite, soils, and sediments in the Rhenish Massif as records of climate and landscape history. Quat. Int. 156–157, 4–12.

Sauer, D., Scholten, T., Felix-Henningsen, P., 2001. Verbreitung und Eigenschaften periglazialer Lagen im östlichen Westerwald in Abhängigkeit von Gestein, Exposition und Relief. Mitteilungen der Deutschen Bodenkundlichen Gesellschaft 96/II, 551–552.

Sazonova, T.S., Romanovsky, V.E., 2003. A model for regional-scale estimation of temporal and spatial variability of active layer thickness and mean annual ground temperatures. Permafrost Periglac. Proc. 14, 125–139.

Schädel, K., Stober, I., 1988. Rezente Großrutschungen an der Schwäbischen Alb. Jahreshefte Geologisches Landesamt Baden-Württemberg 30, 413–439.

Schaetzl, R., Anderson, S., 2005. Soils – Genesis and Geomorphology. Cambridge University Press, Cambridge.

Schäfer, D., Jäger, K.-D., Altermann, M., 1991. Zur Stratigraphie periglazialer Decken im thüringischen Bergland. Archäologisches Korrespondenzblatt Mainz 21, 323–334.

Schaller, M., von Blanckenburg, F., Veit, H., Kubik, P.W., 2002. Influence of periglacial cover beds on in situ-produced cosmogenic ^{10}Be in soil sections. Geomorphology 49, 255–267.

Scheffer, F., Schachtschabel, P., 2010. Lehrbuch der Bodenkunde, 16th ed. Spektrum Verlag, Heidelberg, Berlin.

Scheffer, F., Welte, E., Meyer, B., 1960. Die Rendsinen der mitteldeutschen Berg- und Hügellandschaften (Leine-Weser-Bergland). 1. Mitteilung: Genese und Verbreitungsbedingungen der Rendsinen. J. Plant Nutr. Soil Sci. 90, 18–36.

Scheffer, F., Welte, E., Meyer, B., 1962. Die Rendsinen der mitteldeutschen Berg- und Hügellandschaften (Leine-Weser-Bergland). 2. Mitteilung: Fraktur, Nichtcarbonatgehalt und spezifische Auflösungsgeschwindigkeit des Kalksteins als bestimmende Größen der Bodenentwicklungsgeschwindigkeit. J. Plant Nutr. Soil Sci. 92 (1), 1–17.

Schilling, W., Wiefel, H., 1962. Jungpleistozäne Periglazialbildungen und ihre regionale Differenzierung in einigen Teilen Thüringens und des Harzes. Geologie 11, 428–460.

Schilling, B., Spies, E.-D., 1991. Die Böden Mittel- und Oberfrankens. Bayreuther Bodenkundliche Berichte 17, 67–81.

Schimming, C.G., 1992. Kontamination von Böden. Belastung mit Metallen. In: Blume, H.P. (Ed.), Handbuch des Bodenschutzes. Ecomed Verlag, Landsberg, pp. 277–318.

Schmanke, V., 1999. Untersuchungen zur Hanggefährdung im Bonner Raum – Eine Bewertung mit Hilfe unterschiedlicher Modellansätze. Mainzer Geographische Studien 44.

Schmincke, H.-U., Park, C., Harms, E., 1999. Evolution and environmental impacts of the eruption of Laacher See Volcano (Germany) 12,900 a BP. Quat. Int. 61, 61–72.

Schneider, G., Moldenhauer, K.-M., Nagel, G., 1997. Oberflächennaher Wasseraustrag durch Interflow aus einem bewaldeten Kleineinzugsgebiet (Hoher Taunus). Mitt. Dt. Bodenkdl. Ges. 85, 159–162.

Schneider, R., Schobel, S., Niebes, D., Schröder, D., 2001. Untersuchungen zur Hochwasserentstehung im Labor und Gelände auf unterschiedlichen Skalenniveaus. Mitteilungen der Deutschen Bodenkundlichen Gesellschaft 96/II, 653–654.

Scholten, T., 2003. Verbreitungssystematik und Eigenschaften pleistozäner periglaziärer Lagen in deutschen Mittelgebirgen. Relief, Boden, Paläoklima 19.

Scholten, T., Altermann, M., Schwanecke, W., Felix-Henningsen, P., 1999. Die Bedeutung periglazialer Lagen für Funktionen des Bodens im Ostharz. Mitteilungen der Deutschen Bodenkundlichen Gesellschaft 91/II, 1096–1099.

Scholten, T., Behrens, T., Felix-Henningsen, P., 2001. Modelling geo-spatial soil properties in mid-altitude landscapes. In: Belward, A., Binaghi, E., Brivio, P.A., Lanzarone, G.A., Tosi, G. (Eds.), Geo-spatial Knowledge Processing for Natural Resource Management. Varese, Italy, pp. 324–328.

Scholten, T., Behrens, T., Szibalski, M., Felix-Henningsen, P., 2002. Prediction model Lsys + – spatial modelling of properties of periglacial slope deposits. In: Rubio, J.L., Morgan, R.P.C., Asins, S., Andreu, V. (Eds.), Man and Soil at the Third Millennium, vol. 2. Geoforma Ediciones, Logrono, Spain, pp. 2091–2097.

Schönhals, E., 1957. Ein äolisches Sediment der Jüngeren Dryas-Zeit auf dem Laacher-See-Tuff. Fortschr. Geol. Rheinland u. Westfalen 4, 37–340.

Schönhals, E., 1996. Ergebnisse bodenkundlicher Untersuchungen in der hessischen Lößprovinz mit Beiträgen zur Genese des Würm-Lösses. Justus-Liebig Universität, Gießen.

Schönhals, E., Rohdenburg, H., Semmel, A., 1964. Ergebnisse neuerer Untersuchungen zur Würmlößgliederung in Hessen. Eiszeitalter und Gegenwart 15, 199–206.

Schwanecke, W., 1966. Für die Bodensystematik wichtige Ergebnisse der forstlichen Standortserkundung im Hügelland und Mittelgebirge der Deutschen Demokratischen Republik. Sitzungsberichte DAL Berlin 15, 79–95.

Schwanecke, W., 1970. Die periglaziären Umlagerungszonen im Hügelland und Mittelgebirge der DDR und ihre bodenkundliche Bedeutung. In: Anonymus, (Ed.), Beiträge zur Bodensystematik unter besonderer Berücksichtigung reliktischer und rezenter Merkmale. Tagungsberichte DADL, Berlin, pp. 83–108.

Schwarz, O., 1985. Beiträge zu Oberflächenabfluss und Stoffabtrag bei künstlichen Starkniederschlägen. DVWK-Schriften 71, 185–230.

Scott, W.E., Shroba, R.R., 1980. Stratigraphic significance and variability of soils buried by deposits of the last glacial cycle of Lake Bonneville. Geol. Soc. Am. Abstr. Programs 12, 304.

Scott, W.E., McCoy, W.D., Shroba, R.R., Meyer, R., 1983. Reinterpretation of the exposed record of the last two cycles of Lake Bonneville, Western United States. Quat. Res. 20, 261–285.

Seeger, T., 1990. Abfluss- und Stofffrachtseparation im Buntsandstein des Nordschwarzwaldes. Tübinger Geowissenschaftliche Arbeiten C 6, Tübingen.

Seidel, G. (Ed.), 1995. Geologie von Thüringen. Schweizerbart'sche Verlagsgesellschaft, Stuttgart.

Seidel, G., Steinmüller, A., 1993. Erläuterungen zur Geologischen Karte 1:25.000 von Thüringen Blatt 4936 Camburg. Thüringer Landesanstalt für Geologie, Weimar.

Selby, M.J., 1993. Hillslope Materials and Processes, Second ed. Oxford University of Press, Oxford.

Semmel, A., 1964. Junge Schuttdecken in hessischen Mittelgebirgen. Notizblatt des hessischen Landesamtes für Bodenforschung 92, 275–285.

Semmel, A., 1966. Erläuterungen zur Bodenkarte von Hessen 1:25 000, Blatt Nr. 5224 Eiterfeld. Hessisches Geologisches Landesamt, Wiesbaden.

Semmel, A., 1968. Studien über den Verlauf jungpleistozäner Formung in Hessen. Frankfurter Geogr. H. 45.

Semmel, A., 1977. Grundzüge der Bodengeographie. Teubner, Stuttgart.

Semmel, A., 1980. Quartär. In: Hessisches Geologisches Landesamt, (Ed.), Erläuterungen zur Geologischen Karte 1:25 000, Bl. 5917, Kelsterbach. Geologisches Landesamt Hessen, Wiesbaden, pp. 25–49.

Semmel, A., 1985. Periglazialmorphologie. Wiss. Buchges, Darmstadt.

Semmel, A., 1986. Angewandte konventionelle Geomorphologie – Beispiele aus Mitteleuropa und Afrika. Frankfurter Geowissenschaftliche Arbeiten D 6.

Semmel, A., 1987. Bodenbewegungen im Rhein-Main-Gebiet. Ursachen und Auswirkungen. Forschung Frankfurt 2/3, 18–24.

Semmel, A., 1989. The importance of loess in the interpretation of geomorphological processes and for dating in the FRG. Catena (Suppl.), 15, 179–188.

Semmel, A., 1990. Periglaziale Formen und Sedimente. In: Liedtke, H. (Ed.), Eiszeitforschung. Wissenschaftliche Buchgesellschaft, Darmstadt, pp. 250–260.

Semmel, A., 1991a. Schuttdecken und ihre Bedeutung für den Landschaftshaushalt in hessischen Mittelgebirgen. Geogr. Rundsc. 43, 299–302.

Semmel, A., 1991b. Neotectonics and geomorphology in the Rhenish Massif and the Hessian basin. Tectonophysics 195, 291–297.

Semmel, A., 1991c. Relief, Gestein, Boden, Grundlagen der Physischen Geographie I. Wissenschaftliche Buchgesellschaft, Darmstadt.

Semmel, A., 1992. Gab es während der Jüngeren Tundrenzeit Permafrost im Rheinischen Schiefergebirge? In: Fachbereich Geographie der Humboldt-Universität (Ed.). 18. Tagung des deutschen Arbeitskreises für Geomorphologie, Vortragskurfassungen. Selbstverlag, Berlin: unpaginated.

Semmel, A., 1993a. Bodenerosionsschäden unter Wald – Beispiele aus dem Kristallinen Odenwald und dem Taunus. Jahresberichte der wetterauischen Gesellschaft für Naturkunde 144/145, 5–15.

Semmel, A., 1993b. Grundzüge der Bodengeographie, 3rd ed. Teubner Verlag, Stuttgart.

Semmel, A., 1994a. Zur umweltgeologischen Bedeutung von Hangsedimenten in deutschen Mittelgebirgen. Zeitschrift der Deutschen Geologischen Gesellschaft 145, 225–232.

Semmel, A., 1994b. Gliederung der Solifluktionsdecken in deutschen Mittelgebirgen – noch ein Problem der Wiedervereinigung? Mitteilungen der Deutschen Bodenkundlichen Gesellschaft 74, 29–33.

Semmel, A., 1996a. The historic-genetic approach in applied geomorphology. Z. Geomorph. N.F. 40 (Suppl.), 289–303.

Semmel, A. (Ed.), 1996. Pleistozäne und holozäne Böden aus Lößsubstraten am Nordrand der Oberrheinischen Tiefebene. Frankfurter Geowissenschaftliche Arbeiten D20.

Semmel, A., 1998a. Anmerkung zur Schuttdeckengliederung im Thüringer Wald und in hessischen Bergländern. Geowiss. Mitt. Thüringen Beiheft 7, 62–78.

Semmel, A., 1998b. Lokalitäten von Stratotypen jungpleistozäner äolischer und gelifluidaler Sedimente im Rhein-Main-Gebiet. GeoArchaeoRhein 2, 35–45.

Semmel, A., 2000. Relief und Baugrund an der ICE-Neubaustrecke Köln-Rhein/Main in Hessen. Berichte zur Deutschen Landeskunde 74/4, 329–344.

Semmel, A., 2001a. In memoriam: Die Phäno-Parabraunerde. Mitteilungen der Deutschen Bodenkundlichen Gesellschaft 96, 561–562.

Semmel, A., 2001b. Der oberflächennahe Untergrund in der Rhein-Main-Landschaft – ein Exkursionsführer. Rhein-Mainische Forschungen 121.

Semmel, A., 2002a. Hauptlage und Oberlage als umweltgeschichtliche Indikatoren. Z. Geomorph. N.F. 46, 167–180.

Semmel, A., 2002b. Die periglaziale Hauptlage – Genese, Alter und anthropogene Veränderung. Terra Nostra 2002/6, 342–348.

Semmel, A., 2003. Der Laacher Bimstuff als Zeitmarke der Landschaftsentwicklung in der Wiesbadener Umgebung. Jahrbuch des Nassauischen Vereins für Naturkunde 124, 95–109.

Semmel, A., 2005. Periglaziale Vergangenheit und bodenkundliche Gegenwart in den deutschen Mittelgebirgen. Wald in Sachsen-Anhalt 15, 35–60.

Semmel, A., Plass, W., 1965. Spätkaltzeitliche Umlagerungen in Parabraunerde-Profilen. Mitteilungen der Deutschen Bodenkundlichen Gesellschaft 4, 33–40.

Semmel, A., Terhorst, B., 2010. The concept of the Pleistocene periglacial cover beds in central Europe: a review. Quat. Int 222, 120–128.

Sharp, R.P., 1938. Pleistocene glaciation in the Ruby-East Humboldt Range, northeastern Nevada. J. Geomorphol. 1, 296–323.

Sharp, R.P., 1939. Basin range structure of the Ruby-East Humboldt Range, northeastern Nevada. Geol. Soc. Am. Bull. 50, 881–920.

Sharp, W.D., Ludwig, K.R., Chadwick, O.A., Amundson, R., Glaser, L.L., 2003. Dating fluvial terraces by ^{230}Th/U on pedogenic carbonate, Wind River Basin, Wyoming. Quat. Res. 59, 139–150.

Shroder, J.F., Sewell, R.E., 1985. Mass movements in the La Sal Mountains, Utah. Utah Geol. Mineralog. Surv. Spec. Stud. 64, 48–85.

Silberstein, R.P., Bartle, G.A., Salama, R.B., Hatton, T.J., Reggiani, P., Hodgson, G., Williamson, D.R., Lambert, P., 2002. Mechanisms and control of water logging and groundwater flow in the "Ucarro" sub-catchment. Agric. Water Manage. 53, 227–257.

Simonson, R.W., 1959. Outline of a generalized theory of soil genesis. Soil Sci. Soc. Am. Proc. 23, 152–156.

Simonson, R.W., 1978. A multiple-process model of soil genesis. In: Mahaney, W.C. (Hrsg.), Quaternary Soils. GeoAbstracts, Norwich, pp. 1–25.

Singarayer, J.S., Bailey, R.M., Ward, S., Stokes, S., 2005. Assessing the completeness of optical resetting of quartz OSL in the natural environment. Radiat. Meas. 40, 13–25.

Sivapalan, M., Takeuchi, K., Franks, S.W., Gupta, V.K., Karambiri, H., Lakshmi, V., Liang, X., McDonnell, J.J., Mendiondo, E.M., O'Connell, P.E., Oki, T., Pomeroy, J.W., Schertzer, D., Uhlenbrook, S., Zehe, E., 2003. IAHS decade on Predictions in Ungauged Basins (PUB), 2003–2012: shaping an exciting future for the hydrological sciences. Hydrol. Sci. J. 48, 857–880.

Smalley, I., 1990. Possible formation mechanisms for the modal coarse-silt quartz particles in loess deposits. Quat. Int. 7–8, 23–27.

Smolikova, L., Lozek, V., 1962. Zur Altersfrage der mitteleuropäischen Terra calcis. Eiszeitalter und Gegenwart 13, 101–126.

Sokolov, I.A., 1989. Genesis, identification, and classification of soils with a texturally differentiated profile. Soviet Soil Sci. 21 (3), 1–14.

Spackman, L.K., Munn, L.C., 1984. Genesis and morphology of soils associated with formation of Laramie Basin (mima-like) mounds in Wyoming. Soil Sci. Soc. Am. J. 48, 1384–1392.

Spaulding, W.G., 1990a. Vegetation dynamics during the last deglaciation, southeastern Great Basin, USA. Quat. Res. 33, 188–203.

Spaulding, W.G., 1990b. Vegetational and climatic development of the Mojave Desert: the last glacial maximum to the present. In: Betancourt, J.L., Van Devender, T.R., Martin, P.S. (Eds.), Packrat Middens, The Last 40.000 Years of Biotic Change. University of Arizona Press, Tucson, AZ, pp. 166–199.

Spaulding, W.G., Graumilch, L.J., 1986. The last pluvial climatic episodes in the deserts of southwestern North America. Nature 320, 441–444.

Spaulding, W.G., Leopold, E.B., Van Devender, T.R., 1983. Late Wisconsin paleoecology of the American Southwest. In: Porter, S.C. (Ed.), Late Quaternary Environments of the United States, 1: The Late Pleistocene. University of Minnesota Press, Minneapolis, pp. 259–293.

Spell, T.L., Harrison, T.M., Wolff, J.A., 1990. 40Ar/39Ar dating of the Bandelier Tuff and San Diego Canyon ignimbrites, Jemez Mountains, New Mexico: temporal constraints on magmatic evolution. J. Volcanol. Geotherm. Res. 43, 175–193.

Spektor, V.B., Spektor, V.V., 2009. Karst processes and phenomena in the perennially frozen carbonate rocks of the middle Lena River Basin. Permafrost Periglac. Process. 20, 71–78.

Spies, E.-D., 1991. Böden und Deckschichten der Münchberger Hochfläche und des Frankenwaldes. Mitteilungen der Deutschen Bodenkundlichen Gesellschaft 64, 139–210.

SSS (Soil Survey Staff), 1994. Keys to Soil Taxonomy, 6th ed. Pocahontas Press, Blacksburg.

SSS (Soil Survey Staff), 2010. Keys to Soil Taxonomy, 11th ed. USDA-Natural Resources Conservation Service, Washington, D.C.

Stahr, K., 1979. Die Bedeutung periglazialer Deckschichten für Bodenbildung und Standortseigenschaften im Südschwarzwald. Freiburger Bodenkundliche Abhandlungen 9.

Stahr, A., 2000. Zur Differenzierung periglazialer Deckschichten der montanen und subalpinen Höhenstufen in den Berchtesgadener Alpen. Frankfurter Geowissenschaftliche arbeiten 26, 155–172.

Steinmann, S., 1978. Postglaziale Reliefgeschichte und gegenwärtige Vegetationsdifferenzierung in der alpinen Stufe der Südtiroler Dolomiten (Puez- u. Sellagruppe). Catena-Verlag, Cremlingen.

Steinmüller, A., 1996. Quartär. In: Thüringer Landesanstalt für Geologie (Ed.), Erläuterungen zur Geologischen Karte 1:25 000 von Thüringen. Blatt Weimar Ost, Nr. 5034. Weimar, pp. 49–100.

Stendel, M., Romanovsky, V.E., Christensen, J.H., Sazonova, T., 2007. Using dynamical downscaling to close the gap between global change scenarios and local permafrost dynamics. Glob. Planet. Change 56, 203–214.

Stephan, S., 2000. Bt-Horizonte als Interglazial-Zeiger in den humiden Mittelbreiten: Bildung, Mikromorphologie, Kriterien. Eiszeitalter und Gegenwart 50, 95–106.

Stevens, T., Lu, H., 2009. Optically stimulated luminescence dating as a tool for calculating sedimentation rates in Chinese loess: comparisons with grain-size records. Sedimentology 56, 911–934.

Stocker, E., 1984. Ergebnisse elfjähriger Messungen der Bodenbewegung in der alpinen Stufe der Kreuzeckgruppe (Kärnten). Wiener Geographische Schriften 59/60, 27–35.

Stöhr, W.T., 1963. Der Bimstuff (Trachyttuff), seine Verlagerung, Verlehmung und Bodenbildung (Lockerbraunerden) im südwestlichen Rheinischen Schiefergebirge. Notizblatt Hessisches Landesamt Bodenforschung 91, 318–337.

Stolz, C., Grunert, J., 2010. Quaternary landscape development in Palatinate Forest (Pfälzerwald, south-western Germany). Quat. Int. 222, 129–142.

Stolz, C., Grunert, J., 2012. Observation on the inclination of clasts in perglacial cover-beds in the Rhenish Massif (Germany). Z. Geomorph. N.F. (Suppl.-Iss.) http://dx.doi.org/10.1127/0372-8854/2012/S-00081.

Stückrad, S., Sabel, K.-F., Wilcke, W., 2008. Periglacial transport distance of Pb derived from small-scale ore veins in the Rhenish Slate Mountains. Geoderma 148, 232–239.

Sudakova, N., 1990. Paleogeographicheskie zakonomernosti lednikovogo litogeneza. Lomonossov-University, Moscow.

Suggate, R.P., Almond, P.C., 2005. The Last Glacial Maximum (LGM) in western South Island, New Zealand: implications for the global LGM and MIS 2. Quat. Sci. Rev. 24, 1923–1940.

Svendsen, J.I., Astakhov, V.I., Bolshiyanov, D.Yu., Demidov, I., Dowdeswell, J.A., Gataullin, V., Hjort, C., Hubberten, H.W., Larsen, E., Mangerud, J., Melles, M., Möller, P., Saarnisto, M., Siegert, M.J., 1999. Maximum extent of the Eurasian ice sheets in the Barents and Kara Sea region during the Weichselian. Boreas 28, 234–242.

Svendsen, J.I., Alexanderson, H., Astakhov, V.I., Demidov, I., Dowdeswell, J.A., Funder, S., Gataullin, V., Henriksen, M., Hjort, C., Houmark-Nielsen, M., Hubberten, H.W., Ólafur Ingólfsson, M., Jakobsson, K.H., Kjær, E., Larsen, H., Lokrantz, J. Pekka, Lunkka, A., Lysa, J., Mangerud, A., Matiouchkov, A., Murray, P., Möller, F., Niessen, O., Nikolskaya, L., Polyak, M., Saarnisto, C., Siegert, M.J., Siegert, R.F. Spielhagen, Stein, R., 2004. Late Quaternary ice sheet history of northern Eurasia. Quat. Sci. Rev. 23, 1229–1271.

Tandarich, J.P., Darmody, R.G., Follmer, L.R., Johnson, D.L., 2002. Historical development of soil and weathering profile concepts from Europe to the United States of America. Soil Sci. Soc. Am. J. 66 (2), 335–346.

Tedrow, J., 1977. Soils of the polar landscapes. Rutgers University Press, New Brunswick.

Terhorst, B., 1997. Formenschatz, Alter und Ursachenkomplexe von Massenverlagerungen an der schwäbischen Juraschichtstufe unter besonderer Berücksichtigung von Boden- und Deckschichtenentwicklung. Tübinger Geowissenschaftliche Arbeiten D2, 1–212.

Terhorst, B., 2001. Mass movements of various ages on the Swabian Jurassic escarpment: geomorphologic processes and their causes. Z. Geomorph. N.F. 125 (Suppl.), 65–87.

Terhorst, B., 2007. Periglacial cover beds and soils in landslide areas of SW-Germany. Soil distribution and periglacial cover beds in the Jurassic Cuesta scarp in SW-Germany. Catena 71, 467–476.

Terhorst, B., Damm, B., 2009. Slope stability in the Flysch zone of the Vienna Forest (Austria). J. Geol. Res. http://dx.doi.org/10.1155/2009/589037.

Terhorst, B., Felix-Henningsen, P., 2010. Paläoböden in periglazialen Lagen der Mittelgebirge. In: Blume, H.-P., Felix-Henningsen, P., Fischer, W.R., Frede, H.-G., Horn, R., Stahr, K. (Eds.), Handbuch der Bodenkunde. Wiley-VCH, Weinheim (Chapter 4.5.3.3.6).

Terhorst, B., Kreja, R., 2009. Slope stability modelling with SINMAP in a settlement area of the Swabian Alb. Landslides 6, 309–319.

Terhorst, B., Damm, B., Peticzka, R., Köttritsch, E., 2009. Reconstruction of Quaternary landscape formation as a tool to understand present geomorphological processes in the Eastern Prealps (Austria). Quat. Int. 209, 66–78.

Terzaghi, K., 1950. Mechanism of landslides. Geol. Soc. Am. Rev. Eng. Geol., Berkey Volume, 83–122.

Thackray, G.D., 2008. Varied climatic and topographic influences on Late Pleistocene mountain glaciation in the western United States. J. Quat. Sci. 23, 671–681.

Thalheim, K., Fiedler, H.-J., 1990. Granulometrische und mineralogische Zusammensetzung von Basissedimenten der pleistozänen Deckschichten auf metamorphen und magmatischen Grundgesteinen im Osterzgebirge. Abhandlungen des Staatlichen Museums für Mineralogie und Geologie Dresden 37, 143–154.

Thein, S., 1998. GIS-gestützte Ausweisung rutschungsgefährdeter Gebiete. Standort 22, 13–21.

Thiemeyer, H., 1988. Bodenerosion und holozäne Dellenentwicklung in hessischen Lößgebieten. Rhein-Mainische Forschungen 105.

Thiemeyer, H., 2007. Stark podsolige Terra Fusca-Parabraunerden und Parabraunerde-Rendzinen des östlichen Riesrandes. Mitteilungen der Deutschen Bodenkundlichen Gesellschaft 110 (I), 545–546.

Thiemeyer, H., Veit, H., 1993. Bodenkundliche und schwermineralogische Untersuchungen an ausgewählten Deckschichtenprofilen in NE-Bayern. Berliner Geogr. Arb. 78, 265–286.

Thode, H.G., 1991. Sulphur isotopes in nature and the environment: an overview. In: Krouse, H.R., Grinenko, V.A. (Eds.), Stable Isotopes: Natural and Anthropogenic Sulphur in the Environment. Wiley, Chichester, pp. 1–26.

Thomas, H.R., Cleall, P., Li, Y.-C., Harris, C., Kern-Luetschg, M., 2009. Modelling of cryogenic processes in permafrost and seasonally frozen soils. Géotechnique 59, 173–184.

Thompson, R.S., 1990. Late Quaternary vegetation and climate in the Great Basin. In: Betancourt, J.L., Van Devender, T.R., Martin, P.S. (Eds.), Packrat Middens, The Last 40.000 Years of Biotic Change. University of Arizona Press, Tucson, AZ, pp. 200–239.

Thompson, R.S., 1992. Late Quaternary environments in Ruby Valley, Nevada. Quat. Res. 37, 1–15.

Thompson, R.S., Mead, J.I., 1982. Late Quaternary environments and Biogeography in the Great Basin. Quat. Res. 17, 39–55.

Þórarinsson, S., 1979. Tephrochronology and its application in Iceland. Jökull 29, 33–36.

Tilch, N., Uhlenbrook, S., Didszun, J., Wenninger, J., Kirnbauer, R., Zillgens, B., Leibundgut, C., 2006. Hydrologische Prozessforschung zur Hochwasserentstehung im Löhnersbach-Einzugsgebiet (Kitzbüheler Alpen, Österreich). Hydrologie und Wasserbewirtschaftung 50 (2), 67–78.

Tzedakis, C., 2009. Cenozoic climate and vegetation change. In: Woodward, J. (Ed.), The Physical Geography of the Mediterranean. Oxford University Press, Oxford, pp. 89–137.

Uhlenbrook, S., 2005. Von der Abflussbildungsforschung zur prozessorientierten Modellierung – ein Review. Hydrologie und Wasserbewirtschaftung 49 (1), 13–24.

Utah Geological and Mineral Survey, 1987. Geology and Grand County. Artistic Printing, Salt Lake City, Utah.

Van Devender, T.R., 1990a. Late Quaternary vegetation and climate of the Chihuahuan Desert, United States and Mexico. In: Betancourt, J.L., Van Devender, T.R., Martin, P.S. (Eds.), Packrat Middens, the Last 40.000 Years of Biotic Change. University of Arizona Press, Tucson, AZ, pp. 104–133.

Van Devender, T.R., 1990b. Late Quaternary vegetation and climate of the Sonoran Desert, United States and Mexico. In: Betancourt, J.L., Van Devender, T.R., Martin, P.S. (Eds.), Packrat Middens, the Last 40.000 Years of Biotic Change. University of Arizona Press, Tucson, AZ, pp. 134–165.

Van Loon, A.J., 2006. Lost loesses. Earth Sci. Rev. 74, 309–316.

Van Steijn, H., Bertran, P., Francou, B., Hetu, B., Texier, J.P., 1995. Review of models for genetical interpretation of stratified slope deposits. Permafrost Periglac. Proc. 6, 125–146.

Van Vliet-Lanoë, B., Langohr, R., 1981. Correlation between fragipans and permafrost with special reference to silty Weichselian deposits in Belgium and northern France. Catena 8, 137–154.

Van Vliet-Lanoë, B., Magyari, A., Meilliez, F., 2004. Distinguishing between tectonic and periglacial deformations of Quaternary continental deposits in Europe. Glob. Planet. Change 43, 103–127.

Varnes, D.J., 1978. Slope movement and types of processes. In: Schuster, R.L., Krizek, R.J. (Eds.), Landslides. Analysis and Control. Transportation Research Board Special Report, National Academy of Sciences, pp. 11–33.

Varnes, D.J., 1984. Landslide hazard zonation: a review of principles and practice of the United Nations Educational. UNESCO, Paris.

Veit, H., 1987. Some aspects of the genesis of cover sediments and soils in the "Açungui-Mountains" northwest of Curitiba (Southern Brazil). Zentralblatt Geologie Paläontologie T. 1 (1987), 853–861.

Veit, H., 1988. Fluviale und solifluidale Morphodynamik des Spät- und Postglazials in einem zentralalpinen Flußeinzugsgebiet (südliche Hohe Tauern, Osttirol). Bayreuther Geowissenschaftliche Arbeiten 13.

Veit, H., 1991. Das Bindlacher Schotterfeld und seine Deckschichten. Ein Beitrag zur Quartärstratigarophie in Oberfanken. Geologische Blätter für Nordost-Bayern 41, 1–14.

Veit, H., 1993. Holocene solifluction in the Austrian and southern Tyrolean Alps: dating and climatic implications. Paläoklimaforschung 11, 23–32.

Veit, H., 2002. Die Alpen – Geoökologie und Landschaftsentwicklung. Eugen Ulmer, Stuttgart.

Veit, H., Höfner, T., 1993. Permafrost, gelifluction and fluvial sediment transfer in the alpine/subnival ecotone, central Alps, Austria: present, past and future. Z. Geomorph. N.F. 92 (Suppl.), 71–84.

Veit, H., Stingl, H., Emmerich, K.-H., John, B., 1995. Zeitliche und räumliche Variabilität gelifluidaler Prozesse und ihre Ursachen. Z. Geomorph. N.F. 99 (Suppl.), 107–122.

Veit, H., Mailänder, R., Vonlanthen, C., 2002. Periglaziale Deckschichten im Alpenraum: Bodenkundliche und landschaftsgeschichtliche Bedeutung. Petermanns Geographische Mitteilungen 146 (4), 6–14.

Veit, H., Marti, T., Winiger, L., 2011. Environmental changes in northern Iceland since theYounger Dryas inferred from periglacial slope deposits. The Holocene, http://dx.doi.org/10.1177/0959683611423695.

Velichko, A.A., Andreev, A.A., Klimanov, V.A., 1997. Climate and vegetation dynamics in the tundra and forest zone during the Late Glacial and Holocene. Quat. Int. 41/42, 71–96.

Velichko, A.A., Bogucki, A.B., Morozova, T.D., Udartsev, V.P., Khalcheva, T.A., Tsatskin, A.I., 1984. Periglacial landscapes of the East European Plain. In: Velichko, A.A., Wright Jr., H.W., Barnosky, C.W. (Eds.), Late Quaternary Environments of the Soviet Union. University of Minnesota Press, Minneapolis, pp. 95–118.

Velichko, A.A., Morozova, T.D., Nechaev, V.P., Rutter, N.W., Dlusskii, K.G., Little, E.C., Catto, N.R., Semenov, V.V., Evans, M.E., 2006. Loess/paleosol/cryogenic formation and structure near the northern limit of loess deposition, East European Plain, Russia. Quat. Int. 152/153, 14–30.

Verstraeten, G., Lang, A., Houben, P. (Eds.), 2009. Human impact on sediment dynamics – quantification and timing. Catena Spec. Iss. 77.

Völkel, J., 1993. Periglacial overlying strata in the Bavarian Forest – methods of their stratigraphic division and questions about their ecological significance. Z. Geomorph. N.F. 92 (Suppl.), 113–125.

Völkel, J., 1994. Zur Frage der Merkmalcharakteristik und Gliederung periglazialer Deckschichten am Beispiel des Bayerischen Waldes. Petermanns Geographische Mitteilungen 138 (4), 207–217.

Völkel, J., 1995. Periglaziale Deckschichten und Böden im Bayerischen Wald und seinen Randgebieten. Z. Geomorph. N.F. 96 (Suppl.), pp. 301.

Völkel, J., Leopold, M., 2001. Zur zeitlichen Einordnung der jüngsten periglazialen Ativitätsphase im Hangrelief zentraleuropäischer Mittelgebirge. Z. Geomorph. N.F. 45, 273–294.

Völkel, J., Mahr, A., 1997. Neue Befunde zum Alter der periglazialen Deckschichten im Vorderen Bayerischen Wald. Z. Geomorph. N.F. 41, 115–121.

Völkel, J., Mahr, A., 2001. Die IRSL-Datierung von periglazialen Hangsedimenten – Ergebnisse aus dem Bayerischen Wald. Z. Geomorph. N.F. 45, 285–305.

Völkel, J., Raab, T., 1999. Zur Differenzierung periglazialer, glazigener und kolluvialer Sedimente im Hangrelief von Mittel- und Hochgebirgen. Zentralblatt für Geologie und Paläontologie, Teil I 5–6, 305–318.

Völkel, J., Leopold, M., Roberts, M.C., 2001. The radar signatures and ages of periglacial slope deposits in the Central European highlands of Germany. Permafrost Periglac. Proc. 12, 379–387.

Völkel, J., Leopold, M., Mahr, A., Raab, T., 2002a. Zur Bedeutung kaltzeitlicher Hangsedimente in zentraleuropäischen Mittelgebirgslandschaften und zu Fragen ihrer Terminologie. Petermanns Geographische Mitteilungen 146, 50–59.

Völkel, J., Zepp, H., Kleber, A., 2002b. Periglaziale Deckschichten in Mittelgebirgen – ein offenes Forschungsfeld. Berichte zur deutschen Landeskunde 76 (2/3), 101–114.

Völkel, J., Huber, J., Leopold, M., Dethier, D.P., 2010. Young Quaternary slope sediments and paleosoils in the Colorado Front Range, process and age. Geol. Soc. Am. Abstr. Progr. 42 (5), 177553.

Von Horstig, G., Stettner, G., 1976. Geologische Karte von Bayern 1:25000. Erläuterungen zu Blatt 5735 Schwarzenbach am Wald. Bayerisches Geologisches Landesamt, München.

Vonlanthen, C., 2000. Der Einfluss von Relief und Deckschichten auf Böden und Vegetation in der subalpinen und alpinen Stufe (Furka/CH). Unpubl. Diploma Thesis, University of Berne.

Vucetich, C.G., Pullar, W.A., 1969. Stratigraphy and chronology of Late Pleistocene volcanic tephra beds in Central North Island, New Zealand. New Zealand J. Geol. Geophys. 12, 784–837.

Walk, H., 1982. Die Gehalte der Schwermetalle Cd, Tl, Pb, Bi und weiterer Spurenelemente in natürlichen Böden und ihren Ausgangsgesteinen Südwestdeutschlands. PhD-Dissertation, Univ. Karlsruhe.

Wang, C., Arnold, R.W., 1973. Quantifying pedogenesis for soils with discontinuities. Soil Sci. Soc. Am. Proc. 37, 271–278.

Ward, R., 1978. Floods – A geographical perspective. Macmillan Press, London.

Washburn, A.L., 1973. Periglacial processes and environments. Arnold, London.

Washburn, A.L., 1979. Geocryology. Arnold, London.

Wayne, W.J., 1984a. Glacial chronology of the Ruby Mountains-east Humboldt Range, Nevada. Quat. Res. 21, 286–303.

Wayne, W.J., 1984b. Relative dating techniques to distinguish late Pleistocene-Holocene continental sediments. Chemical Geol. 44, 337–348.

Weise, O.R., 1983. Das Periglazial. Gebr. Bornträger, Berlin, Stuttgart.

Wells, P.V., 1979. An equable glaciopluvial in the west: pleniglacial evidence of increased precipitation on a gradient from the Great Basin to the Sonoran and Chihuahuan Desert. Quat. Res. 12, 311–325.

Wells, S.G., McFadden, L.D., Dohrenwend, J.C., 1987. Influence of late Quaternary climatic changes on geomorphic and pedogenic processes on a desert piedmont, eastern Mojave Desert, California. Quat. Res. 27, 130–146.

Wells, S.G., Dohrenwend, J.C., McFadden, L.D., Turrin, B.D., Mahrer, K.D., 1985. Late Cenozoic landscape evolution on lava flow surfaces of the Cima Volcanic Field, Mojave Desert, California. Geol. Soc. Am. Bull. 96, 1518–1529.

Wenk, G., 2000. Zur Bedeutung unterirdischer Abflusskomponenten in hydrologischen Untersuchungsgebieten des Unterharzes. Erfurter Geographische Studien 9, 109–137.

Wenninger, J., Uhlenbrook, S., Tilch, N., Leibundgut, C., 2004. Experimental evidence of fast groundwater responses in a hillslope/floodplain area in the Black Forest Mountains, Germany. Hydrol. Proc. 18, 3305–3322.

Wenninger, J., Uhlenbrook, S., Lorentz, S., Leibundgut, C., 2008. Identification of runoff generation processes using combined hydrometric, tracer and geophysical methods in a headwater catchment in South Africa. Hydrol. Sci. J. 53, 65–80.

Werner, J., 1958. Zur Kenntnis der braunen Karbonatböden (Terra fusca) auf der Schwäbischen Alb. Arbeiten des Geologisch-Paläontologischen Instituts der Technischen Hochschule Stuttgart. N. F. 16.

Wessely, G., 2006. Geologie der Österreichischen Bundesländer – Niederösterreich. Geologische Bundesanstalt, Wien.

Wilcke, W., Amelung, W., 1996. Small-scale heterogeneity of aluminum and heavy metals in aggregates along a climatic transect. Soil Sci. Soc. Am. J. 60, 1490–1495.

Wilcke, W., Kaupenjohann, M., 1994. Small scale heterogeneity of soil chemical properties. II. Fractions of aluminum and heavy metals. Zeitschrift für Pflanzenernährung und Bodenkunde 157, 459–465.

Williams, P.J., 1957. Some investigations into solifluction features in Norway. Geograph. J. 123, 42–58.

Williams, P., Smith, M., 1989a. The frozen earth. Cambridge University Press, Cambridge.

Wittmann, O., 1991. Berichte aus dem Arbeitskreis für Bodensystematik der DBG. Mitteilungen der Deutschen Bodenkundlichen Gesellschaft 66, 69–74.

Wohlfarth, B., Tarasov, P., Bennike, O., Lacourse, T., Subetto, D., Torssander, P., Romanenko, F., 2006. Late glacial and Holocene palaeoenvironmental changes in the Rostov-Yaroslavl' area, West Central Russia. J. Paleolimnol. 35, 543–569.

Wollschläger, U., Gerhards, H., Yu, Q., Roth, K., 2008. Application of multi-channel ground- penetrating radar to determine thaw depth and moisture content of the active layer. American Geophysical Union, Fall Meeting 2008, abstract #C31E-0545. http://adsabs.harvard.edu/abs/2008AGUFM.C31E0545W (accessed Aug. 7, 2010).

WRB (IUSS Working Group WRB), 2006. World reference base for soil resources. World Soil Resources Reports No. 103, FAO, Rome.

Yaalon, D.H., 1997. Soils in the Mediterranean region: what makes them different? Catena 28, 157–169.

Yaalon, D.H., Ganor, E., 1973. The influence of dust on soils during the Quaternary. Soil Sci. 116, 146–155.

Zalasiewicz, J., Williams, M., Smith, A., Barry, T.L., Coe, A.L., Bown, P.R., Brenchley, P., Cantrill, D., Gale, A., Gibbard, P., Gregory, F.J., Hounslow, M.W., Kerr, A.C., Pearson, P., Knox, R., Powell, J., Waters, C., Marshall, J., Oates, M., Rawson, P., Stone, P., 2008. Are we now living in the Anthropocene? GSA Today 18 (2), 4–8.

Zaruba, Q., Mencl, V., 1982. Landslides and their control. Elsevier, Amsterdam.

Zech, W., Cepel, N., 1977. Anatolien—ein bodengeographischer Streifzug. Mitteilungen der Geographischen Gesellschaft München 62, 155–166.

Zeien, H., Brümmer, G.W., 1989. Chemische Extraktion zur Bestimmung von Schwermetallbindungsformen in Böden. Mitteilungen der Deutschen Bodenkundlichen Gesellschaft 59, 505–510.

Zink, A., 2008. Uncertainties on the luminescence ages and anomalous fading. Geochronometria 32, 47–50.

Zobeck, T.M., Gill, T.E., Popham, T.W., 1999. A two-parameter Weibull function to describe airborne dust particle size distributions. Earth Surf. Process. Landforms 24, 943–955.

Zöller, L., 1980. Über Hangschuttbildung, Plateaulehme und junge Erosion im "Hochwald" (westlicher Hunsrück, Rheinisches Schiefergebirge). Catena 7, 153–167.

Zöller, L., Nehring, F., 2002. Solifluktions-, Löss- und Bodenbildungszyklen seit dem letzten Interglazial im Niederwesterwald. Berichte zur deutschen Landeskunde 76 (2/3), 115–130.

Zufall, M.J., Dai, W.P., Davidson, C.I., Etyemezian, V., 1999a. Dry deposition of particles to wave surfaces: I. Mathematical modeling. Atmos. Environ. 33, 4273–4281.

Zufall, M.J., Dai, W.P., Davidson, C.I., 1999b. Dry deposition of particles to wave surfaces: II. Wind tunnel experiments. Atmos. Environ. 33, 4283–4290.

Index

Note: Page numbers followed by "*f*" indicate figures, and "*t*" indicate tables.

D

E

F

G

H

N

P

R

S

T

W

MIX
Papier | Fördert gute Waldnutzung
FSC® C083411

Kolibri 360 GmbH
Ferdinand-Jühlke-Straße 7
99095 Erfurt, Deutschland
produktsicherheit@kolibri360.de